PHYSICS OF THE SUN
Volume I: The Solar Interior

GEOPHYSICS AND ASTROPHYSICS MONOGRAPHS

PHYSICS OF THE SUN

Volume I: The Solar Interior

Edited by

PETER A. STURROCK

Center for Space Science and Astrophysics,
Stanford University, Stanford, California, U.S.A.

Associate Editors:

THOMAS E. HOLZER

High Altitude Observatory,
National Center for Atmospheric Research, Boulder, Colorado, U.S.A.

DIMITRI M. MIHALAS

High Altitude Observatory,
National Center for Atmospheric Research, Boulder, Colorado, U.S.A.

ROGER K. ULRICH

Astronomy Department, University of California,
Los Angeles, California, U.S.A.

D. REIDEL PUBLISHING COMPANY

A MEMBER OF THE KLUWER ACADEMIC PUBLISHERS GROUP

DORDRECHT / BOSTON / LANCASTER / TOKYO

Library of Congress Cataloging-in-Publication Data
Main entry under title:

Physics of the sun.

  (Geophysics and astrophysics monographs)
  Includes bibliographies and index.
  Contents: v. 1. The solar interior.
    1. Sun. I. Sturrock, Peter A. (Peter Andrew) II. Series.
QB521.P48   1985   523.7   85–20996
ISBN 90-277-1823-7 (set)
ISBN 90-277-1860-1 (v. 1)

Published by D. Reidel Publishing Company,
P.O. Box 17, 3300 AA Dordrecht, Holland.

Sold and distributed in the U.S.A. and Canada
by Kluwer Academic Publishers
190 Old Derby Street, Hingham, MA 02043, U.S.A.

In all other countries, sold and distributed
by Kluwer Academic Publishers Group,
P.O. Box 322, 3300 AH Dordrecht, Holland.

Printed in The Netherlands

TABLE OF CONTENTS

PREFACE

This volume, together with its two companion volumes, originated in a study commissioned by the United States National Academy of Sciences on behalf of the National Aeronautics and Space Administration. A committee composed of Tom Holzer, Dimitri Mihalas, Roger Ulrich and myself was asked to prepare a comprehensive review of current knowledge concerning the physics of the Sun. We were fortunate in being able to persuade many distinguished scientists to gather their forces for the preparation of 21 separate chapters covering not only solar physics but also relevant areas of astrophysics and solar-terrestrial relations.

It proved necessary to divide the chapters into three separate volumes that cover three different aspects of solar physics. Volumes II and III are concerned with 'The Solar Atmosphere' and with 'Astrophysics and Solar-Terrestrial Relations'. This volume is devoted to 'The Solar Interior', except that the volume begins with one chapter reviewing the contents of all three volumes. Our study of the solar interior includes a review of nuclear, atomic, radiative, hydrodynamic and hydromagnetic processes, together with reviews of three areas of active current investigation: the dynamo mechanism, internal rotation and magnetic fields, and oscillations. The last topic, in particular, has emerged in recent years as one of the most exciting areas of solar research.

In preparing our material, the authors and editors benefited greatly from the efforts of a number of scientists who generously agreed to review individual chapters. I wish therefore to take this opportunity to thank the the following individuals for this valuable contribution to our work: S. K. Antiochos, E. H. Avrett, J. N. Bahcall, C. A. Barnes, G. Bicknell, D. Black, M. L. Blake, P. Bodenheimer, F. H. Busse, R. C. Canfield, T. R. Carson, J. I. Castor, J. Christensen-Dalsgaard, E. C. Chupp, A. N. Cox, L. E. Cram, P. R. Demarque, L. Fisk, W. A. Fowler, D. O. Gough, L. W. Hartmann, J. W. Harvey, R. F. Howard, P. Hoyng, H. S. Hudson, G. J. Hurford, C. F. Kennel, R. A. Kopp, A. Krueger, R. M. Kulsrud, R. B. Larson, H. Leinbach, R. E. Lingenfelter, J. L. Linsky, D. B. Melrose, M. J. Mitchell, A. G. Newkirk, F. W. Perkins, R. Roble, R. T. Rood, R. Rosner, B. F. Rozsynai, S. Schneider, E. C. Shoub, B. Sonnerup, H. Spruit, R. F. Stein, M. Stix, J. Tassoul, G. Van Hoven, G. S. Vaiana, A. H. Vaughan, S. P. Worden, R. A. Wolf, and J. B. Zirker.

On behalf of the editors of this monograph, I wish to thank Dr Richard C. Hart of the National Academy of Sciences, Dr David Larner of Reidel Publishing Company, and Mrs. Louise Meyers of Stanford University, for the efficient and good-natured support that we received from them at various stages of the preparation of this volume.

Stanford University,   P. A. STURROCK
April, 1985

Peter A. Sturrock (ed.), Physics of the Sun, Vol. I, · p. ix.
©1986 *by D. Reidel Publishing Company.*

CHAPTER 1

INTRODUCTION AND SUMMARY

PETER A. STURROCK

1. Preliminary Remarks

In has long been recognized that the Sun plays a key role in the solar system in influencing the interplanetary medium and the atmospheres of planets, so that the study of solar–terrestrial relations is one of the most important aspects of solar physics. The significance of plasma physics in the solar atmosphere — and indeed in the entire heliosphere — is now widely accepted. A more recent development is that the relationship of solar physics to astrophysics in general, and to stellar physics in particular, is also receiving close attention. For instance, an understanding of the long-term evolution of the Sun and solar variability can be obtained only partially from solar observations, paleoclimatic studies, and analysis of meteoritic and lunar samples. A thorough understanding requires the use of additional information that could be obtained from observations of a large number of stars similar to the Sun. A carefully chosen set of such observations would permit the development of a statistical picture of the time history of solar luminosity variations, cycle morphology, and magnetic activity (including spots and flares) over the Sun's entire lifetime. The existence of such information would make possible realistic estimates of the variation of the Sun's radiation from its early main sequence phase to the present, and would allow predictions of its future behavior.

Space missions of the last decade have changed our view of the Sun in many ways. The most visually dramatic changes came from Skylab. The X-ray telescopes gave us new insight into the structure of the corona, showing that magnetic loops are the fundamental structure of active regions and flares, and that 'coronal holes' are the source of high-speed solar wind streams. In addition, the coronograph on Skylab identified 'coronal transients' — huge masses of plasma ejected by flares and erupting prominences which may travel to the Earth's orbit and cause geomagnetic disturbances.

Skylab set the stage for more specialized physical studies by OSO-8, ISEE and SMM. As a result of these studies, we have learned that the Sun's corona is not heated purely by sound waves, and that the magnetic field must play a crucial role in coronal heating. We have learned that the photospheric magnetic field is not distributed smoothly but is concentrated in small tubes of high field strength, with important implications for all magnetic structures and solar activity. Concerning solar flares, we have learned to distinguish between the primary energy release which occurs in the corona, and secondary processes in the chromosphere and elsewhere which involve energy exchange by electron beams, ion beams, heat conduction and X-rays. Our earlier concept that ion acceleration

Peter A. Sturrock (ed.), Physics of the Sun, Vol. I, pp. 1–14.

occurs at a later stage than electron acceleration has proved to be incorrect, leading to the search for one process or two closely related processes for accelerating both species of particles. More recently, we have learned from SMM data that one of the fundamental 'constants' of solar–terrestrial physics is not a constant: the solar luminosity has been found to vary on a time-scale of one day. This new knowledge comes at a time of renewed interest in the relationship between the Sun and the Earth's climate. It also poses new and difficult questions concerning the propagation and storage of energy through the convection zone.

For all of these and for many other reasons, it is clear that space missions have brought about profound changes in our perception of the Sun, both as a star and as the central engine of the solar-planetary system.

The above points will be developed in more detail in the remaining sections of this summary. We first discuss the Sun's interior, then proceed to various layers of the Sun's atmosphere and to a discussion of solar activity. We then discuss the relationship of solar physics to other areas of astrophysics, and conclude with a discussion of solar-terrestrial relations. In discussing each topic, we attempt to point out significant recent advances, and problems that appear deserving of special attention for future research. We shall also point out the great potential of space observations in providing critical data for use in answering these questions.

2. The Solar Interior

The essential test of a model of the Sun rests on a comparison with the basic macroscopic properties, namely the mass, radius, luminosity, and composition. Until recently, we thought that the relevant physics was well known except for one deficiency — we have no theory of convection valid for application to stellar structure. For this reason, standard models necessarily involve arbitrary parameters such as the ratio of the mixing length to the scale height. We may therefore note that an important outstanding problem facing solar physicists, and indeed astrophysicists in general, is the development of a reliable theory of convection in stellar envelopes.

In recent years, however, confidence in our understanding of stellar interiors has been profoundly shaken by the conflict between the actual measurement of the neutrino flux from the Sun and theoretical estimates of this flux. In attempts to resolve this discrepancy, many questions have been analyzed in detail that previously received only cursory examination as discussed in Chapter 2, *Thermonuclear Reactions in the Solar Interior* by P. D. M. Parker, and in Chapter 17, by M. J. Newman, *The Solar Neutrino Problem: Gadfly for Solar Evolution Theory*. As Parker comments, "the observation of solar neutrinos is a crucial test of our understanding of the solar interior, and until we can understand the disagreement between the current model predictions and the current experimental results, we are forced to conclude that after 50 years we still have only circumstantial and indirect evidence for thermonuclear reactions in the solar interior."

Physicists and astrophysicists have considered many possible changes in the physical processes involved in producing the solar neutrino flux. It has recently been suggested that the neutrino may have nonzero mass, in which case each electron neutrino emitted from the Sun would cycle through three states, of which only one would be detected, so reducing the count rate by a factor of three. It has also been noted that the plasma in the Sun's deep interior does not satisfy a basic requirement for the applicability of plasma collision theory (that there be many particles per Debye cube): this calls into question the applicability of standard calculations of thermonuclear reaction rates. Another important outstanding question concerns convection in the Sun's interior: if some mechanism such as convective overshoot or rotationally driven circulation were sufficiently effective for the Sun's interior to be thoroughly mixed, it might be possible to reconcile calculations of the neutrino flux with the data. Since Davis's current experiment (Bahcall and Davis, 1981) is sensitive only to the high-energy tail of the solar neutrino spectrum, there appears a strong case for developing deterctors, such as the proposed gallium detector, which are capable of measuring the bulk of the neutrino flux.[1]

One of the important unknowns concerning the Sun's interior is its state of rotation. In particular, Dicke and Goldenberg (1967) drew attention to this problem and carried out a series of observations on the figure of the Sun from which they were able to extract information relevant to the Sun's internal rotation. The question is still unresolved, and there is no doubt that valuable data, which could perhaps resolve the issue, can be obtained by means of an accurately tracked spacecraft which makes a close encounter with the Sun, or by means of accurate measurements of solar oscillations, over a period of years, which can best be made from space.

Calculation of the neutrino flux depends sensitively on estimates of the internal temperature of the Sun. As W. F. Huebner states in Chapter 3, devoted to *Atomic and Radiative Processes in the Solar Interior*, "the relationship between pressure, density, temperature, and internal energy of the plasma is determined by the equation of state (but) the relationship of these quantities to the radial distance in the Sun is determined primarily by the opacity". Hence, directly or indirectly, experimental data on the Sun's neutrino flux has stimulated a careful re-examination of atomic and radiative processes in stellar interiors.

Quite recently, a powerful new tool for examining the interior of the Sun has become available to solar astronomers. This is the study of waves and oscillations, visible on the surface of the Sun, which is reviewed by T. Brown, B. Mihalas, and E. Rhodes in Chapter 7, *Solar Waves and Oscillations*. Since the modal character of the 'five-minute' oscillations was first established by Deubner (1975), increasingly accurate and sophisticated measurements have been made by many investigators. As discussed by Brown, Mihalas, and Rhodes, analysis of waves and oscillations will not only help select among prospective models of the Sun's internal density and temperature, but also provide information about the Sun's internal rotation. Observations made at the South Pole, with a time base of several days (Grec *et al.*, 1983), suggest that the interior of the Sun rotates significantly faster than its visible surface. Pursuit of this new area of investigation, which is coming to be known as 'solar seismology', may eventually require a dedicated spacecraft with equipment capable of determining the amplitude and spatial structure of the oscillations.

Some of the above topics concerning the solar interior are also touched upon by W. H.

---

[1] Since extensive lists of references are provided in each of the chapters referred to, we make no attempt to provide this chapter with comprehensive citations to all work mentioned in the text. However, it does seem appropriate to provide a reference where we refer to a very specific contribution.

Press in Chapter 4, dealing with *Hydrodynamic and Hydromagnetic Phenomena in the Deep Solar Interior*. Press stresses that the conventional view, that hydrodynamic and hydromagnetic effects are unimportant in the 'deep' solar interior (below the region where convection is expected), may need modification. Observational data leading one to this opinion includes a possible light-element depletion, the apparent neutrino discrepancy, tentative indications of a rapidly rotating core, and the 160 minute oscillations. One mechanism that would lead to flows in the deep interior is deep convective overshoot below the convection zone. Another mechanism, which would operate in the solar core itself, is nucleothermal destabilization of *g*-modes — that is, the spontaneous excitation, by a thermodynamic mechanism, of waves which are more closely related to water waves than to sound waves. It is clear that current solar observational data provide a strong challenge to the long-held view that the interior of a star can be divided into a convection zone and a radiative zone free from radial motion. A more detailed understanding of internal flows in the Sun, and the mechanisms that drive them, would clearly have great significance for our understanding of stellar structure in general.

We referred, early in this chapter, to the pressing need to improve our understanding of convection in the Sun. This is critical for understanding the observational data referred to in the preceding paragraph. It is also critical for understanding the solar dynamo, a topic discussed by P. A. Gilman in Chapter 5, *The Solar Dynamo: Observations and Theories of Solar Convection, Global Circulation, and Magnetic Fields*. It is widely accepted that the magnetic field visible at the Sun's surface is due to dynamo action in which the interplay of differential rotation, convection, and magnetic field leads to the enhancement of a weak field. A number of models of dynamo action have been developed in recent years; in most of these, turbulence is regarded as a small-scale random process, the effects of which may be described by a suitable averaging procedure. A typical dynamo model may show that a certain combination of differential rotation and turbulence will lead to the enhancement of a weak magnetic field, and nonlinear calculations may yield typical forms for the time-evolution of magnetic field in that model.

Until recently, the key test for a dynamo model has been comparison with the Maunder 'butterfly diagram' representing the time variation of sunspot activity in latitude during the course of a solar cycle. Quite recently, important additional information concerning the sunspot cycle has been obtained by However and La Bonte (1980). They find evidence for a 'torsional oscillation' propagating in 22 years from the poles to the equator, the velocity amplitude of the oscillation being only 3 m s^{-1}(!). This wave-like motion is closely related to the magnetic sunspot cycle and constitutes an additional fact that a successful dynamo theory must explain.

An important question concerning the internal structure of the Sun is the possible existence of high internal rotation and strong magnetic fields. These questions, and their significance for solar physics and astrophysics, are discussed by R. Ulrich in Chapter 6 on *Solar Internal Stresses: Rotation and Magnetic Fields*. Knowledge of the internal rotation and internal magnetic field is important for questions of cosmogony, since the initial rotation rate of the Sun must be at least as rapid as the present-day rotation rate of the core, and a magnetic field in the core must be a relic of the magneitc field in the primeval nebula.

Another reason for current interest in the internal stresses of the Sun is that rapid internal rotation should lead to circulation which would mix different layers of the Sun, tending to homogenize the element abundances. This process is important for our understanding of the observed abundance of light elements such as lithium on the Sun's surface. Furthermore, variation of the rotation rate with radius plays an important part in dynamo theory, discussed by Gilman, and may also have some importance in theoretical estimates of the neutrino production rate, discussed by Parker.

In recent years, a variety of techniques have evolved in an attempt to obtain information about the internal stresses of the Sun. Dicke and Goldberg (1967), and more recently Hill and his collaborators (Hill and Stebbins, 1975), have made accurate ground-based measurements of the figure of the Sun, with the goal of measuring the solar oblateness. There is as yet no agreement on the observational results, and it is important that further observations should be made, either from the ground or from space. As far as tests of general relativity are concerned, the crucial quantity to be determined is the gravitational quadrupole moment which could perhaps be inferred from an accurate oblateness measurement, but can be determined more precisely and unambiguously by study of the orbit of a spacecraft flying close to the Sun.

Ulrich, and also Brown, Mihalas, and Rhodes, stress that accurate measurements of global oscillations of the Sun — such as could be made from space — hold great promise for study of the Sun's internal rotation and may possibly yield information about the Sun's internal magnetic field of this field is sufficiently strong.

We see, from the preceding discussion, that solar physicists are now concerned with, and seeking ways to determine, many aspects of the internal structure of the sun: its rotation, magnetic field, convection, oscillations, etc. Further advance in our understanding will depend critically on new observational data, such as a determination of the internal rotation rate as a function of radius and latitude. As Gilman points out, however, a critical need is for information concerning the interaction between velocity fields and magnetic fields on very small spatial scales, smaller than that of the basic convection element near the photosphere, which is a 'granule'. To quote Gilman: "The necessary spatial resolution can only be obtained from a space platform such as the Shuttle, using the Solar Optical Telescope".

3. Atmosphere

The atmosphere of the Sun may be observed over a wide range of the electromagnetic spectrum ranging from radio to gamma-ray frequencies, sometimes with spatial resolution better than 1 arcsec, and with a variety of spectroscopic techniques that yield information on the line-of-sight velocity field and the line-of-sight magnetic field. The problem of inferring the structure of the atmosphere from observed radiation is discussed by R. G. Athay in Chapter 8 on *Radiation Output*. Direct inference of atmospheric structure from radiation output is not possible; the procedure must be based upon models of the atmosphere or of components of the atmosphere. A detailed discussion of the complexities of *Chromospheric Fine Structure* is presented by Athay in Chapter 9. A phenomenological review of our knowledge of the upper atmosphere is presented in Chapter 10, *Structure, Dynamics, and Heating of the Solar Atmosphere*, written by F. Q. Orrall and G. W. Pneuman. A theoretical discussion of *Physical Processes in the Solar Corona* is presented in Chapter 11, prepared by R. Rosner, B. C. Low, and T. Holzer.

From these four chapters, it is clear that much has been learned in recent years about the Sun's atmosphere, yet fundamental questions remain unanswered. A few years ago, there was widespread belief that the corona is heated by a flux of acoustic waves generated by convective motion at the photosphere. Spectroscopic observations made by means of the OSO–8 spacecraft, however, indicate that the needed acoustic wave flux is not present in the upper chromosphere. We learned from X-ray observations made by telescopes on Skylab that, at least in active regions, the structure of the solar corona is dominated by loop-like structures that may be attributed to a magnetic field. It is therefore considered possible, and perhaps likely, that coronal heating in these regions is caused by the dissipation of magnetic energy.

Although such theories appear promising for active regions, it is not clear that coronal heating in coronal holes (which have predominantly open magnetic structure) can be attributed to the dissipation of magnetic energy. Although it appears likely that photospheric motions can slowly twist closed magnetic flux tubes in active regions, to build up magnetic free energy in the form of field-aligned currents which can then dissipate and so heat the corona, photospheric motions cannot lead to an accumulation of free energy in the open magnetic field configurations of coronal holes. They can produce outward propagating torsional Alfvén waves but these travel through the inner and middle corona without contributing significantly to coronal heating.

The challenge to our understanding of the solar atmosphere is, however, much greater than simply understanding the mechanism for coronal heating. One of the most important areas of research concerns the origin of the solar wind. According to early theoretical ideas, the solar wind is a direct consequence of the high temperature of the corona. Although it is clear that the corona can generate a wind without any other mechanism being involved, this simple theory does not lead to a model which explains either the parameters characterizing the steady state of the solar wind or the great variability of these parameters. Hence, current theoretical research involves the investigation of possible nonthermal processes which lead to the injection of additional energy and momentum into the solar wind, and there is a manifest need for observational evidence related to these questions, such as could be obtained by means of an appropriate space mission.

The atmosphere involves a wide range of structures and dynamical phenomena including sunspots, fibrils, prominences, spicules, and 'bright points', to mention only a few, all of which need to be explained if we are to say that we 'understand' the solar atmosphere. Improved understanding depends crucially on the acquisition of improved diagnostic data with high spectral and spatial resolution, especially in the UV part of the spectrum. Holzer, Low, and Rosner also point out the need for more detailed MHD models of solar phenomena, and stress that it is unlikely that these models can be constructed using analytical techniques alone. The current trend is towards the development of time-dependent, multidimensional MHD models with a level of complexity similar to that of codes being developed as part of the CTR (controlled themonuclear reactor) program.

Recent discoveries have not been limited to the upper chromosphere and corona. Much has been learned also about the photosphere and low chromosphere. One of the most important discoveries has been the determination that much of the magnetic flux threading the photosphere is confined in small flux tubes, of diameter of order 400 km, in which the field strength is between 1000 and 2000 gauss. These structures may well be the basic 'building blocks' of magnetic structures in the Sun. If so, our models of the coronal magnetic field should consist of a number of discrete flux tubes rather than a distributed field configuration. The significance of this and related problems is emphasized by Athay in his chapter on *Chromospheric Fine Structure*. Observations of the required resolution will be possible only with the advent of an optical solar telescope in space. Stereoscopic observations would be invaluable in resolving the three-dimensional structure of the atmosphere.

As we have noted above, study of the Sun's atmosphere does not end with the corona: the atmosphere extends outward into the solar wind, and the structure of the solar wind depends critically on the structure of the magnetic field and also on various nonthermal processes that may lead to the exchange of energy and momentum between various layers of the Sun's atmosphere and the solar wind. Thus, the upper regions of the solar atmosphere and the solar wind must be viewed as a tightly coupled system, and this close coupling emphasizes the importance of obtaining data simultaneously from a spacecraft looking at the Sun's corona and from one or more spacecraft measuring the solar wind. It is possible that one day we shall obtain the solar data from a permanent solar observatory in space, equipped with many high-resolution instruments covering a wide range of wavelengths.

Solar activity, which is concerned with the development of active regions, pores, sunspots, prominences, etc., may be viewed as the study of the active role of magnetic fields in the Sun's atmosphere (Priest, 1982). The most complex manifestations of this process is the flare which produces sudden bursts of energetic electromagnetic radiation, radio waves, high-energy particles, and plasma. It is believed that the energy released during a flare is associated with stresses (currents) in the coronal magnetic field, which have built up comparatively slowly as a result of changes and motions of the photospheric magnetic field. The sudden release of the energy stored in the coronal magnetic field is attributed to some form of plasma instability. These concepts are discussed in Chapter 12, *Magnetic Energy Storage and Conversion in the Solar Atmosphere*, by D. Spicer, J. Mariska, and J. Boris.

From an astrophysical point of view, one of the most important properties of a flare is the acceleration of particles to high energy – in the case of a solar flare, sometimes as high as 10 GeV. Our present understanding of the complex problem of acceleration in solar flares is reviewed by M. Forman, R. Ramaty, and E. Zweibel in Chapter 13, The *Acceleration and Propagation of Solar Flare Energetic Particles*. Study of data concerning radio waves, X-ray, gamma rays, and particles emitted from the Sun suggests that there are at least two different acceleration mechanisms at work in a solar flare. Three possible mechanisms discussed by the authors are stochastic acceleration, shock acceleration, and 'direct' acceleration due to a low-frequency electric field parallel to the magnetic field. The energetic particles observed in interplanetary space are probably accelerated in the corona by shocks or by turbulence produced by shocks. The prompt acceleration of nuclei and relativistic electrons, as manifested by gamma-ray lines and continuum radiation, may be due to electric fields associated with current interruption or reconnection in closed magnetic field regions, or to shock or stochastic acceleration in these regions. Clarification of these possibilities requires coordinated high-resolution observations at a variety of wavelengths – especially hard X-rays – and the direct measurement of particle fluxes from the Sun. The latter measurements should provide data on the elemental and isotopic composition of particle fluxes from the Sun.

The detection of gamma-ray lines by spectrometers on the OSO–7, HEAO–1, and HEAO–3 spacecraft, and of gamma-ray lines and high-energy neutrons by the SMM spacecraft, have provided important new information concerning the acceleration of high-energy particles during flares. This topic is disucssed in detail by R. Ramaty in Chapter 14, *Nuclear Processes in Solar Flares*. As Ramaty points out: "Gamma-ray lines are the most direct probe of nuclear processes in the solar atmosphere." One of the remarkable results to come from these observations is that protons and nuclei can be accelerated to energies of tens or even hundreds of MeV in only a few seconds. Solar gamma-ray astronomy has made a dramatic beginning. Future observation of solar gamma-ray lines with improved sensitivity and spectral resolution will provide unique information on particle acceleration during solar flares, particularly such questions as the timing of the acceleration, the beaming of the energetic particles, the temperature of the energetic particle interaction site, and the compositions of the ambient medium and of the energetic particles.

Radio observations provide a quite different channel for observing nonthermal processes occurring in the Sun's atmosphere, especially at the time of solar flares. A selective review of this very extensive topic is given by M. Goldman and D. F. Smith in Chapter 15, *Solar Radio Emission*. Radio emission is due to electrons, but it appears that a variety of mechanisms are at work, including gyrosynchrotron radiation and radiation from plasma oscillations. In addition to observations obtained by dedicated solar observatories, important data has been obtained from the VLA, which provides extremely high spatial resolution, and from low-frequency radio receivers on spacecraft such as those of the IMP series. Spacecraft penetrating into the solar wind (such as ISEE-C) are able to probe the region in which radio waves of very low frequency are believed to be generated. Such observations give direct insight into the way in which waves and oscillations are generated in a plasma by an electron stream, and the nonlinear processes that eventually lead to radio emission. As explained by Goldman and Smith, these observations have confirmed certain concepts but challenged others. These observations demonstrate that it is indeed possible to make *in-situ* measurements of certain plasma processes of astrophysical significance.

4. The Sun in its Astrophysical Context

Although much of the material discussed in earlier chapters of the review is applicable to a wide range of stars, and also, occasionally, to nonstellar objects, it is instructive to focus on the relationship between the Sun and other main sequence stars. This is the theme of Chapters 16 through 19 of the review.

Information concerning the *Formation of the Sun and its Planets* is reviewed by W. M. Kaula in Chapter 16. There is reason to believe that two supernova events preceded the formation of the solar system: one 200 million years before formation, and the other only two million years before. The latter may possibly have influenced the collapse of the primeval gas cloud. Studies of remanent magnetism and inert gases in meteorites indicate that the early Sun was considerably more active than the present Sun. As we shall see later, this appears to be a typical trend for single main sequence stars.

It has been found that, for binaries with periods less than 100 years, the frequency with which a primary is found to have a secondary companion increases as the one-third power of the ratio of the mass of the primary to the mass of the secondary. If this law is assumed to apply to smaller primary masses than the range for which it was established, one infers that almost all stars of mass less than about 1.5. $M_\odot$ have companions, the smallest companions being planets. Observation of Barnard's star gives suggestive, but not conclusive, evidence that there are planets in orbit around that star. Hence, from a strictly observational viewpoint, we still do not know whether or not the Sun is unusual in having planets.

Since, in developing models of the Sun, one of the critical parameters is its assumed age, it is perhaps appropriate to reconsider the usual assumption that the age of the Sun can be inferred from the age of meteorites. In order to resolve this question, we need even more detailed models of the evolution of the primeval nebula into a central object and a surrounding disk, with the central object evolving into a main sequence star, and the disk condensing first into planetesimals and then into planets. Such studies may help to resolve some current problems, such as the fact that chondritic meteorites indicate higher nebula temperatures (of order 1500 K) than are provided by present models.

Some of the content of Chapter 17 by M. J. Newman, *The Solar Neutrino Problem: Gadfly for Solar Evolution Theory*, has been discussed in conjunction with chapters dealing with the solar interior. However, Newman also points out certain general astrophysical questions that have been brought to the attention of the astrophysics community by the neutrino problem. For instance, it has been speculated that the gravitational constant may vary in the course of time, and that there may be a black hole in the center of the Sun. Less dramatic possibilities concern the evolution of the Sun. For instance, it has been suggested that the initial Sun may have been nonuniform in the sense that the composition was a function of radius. There has also been speculation that there was a very strong primordial magnetic field which is now locked in the core, and/or that the core rotates much more rapidly than the surface. Information about the solar interior to be obtained by 'solar seismology', discussed earlier in this chapter, will clarify these possibilities. Other relevant data may be obtained by gravitational tests involving a spacecraft, or by accurate measurements (from the ground or from space) of the figure of the Sun.

It is also possible that some peculiarity in the evolution of the Sun is responsible for the present situation. It has, for instance, been proposed that the surface composition of the Sun may not reflect the primordial composition, since it may have been influenced by accretion. The accreted matter may come from the solar system itself (comets, meteorites, etc.), or from dense interstellar clouds.

Chapter 18 by J. P. Cassinelli and K. B. MacGregor, *Stellar Chromospheres, Coronae, and Winds*, and Chapter 19 by R. W. Noyes, *Solar and Steller Magnetic Activity*, both deal with chromospheric and coronal radiation from the Sun and similar stars, and with related topics. As Cassinelli and MacGregor remark: "The discovery of chromospheres, coronae, and winds associated with stars other than the sun affords a unique opportunity for the fruitful exchange of ideas between solar and stellar physicists. In particular, the observation and interpretation of solar-type phenomena in stars of different spectral types, luminosity classes, and ages should significantly increase our understanding of the 'Sun as a star'."

Since the solar corona is strongly influenced by — and may even own its existence to — the solar magnetic field, and since the chromosphere also is strongly affected by the magnetic field, study of these regions is tightly coupled to the study of magnetic activity in general, and the dynamo process in particular. White light observations of stars give evidence for the existence of spots, rotation, and even differential rotation. The existence of spots is closely coupled to chromospheric activity that may be conveniently monitored by means of Ca II emission. Measured in this way, chromospheric activity in stars is found to be related to star spots in much the same way as it is on the Sun. Surveys of chromospheric activity in main sequence stars have given valuable insight into the way in which dynamo activity depends upon stellar parameters. For stars of a given spectral type, older stars show cyclic behavior similar to the solar cycle, whereas younger stars show a stochastic variation in activity. There is a clear gap between these two regimes. It appears, therefore, that there are at least two distinct regimes of dynamo activity, and that a star may be comparatively quiet for a short time after leaving the chaotic regime and before entering the cyclic regime. Such data, derived from large numbers of stars, provide stringent additional tests for dynamo models, beyond tests derived from solar data alone.

Similarly, analysis of chromospheric and coronal emission from other stars provides an additional test of theories of chromospheric and coronal heating developed originally for the Sun. We find, for instance, that data on stellar coronae disagree with the implications of acoustic heating theory, as is the case for the solar corona. In attempting to understand strong patterns such as the Wilson—Bappu effect, relating the width of Ca II line-emission cores with luminosity, astrophysicists may build upon our understanding of the solar chromosphere, but must also estimate the way in which atmospheric parameters vary with the basic stellar parameters of effective temperature and surface gravity.

Consideration of the problems posed by stellar observations calls for additional information, some of which can best be obtained by further observation of the Sun. For instance, spatially resolved observations of the magnetic and velocity fields on the Sun, such as can best be obtained from an optical solar telescope in space, are required to tighten constraints on the dynamo mechanism and on the mechanisms involved in small-scale processes such as sunspot formation, flares, spicules, etc. We also need further information on the latitude, longitude and solar-cycle-phase dependence of ephemeral active regions that can be observed as coronal bright points. This information could be provided by a suitably instrumented spacecraft. Supplementary diagnostics information on coronal structure can be obtained by high-resolution observations of the 'pinhole' type. In order to obtain further information about the way in which a wind exerts a braking torque on a star, such as the Sun, we need out-of-the-ecliptic observations of the solar wind.

Similarly, our understanding of the Sun could be advanced by increasing our knowledge of stellar atmospheres and stellar activity. It is desirable to have more detailed observations of the chromospheric and coronal radiation to learn more about the way in which the dynamo mechanism varies with basic stellar parameters, and to provide further checks of theories of chromospheric and coronal heating. Some of these observations can be made from the ground, but UV and X-ray observations must be made from space.

5. Solar–Terrestrial Relations

One of the most important reasons for studying the Sun in great detail is to improve our understanding of the various ways in which the Sun affects our terrestrial environment. As time goes by, we are improving our understanding of certain of the well-established effects, such as geomagnetic storms, but we are also obtaining evidence pointing to new and unsuspected effects, possibly even including solar influences on climate and weather.

The very complex manner in which the sun's electromagnetic radiation influencess the Earth's atmosphere, and all life forms on Earth, is outlined by R. Dickinson in Chapter 20, *Effects of Solar Electromagnetic Radiation on the Terrestrial Environment.*

These effects involve a combination of physical, chemical, and biological processes. The sensitivity of the Earth's climate to variation of insolation has recently been underscored by the demonstration that historical major climatic fluctuations may be attributed to variations in the Earth's orbital parameters. In particular, variation of insolation due to orbital effects are believed to have been a contributing cause of the last ice age that peaked 20 000 years ago, as well as earlier ice ages over the last million years.

Recognition of the sensitivity of our climate to fluctuations in solar radiation received on Earth clearly makes it important to understand the stability — or lack of stability — of the total radiative energy input from the Sun, heretofore referred to as the 'solar constant'. The intrinsic fluctuations are extraordinarily difficult to measure from the Earth's surface. Measurements have been made over a period of over two years, however, by the active cavity radiometer on the SMM spacecraft. This instrument has a precision of 100 p.p.m. and can detect changes much smaller than this value. Measurements with this instrument have shown that the Sun's luminosity in far from constant; a variation of 1000 p.p.m. may be measured in one solar rotation.

It will be important to try to determine what changes, if any, in the Earth's atmosphere may be correlated with these changes in the Sun's luminosity. Synoptic observations of the Earth's atmosphere are now being made by spacecraft. Solar physicists, for their part, are trying to understand the origin of these fluctuations. There is certainly a strong component that may be attributed to sunspots, but there are other effects that must be pinned down. The very fact that sunspots can cause a diminution of the total radiative output from the Sun came as a surprise to many solar physicists, who believed that the 'missing flux' from a sunspot would appear elsewhere on the Sun's surface with negligible time delay. It now seems that substantial amounts of energy are being stored in the outer layers of the Sun, and further study of the influence of sunspots and other features on the Sun's luminosity may yield valuable insight into the structure of the Sun's convection zone, the transport of energy through that zone, and perhaps into the nature of stellar convection itself.

In addition to the effects on the Earth's atmosphere of variations of the total radiation from the Sun, we know that certain layers of the Earth's atmosphere — such as the ionosphere — are very sensitive to variations in solar ionizing radiation, including UV and X-ray. Changes in the ionosphere are important for a number of reasons, including the effect of such changes on radio propagation, we will be discussed later.

Changes in the solar wind itself, and in the magnetic field that permeates the solar wind, can both have important terrestrial effects, as discussed by N. Crooker and G. Siscoe in Chapter 21 dealing with *The Effect of the Solar Wind on the Terrestrial Environment.*

For instance, when the interplanetary magnetic field that sweeps past the Earth has a southward component for an hour or longer, there are auroral displays and there may even be disruptions in power distribution systems and in radio and cable communication systems. It is somewhat surprising that the impact of the expanding corona can be so pronounced, since its energy flux is six to seven orders of magnitude smaller than the energy flux radiated by the Sun and the solar wind hardly penetrates the magnetosphere: only about 0.1% of the solar wind mass flux incident on the magnetopause manages to cross it.

It is clear that subtle effects take place in the Earth's magnetic field structure, caused by the interplay of the solar wind with the nagnetosphere. The interaction is sensitive to the direction of the magnetic field permeating the solar wind. A number of characteristic forms of interaction have been identified. For instance, it is known that high-velocity streams have predictable effects. Some of these streams, often associated with coronal holes, are long lived, so that they recur with a period of approximately 27 days, due to the Sun's rotation. Other streams, such as those produced by coronal transients and other flare-related processes, are nonrecurrent. These nonrecurrent streams tend to have higher velocities and so give rise to shock waves that have important geomagnetic effects.

Even in the absence of solar activity, the Earth's magnetosphere is not static. Interaction between the magnetosphere and the solar wind at the magnetopause gives rise to convection of the magnetospheric plasma. This, in turn, gives rise to field-aligned or 'Birkeland' currents that transfer energy from the magnetopause, where it is derived from the solar wind, to the ionospheric layers in the auroral regions. These currents play a key role in ionospheric electrojets, aurorae, and magnetic substorms.

During times of magnetic activity, high-energy electrons from the outer magnetosphere may penetrate to the middle atmosphere, where their presence leads to a depletion in the stratospheric ozone. Because ozone in the middle atmosphere affects the radiation reaching the lower atmosphere, changes in the stratospheric ozone abundance could conceivably influence the Earth's weather.

There have been claims for over 100 years that some component of the Earth's weather fluctuations may be attributed to solar influences. For instance, there is some statistical support for the proposition that droughts in the high plains of the United States tend to occur at every other sunspot minimum. On a shorter time-scale, there is suggestive evidence (Wilcox et al., 1973) that a correlation exists between an atmospheric 'vorticity index' and the passage of 'sector boundaries' in the solar wind, where the predominantly radial component of magnetic field changes sign. This question deserves continued critical study and a vigorous search for physical mechanisms which might explain such effects.

In addition to the steady output of charged particles that comprise the solar wind, and sudden changes in the flow, which comprise 'coronal transients', high-speed streams, etc., the Sun also emits occasional streams of high-energy ions and electrons which influence the Earth's environment. These processes are reviewed by G. Reid in Chapter 22, *Solar Energetic Particles and their Effects on the Terrestrial Environment*. The influence of these particles on the terrestrial environment depends critically on their access to the Earth's atmosphere, which is determined by the structure of the Earth's magnetosphere. Particles may most easily reach the 'polar caps', where the magnetic field couples into the Earth's geomagnetic tail, or even out into interplanetary space.

Hence, solar 'particle events' typically cause enhanced ionization in the polar cap regions of the ionosphere. This gives rise not only to greatly increased absorption of radio waves, which may amount to a radio blackout, but also to a change in the phase of transmitted waves, which can have important adverse influences on radio navigation systems. The flux of solar particles directly into the polar cap regions also gives rise to a faint, diffuse aurora over the entire polar cap, called a 'polar-glow aurora'. This is one of the few observable effects that may be attributed directly to solar particles.

The influx of high-energy particles into the Earth's atmosphere also has important chemical consequences. For instance, energetic particles can dissociate molecules giving rise to an increase in NO concentration, leading in turn to NOX compounds. Furthermore, such chemical changes have an important influence on the ozone abundance, so that solar particle events can lead to a reduction in the stratospheric ozone content. As Reid remarks: "Solar particle events clearly have the potential for modifying the chemical composition of the middle atmosphere."

The terrestrial effects of solar particle events vary greatly from case to case. For instance, the February 1956 event increased the ground level neutron flux by a factor of 90 in some regions, and added 10% of an entire solar cycle's of C^{14}. The events of August 1972 caused a substantial and long-lasting depletion of stratospheric ozone. In historic or prehistoric times, there may will have been much larger events. One must also remember that the Earth's geomagnetic field has gone through reversals. If the Earth's magnetic field dropped to a very low value during a reversal, the Earth would not have been shielded from charged particles as it is now. There is indeed some evidence for correlation of the extinction of marine micro-organisms with polarity reversals, and some evidence for correlation of polarity reversals with changes in the Earth's climate.

As mentioned earlier, understanding of any apparent statistical correlation between solar activity, solar sector structure, etc., on the one hand, and the Earth's climate and weather, on the other, requires that we identify a 'trigger mechanism' by which a phenomenon with little energy content can influence the Earth's global circulation pattern, a system with comparatively high energy content. Proposed trigger mechanisms are electrical, radiative, or dynamical.

Of these possibilities, one of the most interesting is that the global electric field plays a key role. It is known that the electrical conductivity of the Earth's upper atmosphere is strongly influenced by the galactic cosmic-ray flux, and we know that the cosmic-ray flux is strongly influenced by the interplanetary magnetic field. Unfortunately, as Reid points out: "Our understanding of the global electric field is fairly rudimentary." If the case for a causal relationship between a nonradiative solar output and the Earth's weather is firmly established, and if the search for a physical mechanism zeroes in upon the global electric field, it will then become most important to find some mechanism for obtaining global synoptic maps of the atmospheric electric field. Some information is now being obtained by means of tethered balloons and aircraft, but the need for acquiring global data should stimulate a search for some method of measuring electric fields from space. No proposals for such measurements have yet been advanced.

It will be noted that the study of solar—terrestrial relations is important not only for practical reasons; the recognition of such effects also poses searching questions concerning the Sun itself and the complicated environment of the Earth.

It is clear from the material contained in the following chapters, and from the extensive studies on which they are based, that solar physics is now in an exciting state in which observations have sometimes confirmed established concepts, sometimes called for their modification, and sometimes called for the development of new models and theories. Furthermore, solar physics is enriched by the fact that it may be viewed as a testbed of physics, as the point of origin of solar—terrestrial relations, or in the context of stellar astrophysics. The editors and authors hope that this monograph will help stimulate both established scientists and students now embarking on their scientific careers to pursue these fascinating studies and so provide answers to the outstanding questions of solar physics.

References

Bahcall, J. N. and Davis, R., Jr: 1981, in C. A. Barnes, D. D. Clayton, and D. N. Schramm (eds.), *Essays in Nuclear Astrophysics*, Cambridge Univ. Press.
Deubner, F. L.: 1975, *Astron. Astrophys.* **44**, 371.
Dicke, R. H. and Goldenberg, H. M.: 1967, *Phys. Rev. Lett.* **18**, 313.
Grec, G., Fossat, E., and Pomerantz, M. A.: 1983, *Solar Phys.* **82**, 55.
Hill, H. A. and Stebbins, R. T.: 1975, *Astrophys. J.* **200**, 471.
Howard, R. F. and La Bonte, B. J.: 1980, *Astrophys. J.* **239**, L33.
Priest, E. R.: 1982, *Solar Magnetohydrodynamics*, Reidel, Dordrecht.
Wilcox, J. M., Scherrer, P. H., Svalgaard, L., Roberts, W. O., and Olson, R. H.: 1973, *Science* **180**, 185.

Center for Space Science and Astrophysics,
Stanford University,
Stanford, CA 94305,
U.S.A.

CHAPTER 2

THERMONUCLEAR REACTIONS IN THE SOLAR INTERIOR

PETER D. MacD. PARKER

1. Introduction

The probable role of thermonuclear reactions in the solar interior had already begun to be recognized as early as the 1920s by astrophysicists such as Eddington (1926). This recognition was based *not* on any direct observation of these reactions but instead was arrived at more on the basis of default. The realization that the Earth was at least a few billion years old (see Burchfield, 1975, for an interesting history of the development of this idea) coupled to the known mass, radius, and luminosity of the Sun demonstrated the inadequacy of classical (gravitational and chemical) energy sources for the sun. The enormous energy available from the then newly discovered nuclear reactions could, however, readily supply the required luminosity; the recognition by Gamow (1928) of the role of tunneling under the Coulomb barrier at low energies then made it possible for these reactions to occur in the solar interior where kT was most probably equivalent to only $\lesssim$ few keV, well below the Coulomb barrier (Atkinson and Houtermans, 1929). Following this early work which laid the foundations for nuclear astrophysics, by the end of the 1930s Weizsäcker (1937, 1938), Bethe and Critchfield (1938), and Bethe (1939) had laid out the energetics and many of the theoretical details for the specific reactions in the CN cycle and the p–p chain, as processes for carrying out the conversion of hydrogen to helium.

During the late 1940s and the 1950s initial laboratory measurements were made on all of the important reactions in the CN cycle and the p–p chain (except, of course, for the $^1H(p, e^+\nu)^2D$ reaction). (See Sections 3 and 4 for a detailed discussion of these individual reactions.) Of equal significance during this time was the recognition (Davis, 1955; Cameron, 1958; Fowler, 1958) of the possibility and the significance of detecting the neutrinos from some of these reactions as a way of probing the solar interior for direct evidence of the role of the thermonuclear reactions. (For a complete history of the development of the ^{37}Cl Solar Neutrino Experiment, see Bahcall and Davis (1981).) In parallel with the design and implementation of the ^{37}Cl neutrino detector (Davis, 1964; Davis et al., 1968) and motivated by the obvious significance of the solar neutrino measurements, during the decades of the 1960s and 1970s there was substantial effort devoted to the refinement of experimental measurements of the important p–p chain and CN cycle reactions. The net result of all of this effort is that there is now a measured (Davis et al., 1983) yield of ^{37}Ar from the ^{37}Cl deterctor corresponding to 1.8 ± 0.3 SNU (1 SNU = 10^{-36} per ^{37}Cl atom s^{-1}) which is in disagreement with the current theoretical predictions (6.2 ± 1.0 SNU) as determined using standard solar models incorporating the "best" values for the various nuclear reaction rates. (See Bahcall *et al.*, 1982, and Section 5 of this chapter.) The details of our present knowledge and our present uncertainties regarding each of these relevant nuclear reactions are discussed in Sections 3 and 4.

The observation of solar neutrinos is a crucial test of our understanding of the solar interior, and until we can understand the disagreement between the current model predictions and the current experimental results we are forced to conclude that after 50 years we still have only circumstantial and indirect evidence for thermonuclear reactions in the solar interior.

2. Reaction Rate Formalism

In discussing the details of thermonuclear reactions in stellar interiors, the reaction rate between two types of nuclei can be written as (e.g., Burbidge *et al.*, 1957; Parker *et al.*, 1964; Fowler *et al.*, 1967; Clayton, 1968; Barnes, 1971; Fowler *et al.*, 1975; Rolfs and Trautvetter, 1978):

$$P_{12} = \frac{n_1 n_2}{1 + \delta_{12}} \langle \sigma v \rangle_{12} \text{ reactions cm}^{-3} \text{ s}^{-1}, \tag{2.1}$$

where n_1 and n_2 are the number densities of nuclei of type 1 and type 2 (with atomic number Z_1 and Z_2, and atomic mass A_1 and A_2), and where $\langle \sigma v \rangle_{12}$ is related to the product of the reaction cross-section and the flux, averaged over the Maxwell–Boltzmann velocity distribution

$$\langle \sigma v \rangle = \int_0^\infty [\sigma(v) \cdot v] \, \Phi(v) \, dv \text{ cm}^3 \text{ s}^{-1}, \tag{2.2}$$

where,

$$\Phi(v) \, dv = \left(\frac{\mu}{2\pi kT} \right)^{3/2} \exp\left(-\frac{\mu v^2}{2kT} \right) 4\pi v^2 \, dv;$$

therefore,

$$\langle \sigma v \rangle = \left[\frac{8}{\pi \mu (kT)^3} \right]^{1/2} \int_0^\infty E \, \sigma(E) \exp\left(-\frac{E}{kT} \right) dE, \tag{2.3}$$

where μ is the reduced mass, and where v and E are the velocity and kinetic energy in the center-of-mass system. In order to evaluate $\langle \sigma v \rangle$ the energy dependence of the reaction cross-section must be determined. (A number of different cases have been considered for both neutron-induced and charged-particle-induced reactions, including non-resonant and resonant situations, and Fowler *et al.* (1967) and Fowler *et al.* (1975) have tabulated the appropriate expressions for $\langle \sigma v \rangle$ for a wide variety of specific reactions.)

In the study of the solar interior, almost all of the relevant nuclear reactions are nonresonant and charge particle induced. For this class of reactions, the energy dependence of the cross-section at low energies is dominated by the Coulomb barrier and may be written as

$$\sigma(E) = \frac{S(E)}{E} \exp(-2\pi\eta),$$ (2.4)

where the $1/E$ term comes from the λ^2 geometrical factor, in which $\exp(-2\pi\eta)$ is the usual Gamow barrier penetration term,

$$\eta = Z_1 Z_2 e^2 / \hbar v$$

$$2\pi\eta = 31.290 \, Z_1 Z_2 A^{1/2} / E_{cm}^{1/2} \text{ (keV)}; \, A = \mu/(1 \text{ amu}),$$

and where, for nonresonant reactions, $S(E)$ is then only a slowly varying function of energy.

In order to relate the $\sigma(E)$ and $S(E)$ quantities measured in the laboratory to the appropriate cross-sections in the solar interior, a correction (f_0) must be included to take into account the effects of electron screening (Salpeter, 1954). In the weak-screening approximation appropriate for the sun, this can be written as

$$f_0 = \exp(0.188 Z_1 Z_2 \, \zeta \, \rho^{1/2} \, T_6^{-3/2}),$$

where

$$\zeta = \left[\sum_i \left(X_i \frac{Z_i^2}{A_i} + X_i \frac{Z_i}{A_i} \right) \right]^{1/2},$$

where X_i is the mass fraction of nuclei of type i

$$X_i = n_i A_i / (6.022 \times 10^{23} \, \rho)$$

and where T_6 is the temperature in units of 10^6 K, T_9 in units of 10^9 K, etc. Substituting (2.4) into (2.2),

$$\langle \sigma v \rangle = \left(\frac{8}{\pi \mu (kT)^3} \right)^{1/2} f_0 \int_0^\infty S(E) \exp(-2\pi\eta) \exp(-E/kT) \, dE.$$ (2.5)

The energy dependence of this integrand is shown schematically in Figure 1. In order to evaluate this integral we expand $S(E)$ in a Taylor series,

$$S(E) = S(0) + ES'(0) + \tfrac{1}{2} E^2 S''(0) + \ldots$$ (2.6)

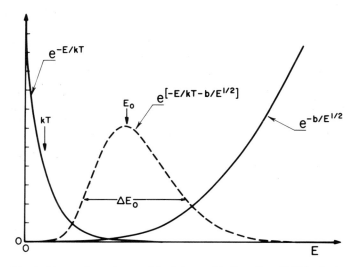

Fig. 1. Energy dependence of the integrand in Equation (2.5) for $\langle \sigma v \rangle$.

and substitute this into (2.5) to yield (Bahcall, 1966; Barnes, 1971),

$$\langle \sigma v \rangle = \left(\frac{2}{\mu(kT)^3} \right)^{1/2} (\Delta E_0) (f_0 S_{\text{eff}}) \exp(-3E_0/kT)$$

$$= 1.3006 \times 10^{-14} \left(\frac{Z_1 Z_2}{A} \right)^{1/3} f_0 S_{\text{eff}} T_9^{-2/3} \exp(-3E_0/kT), \tag{2.7}$$

where

$$E_0 = [\pi(Z_1 Z_2 e^2/hc)kT(\mu c^2/2)^{1/2}]^{2/3}$$

$$= 1.2204(Z_1^2 Z_2^2 A T_6^2)^{1/3} \text{ keV}$$

$$= 0.12204(Z_1^2 Z_2^2 A T_9^2)^{1/3} \text{ MeV}, \tag{2.7a}$$

$$\Delta E_0 = 4[E_0 kT/3]^{1/2} = 0.74889(Z_1^2 Z_2^2 A T_6^5)^{1/6} \text{ keV}$$

$$= 0.23682(Z_1^2 Z_2^2 A T_9^5)^{1/6} \text{ MeV}, \tag{2.7b}$$

$$k = 0.086171 \text{ keV}/10^6 \text{ K} = 0.086171 \text{ MeV}/10^9 \text{ K},$$

$$S_{\text{eff}} = S(0)\left[1 + \frac{5kT}{36E_0}\right] + S'(0)E_0\left[1 + \frac{35kT}{36E_0}\right] + \frac{1}{2} S''(0)E_0^2\left[1 + \frac{89kT}{36E_0}\right]$$

$$= S(0) [1 + 0.098068(Z_1^2 Z_2^2 A)^{-1/3} T_9^{1/3}] +$$

$$+ S'(0) [0.12204(Z_1^2 Z_2^2 A)^{1/3} T_9^{2/3} + 0.083778 T_9] + \tag{2.7c}$$

$$+ \tfrac{1}{2} S''(0) [0.014894(Z_1^2 Z_2^2 A)^{2/3} T_9^{4/3} + 0.025999(Z_1^2 Z_2^2 A)^{1/3} T_9^{5/3}].$$

(Fowler *et al.*, (1975) tabulate the quantity $N_A \langle \sigma v \rangle / f_0$ for a wide variety of nuclear reactions.) E_0 corresponds to the maximum of the integrand, as shown in Figure 1, and ΔE_0 corresponds to the full width of the integrand at $1/e$ of its maximum value. (In the solar interior at a typical temperature of $T_6 = 16$, for the ^{3}He $(\alpha, \gamma)^7$ Be reaction in the proton–proton chain, $E_0 = 23.4$ keV and $\Delta E_0 = 13.1$ keV.)

On the basis of this formalism numerous laboratory measurements have been carried out to determine $S(0)$, $S'(0)$ and $S''(0)$ for the various nuclear reactions in the proton–proton chain and the CNO cycle. Due to the effects of the Coulomb barrier, at the present time it is not technically feasible to measure the cross-sections for most of these reactions in the region of E_0, and, therefore, the determination of S_{eff} must be based on extrapolation from measurements at higher energies, typically $E \gtrsim 100$ to 200 keV. In the next sections we discuss the present status of these experimental measurements and extrapolations for each of the various nuclear reactions in the CN cycle and the proton–proton chain.

3. Carbon–Nitrogen Cycle

Following the original work of Bethe (1939) outlining the role of ^{12}C as a nuclear catalyst in the conversion of hydrogen to helium via the carbon–nitrogen cycle, more detailed and sensitive measurements have been made of the individual nuclear reactions in the cycle. The whole process is now described as a tricycle (e.g. Caughlan and Fowler, 1962; Rolfs and Rodney, 1974a; Rolfs and Rodney, 1975) involving two additional side chains which remove material from the main CN cycle at one point only to return it at another; see Figure 2. These side chains do not significantly affect energy generation in the CN

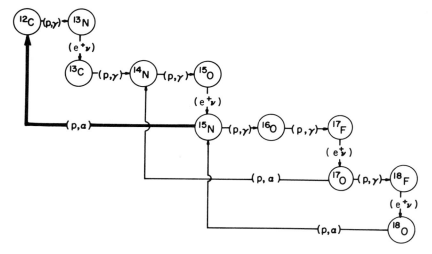

Fig. 2. The CNO tri-cycle.

cycle, but they can significantly affect the isotopic abundance of ^{17}O and ^{18}O in the interiors of stars operating on the CN cycle, which could then be important in the production of neutrons in later stages of the evolution of such stars. According to current, standard (traditional) solar models (e.g., Bahcall *et al.*, 1980) the CN cycle plays only a minor role (~1.2%) in the generation of energy in the Sun. It has also been argued (e.g., Bahcall *et al.*, 1968; Bahcall, 1979) that the results of the solar neutrino experiment place an observational limit (<10%) on the role of the CN cycle in the Sun; *however*, we should keep an open mind on this situation until we better understand the results of Davis's measurements (Davis, 1978) and the disagreement between those experimental results and the results of the standard solar models.

Our knowledge of the low-energy cross-sections for the nuclear reactions in the main CN cycle (>99.9% of the CNO tri-cycle) seems to be fairly well established experiment-ally. Measurements of the cross-sections for the reactions $^{12}C(p, \gamma)$ ^{13}N (Bailey and Stratton, 1950; Hall and Fowler, 1950; Lamb and Hester, 1957a; Hebbard and Vogl, 1960; Vogl, 1963; Fowler and Vogl, 1964; Rolfs and Azuma, 1974), $^{13}C(p, \gamma)$ ^{14}N (Woodbury and Fowler, 1952; Hebbard and Vogl, 1960; Hester and Lamb, 1961; Vogl, 1963; Fowler and Vogl, 1964), $^{14}N(p, \gamma)$ ^{15}O (Lamb *et al.*, 1957b; Hebbard and Bailey, 1963; Hensley, 1967), and $^{15}N(p, \alpha_0)$ ^{12}C (Schardt *et al.*, 1952; Zyskind and Parker, 1979) are generally in good agreement for each reaction. For all four of these reactions, the cross-sections at low energies are very strongly influenced by broad resonances located only a few hundred keV above threshold; in the energy range $0 \leq E_{cm} \leq 100$ keV the S-factors for these reactions can be expanded in a Taylor series (Equation (2.6)) with the parameters listed in Table I. The $^{14}N(p, \gamma)$ ^{15}O reaction is by far the slowest

TABLE I

Reaction	$S(0)$ (MeV-barn)	$S'(0)$ (barn)	$S''(0)$ (barn Mev^{-1})	Reference
$^{12}C(p, \gamma)$ ^{13}N	1.45×10^{-3}	2.45×10^{-3}	6.80×10^{-2}	Rolfs and Azuma (1974).
$^{13}C(p, \gamma)$ ^{14}N	5.50×10^{-3}	1.34×10^{-2}	9.87×10^{-2}	Fowler *et al.* (1967).
$^{14}N(p, \gamma)$ ^{15}O	3.32×10^{-3}	-5.91×10^{-3}	9.06×10^{-3}	Fowler *et al.* (1975).
$^{15}N(p, \alpha)$ ^{12}C	78.0	351.0	1.11×10^4	Zyskind and Parker (1979).

reaction in the CN cycle (typically more than 100X slower than the other reactions for $T_6 \sim 15-20$), and, therefore, its rate determines the role of the CN cycle in the Sun in competition with the p–p chain. Because of the importance of the rate of this reaction, the possible role of narrow states near threshold or below threshold has been studied carefully and shown to have a <1% effect on the low-energy cross-section for the $^{14}N(p, \gamma)$ ^{15}O reaction (e.g. Hensley, 1967).

More recently, remeasurements of the (p, α) and (p, γ) reactions on ^{15}N (Rolfs and Rodney, 1974b; Zyskind and Parker, 1979), ^{17}O (Rolfs and Roaney, 1975; Kieser *et al.*, 1979), and ^{18}O (Lorenz-Wirzba *et al.*, 1979; Wiescher *et al.*, 1980) have better defined the roles of the side branches of the CNO tri-cycle. These results indicate that $\leq 10^{-3}$ of the ^{15}N in the main CN cycle is diverted to form ^{16}O; at ^{17}O some of this material is returned to the main CN cycle, but, depending on the temperature, a substantial fraction of this will continue to form ^{18}F; at ^{18}O all but about 6×10^{-3} of this is returned to the main CN cycle. The net effect is thus only a very small (~10^{-6}) leak of material out of the CN cycle (most of which is most probably returned via the ^{19}F(p, α) ^{16}O reaction) which does not have any substantial impact on the energy generation or the neutrino spectrum [$E_\nu(^{17}$F$) \leq 1.74$ MeV; $E_\nu(^{18}$F$) \leq 0.63$ MeV] of the CN cycle, but can affect the ^{17}O and ^{18}O abundances in evolving stars.

4. Proton–Proton Chain

The proton–proton chain (Figure 3) is the dominant series of thermonuclear reactions

$4\,^1\text{H} \longrightarrow\,^4\text{He}+2e^+ +2\nu$ $Q = +26.731$ MeV

$^1\text{H}+\,^1\text{H}+e^- \rightarrow\,^2\text{D}+\nu$ 1.442 MeV $E_\nu = 1.442$ MeV

$^1\text{H}+\,^1\text{H} \longrightarrow\,^2\text{D}+e^+ +\nu$ 1.442 MeV $E_{\nu max} = 0.420$ MeV

    $^2\text{D}+\,^1\text{H} \longrightarrow\,^3\text{He}+\gamma$ 5.493 MeV

   a) $^3\text{He}+\,^3\text{He} \rightarrow\,^4\text{He}+2\,^1\text{H}$ 12.859 MeV

   b) $^3\text{He}+\,^4\text{He} \rightarrow\,^7\text{Be}+\gamma$ 1.587 MeV

    b1) $^7\text{Be}+e^- \rightarrow\,^7\text{Li}+\nu$ 0.862 MeV $E_\nu = \begin{cases} 0.862 \text{ MeV } 89.7\% \\ 0.384 \text{ MeV } 10.3\% \end{cases}$

       $^7\text{Li}+\,^1\text{H} \rightarrow 2\,^4\text{He}$ 17.347 MeV

    b2) $^7\text{Be}+\,^1\text{H} \rightarrow\,^8\text{B}+\gamma$ 0.135 MeV

       $^8\text{B} \rightarrow\,^8\text{Be}^* +e^+ +\nu$ 15.079 MeV $E_{\nu max} \approx 14.02$ MeV

       $^8\text{Be}^* \rightarrow 2\,^4\text{He}$ 2.995 MeV

Fig. 3. The proton–proton chain, showing the three most important terminations – (a), (b1), and (b2).

in the solar interior, accounting for $\approx 99\%$ of the energy produced in the Sun, according to standard solar models (e.g. Bahcall *et al.*, 1980). The $^1\text{H} +\,^1\text{H} \rightarrow\,^2\text{D} + e^+ + \nu$ reaction is by far the slowest one in this sequence, and, therefore, its cross-section determines the overall rate of the p–p chain and hence the temperature of the solar interior. Two important branch points occur in the p–p chain, one involving the burning of ^{3}He through either the ^{3}He(^{3}He, 2p) ^{4}He or ^{3}He(α, γ) ^{7}Be reactions and the other involving subsequent burning of this ^{7}Be through either the ^{7}Be(e^-, ν) ^{7}Li or ^{7}Be(p, γ) ^{8}B reactions. Although the ^{7}Be(e^-, ν) and ^{7}Be(p, γ) terminations play only relatively minor roles in the generation of energy in the Sun (12.4% for the ^{7}Be(e^-, ν) termination and only 0.02% for the ^{7}Be(p, γ) termination, compared to 86.4% for the ^{3}He(^{3}He, 2p) termination), these two side chains – especially the ^{7}Be(p, γ) ^{8}B branch - play extremely important roles in the ^{37}Cl solar neutrino experiment. The present status of our knowledge of the cross-sections and decay rates for the important reactions in the p–p chain are discussed below. (In contrast to the reactions in the CN cycle, the reactions in the p–p chain are nonresonant at very low energies and are not significantly affected by the tails of low-lying, broad resonances.)

^{1}H(p, $e^+\nu$) ^{2}D

Because of its very small cross-section, this reaction cannot be measured experimentally in the laboratory, and instead its rate must be calculated on the basis of our knowledge of the weak interaction. Bahcall and May (1969) have derived a formula for $S(E)$ for this reaction (their Equ. (12)) which is proportional to $\Lambda^2(E)$ (the square of the overlap matrix element between the deuteron and the p–p initial state) and is also roughly proportional to the inverse of the free neutron lifetime. There have been four published direct measurements of the free neutron lifetime with individual uncertainties of $\leqslant \pm 3\%$ (Sosnovskii et al., 1959; Christensen et al., 1972; Bondarenko et al., 1978; Byrne et al., 1980), but there is unfortunately a substantial scatter in these results, $\sim 15\%$. In addition, four indirect determinations of the free neutron lifetime have been made by measuring the e–ν angular correlation coefficient (Stratowa et al., 1978) and by measuring the beta-decay asymmetry parameter using polarized neutrons (Krohn et al., 1975; Erozolimskii et al., 1979; and Bopp et al., 1984). These two indirect methods determine the ratio $|g_A/g_V|$ which is related to a comparison of the neutron lifetime to the lifetime for superallowed $0^+ \rightarrow 0^+$ decays. See, for example, Segre (1977; Chap. 9.9) who shows that

$$\frac{2g_V{}^2}{g_V^2 + 3g_A^2} = \frac{(ft)_n}{(ft)_{\text{superallowed } 0^+ \rightarrow 0^+}}$$

These eight independent determinations of the neutron half-life are listed in Table II

TABLE II

Authors	Measurement	Neutron half-life (s)
Sosnovskii et al. (1959)	D	702 ± 18
Christensen et al. (1972)	D	637 ± 10
Krohn and Ringo (1975)	A	629 ± 12.5
Bondarenko et al. (1978)	D	608 ± 11
Stratowa et al. (1978)	C	625 ± 14
Erozolimskii et al. (1979)	A	627 ± 10
Byrne et al. (1980)	D	649 ± 12
Bopp et al. (1984)	A	616 ± 7.5

D = Direct measurement of half-life
C = Measurement of e–ν angular correlation
A = Measurement of beta-decay asymmetry

and are plotted in Figure 4. The weighted mean of these results is $t_{1/2}(n) = 629 \pm 7.5$ s (this is not inconsistent with the average of the direct measurements which gives 649 ± 16 s). Using the value of 629 s for the neutron half-life, together with the most recent calculations (Gari, 1978) of Λ^2 [$\Lambda^2(0) = 6.91$] and of the mesonic-exchange correction term [$(1 + 0.02)^2$] (not included in Bahcall et al., 1969), $S_{11}(E)$ can be evaluated as

$$S_{11}(0) = 4.03 \times 10^{-25} \text{ MeV-barn,}$$

$$S'_{11}(0) = + 4.52 \times 10^{-24} \text{ barn,}$$

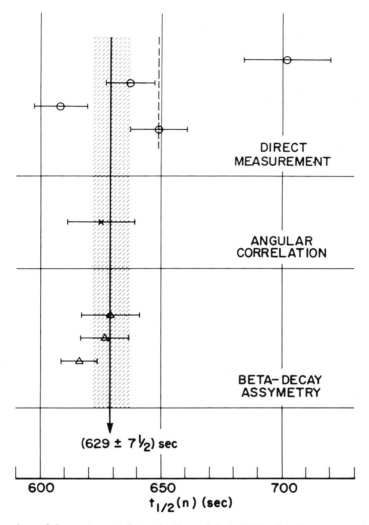

Fig. 4. Comparison of the most recent determinations of the half-life of the free neutron (see Table II).

with an estimated uncertainty in $S_{11}(0)$ of $\pm2\%$, due to the uncertainty in the neutron lifetime ($\pm1.2\%$), the uncertainty in the meson-exchange correction ($\pm0.7\%$), and the uncertainty in Λ^2 ($\pm0.8\%$).

A variation of this reaction is the $p + e^- + p \rightarrow d + \nu$ reaction. Its rate is even slower (typically $\sim 500\times$ slower) than the p–p reaction (Bahcall and May, 1969; Fowler et al., 1975),

$$R_{pep} \simeq 5.51 \times 10^{-5} \, \rho(1 + X_H) \, T_6^{1/2} (1 + 0.02T_6)(1 + 0.0088T_6^{2/3})^{-1} R_{pp}, \quad (4.1)$$

so that it does not play a significant role in solar hydrogen burning. It does, however, play an important role in the ^{37}Cl solar neutrino experiment because the monoenergetic neutrino ($E\nu$ = 1.44 MeV) is above the ^{37}Cl(ν, e$^-$) ^{37}Ar threshold and provides a lower limit of 0.25 SNU (Bahcall and May, 1969) for the theoretical predictions for this experiment, regardless of uncertainties in the opacities or in the relative roles of the various terminations to the p–p chain, assuming only that the Sun is currently generating sufficient thermonuclear energy via the p–p chain in its interior to balance is external luminosity.

^{2}D(p, γ) ^{3}He

The cross-section for this reaction has been measured directly (Griffiths et al., 1963) down to E_{cm} = 15 keV. Its extrapolation to lower energies is straightforward and determines

$$S_{12}(0) = (2.5 \pm 0.4) \times 10^{-4} \text{ keV-barn,}$$
$$S'_{12}(0) = + 7.9 \times 10^{-6} \text{ barn.}$$

Because this is the only important deuterium-burning reaction in the proton–proton chain (e.g., Parker et al., 1964) and because its rate is so much faster than the rate for the ^{1}H(p, e$^+$ ν) ^{2}D reaction, it does not contribute to the determination of the overall rate of the proton–proton chain and does not influence the solar neutrino spectrum.

^{3}He(^{3}He, 2p) ^{4}He

The cross-section for this reaction has been measured in four separate experiments (Neng-Ming et al., 1966; Bacher and Tombrello, 1965; Tombrello, 1967; Dwarakanath and Winkler, 1971; Dwarakanath, 1974) over the energy range 30 keV $\leq$ 10 MeV. The results of all four of those experiments are in agreement over the energy ranges where they overlap. Their results determine

$$S_{33}(0) = 5.5 \pm 0.5 \text{ Mev-barn,}$$
$$S'_{33}(0) = -3.1 \text{ barn,}$$
$$S''_{33}(0) = +2.8 \text{ barn MeV}^{-1}.$$

Only negative results have been obtained in attempts (e.g. Parker et al., 1973; Halbert et al., 1973; Dwarakanath, 1974) to discover low-energy resonances (E_{cm} < 30 keV, including subthreshold resonances) (see also Fetisov and Kopysov, 1975) which might enhance the role of this reaction in the proton–proton chain.

^{3}He(α, γ) ^{7}Be

The ^{37}Cl solar neutrino detector is particularly sensitive to the rates of the ^{3}He(α, γ) ^{7}Be reaction and the ^{7}Be(p, γ) ^{8}Be reaction because these two reactions control the production of high-energy neutrinos in the Sun. Prior to 1979 there had been two independent measurements of $S(E)$ at low energies for the ^{3}He(α, γ) ^{7}Be reaction (Parker and Kavanagh, 1963; Nagatani et al., 1969) which determined a mean value of $S_{34}(0)$ = 0.52 $\pm$ 0.05 keV-barn (see discussion in Bahcall et al., 1982). As the result of a preliminary report of an experimental determination of a substantially smaller value of

$S_{34}(0) = 0.30 \pm 0.03$ keV-barn (Rolfs, 1979; Kräwinkel et al., 1982), five additional independent determinations of $S_{34}(0)$ have recently been made. These include both the direct measurement of the gamma-rays produced in the actual capture reaction (Osborne et al., 1982, 1984; Alexander, et al., 1984) and measurement of the resulting ^{7}Be radioactivity (Osborne et al., 1982, 1984; Volk et al., 1983; and Robertson et al., 1983). The radioactivity technique has an advantage in determining the *overall* rate for this reaction since it completely integrates over corrections for angular distribution effects, efficiency variations, and the gamma-decay branching ratio. All these results are summarized in Table III. There is good agreement between seven of these eight

TABLE III

Authors	Measurement	$S_{34}(0)$ (keV-barn)
Parker and Kavanagh (1963)	D	0.47 ± 0.05
Nagatani et al. (1969)	D	0.57 ± 0.06
Kräwinkel et al. (1982)	D	0.31 ± 0.03
Osborne et al. (1982, 1984)	D	0.52 ± 0.03
Osborne et al. (1982, 1984)	R	0.53 ± 0.04
Robertson et al. (1983)	R	0.63 ± 0.04
Volk et al. (1983)	R	0.56 ± 0.03
Alexander et al. (1984)	D	0.47 ± 0.04

D = Measured direct-capture gamma-rays
R = Measured ^{7}Be radioactivity

measurements.[1] The individual values for $S_{34}(0)$ are determined by using direct-capture calculations (e.g. Tombrello and Parker, 1963; Kim et al., 1981; Liu et al., 1981; Williams and Koonin, 1981; Walliser et al., 1983, 1984; Kajino and Arima, 1984; etc.) to extrapolate from the measured values of $S_{34}(E)$ to $E = 0$. A weighted average of the seven consistent results determines

$$S_{34}(0) = 0.54 \pm 0.02 \text{ keV-barn}$$

with an energy derivative of

$$S'_{34}(0) = -3.1 \times 10^{-4} \text{ barn}$$

which is determined primarily from the energy dependence measured by Kräwinkel et al. (1982) and by Osborne et al. (1982, 1984) and from the energy dependence predicted by the direct-capture calculations.

---

[1] As this chapter goes to press there are also preliminary reports (Fowler, private communication) that a correction may have been found to the density distribution in the gas-jet target used in the Munster–Stuttgart data (Kräwinkel et al., 1982) which will increase their result so that it will be in agreement with the other seven measurements.

$^7Be(e^-, \nu)\ ^7Li$

The decay of ^7Be via electron capture has been well studied in the laboratory. This decay goes to both the ground state and the first excited state of ^7Li (the distinction between these two decays is important to the solar neutrino experiment because of the resulting differences in the neutrino spectrum); the percentage of decays to the first-excited of ^7Li is $(10.32 \pm 0.16)\%$. The measured laboratory half-life for the decay of ^7Be is (53.29 ± 0.07) days. However, in the solar interior the half-life of ^7Be will be substantially longer due to the ionization of the ^7Be; the decay rate for capture of electrons from the continuum will then be proportional to the electron density n_e

$$n_e = \frac{\rho}{m_p} \frac{(1 + X_H)}{2} ,$$

and over the temperature range $10 \leq T_6 \leq 16$ this decay rate can be written (Bahcall and Moeller, 1969) as

$$\lambda_c = \frac{1}{\tau_c} = 4.62 \times 10^{-9} \left(\rho \frac{1 + X_H}{2}\right) T_6^{-1/2} [1 + 0.004 (T_6 - 16)]\ s^{-1}. \qquad (4.2)$$

There are also contributions due to electron capture from bound states, so that

$$\lambda = \lambda_c + P_K \lambda_K + P_L \lambda_L + \ldots,$$

where P_K is the fractional occupation of the K shell, etc. The relative sizes of these additional contributions are functions of temperature and density; by averaging over the appropriate regions of the solar interior Iben et al. (1967) and Bahcall and Moeller (1969) have shown that in almost all solar model calculations it is sufficient to take

$$\lambda \approx 1.2\lambda_c = 5.5 \times 10^{-9} \left(\rho \frac{1 + X_H}{2}\right) T_6^{-1/2} [1 + 0.004(T_6 - 16)]\ s^{-1}. \qquad (4.3)$$

$^7Be(p, \gamma)\ ^8B$

Because of its enormous sensitivity to the high-energy neutrinos from the positron decay of ^8B, the interpretation of the ^{37}Cl solar neutrino experiment is particularly dependent on an accurate measurement of the $^7Be(p, \gamma)\ ^8B$ cross-section at low energy. Six independent experimental studies of this reaction have been reported (Kavanagh, 1960; Parker, 1966, 1968; Kavanagh et al., 1969, 1972; Vaughn et al., 1970; Wiezorek et al., 1977; Filippone et al., 1983). All these measurements (except Wiezorek et al.) determine the amount of ^7Be in their targets by measuring the rate at which ^7Li is being produced in the target from the decay of the ^7Be. The ^7Li density is measured using the 770 keV resonance in the $^7Li(d, p)\ ^8Li$ reaction; the absolute cross-section at the peak of this resonance has been determined from a large number of independent measurements to be 157 ± 10 mb (Filippone et al., 1983). An independent check on this number is provided in a comparison performed by Filippone et al. (1983) to determine the ^7Be areal density in their target using both the ^{7}Li(d, p) ^{8}Li resonance and a mapping of the ^{7}Be activity profile using a tightly collimated gamma-ray detector; the two methods agreed to within 7%, well within their uncertainties.

The results of the six determinations of $S_{17}(0)$ are listed in Table IV; a direct-capture

TABLE IV

Authors	$S_{17}(0)$ (keV-barn)
Kavanagh (1960)	0.016 ± 0.006
Parker (1966, 1968)	0.028 ± 0.003
Kavanagh et al. (1969); Kavanagh (1972)	0.0273 ± 0.0024
Vaughn et al. (1970)	0.0214 ± 0.0022
Wiezorek et al. (1977)	0.045 ± 0.011
Filippone et al. (1983)	0.0221 ± 0.0028

model calculation (e.g. Tombrello, 1965; Williams and Koonin, 1981) has been used to extrapolate the measured $S(E)$ to $E = 0$. The weighted mean of these six results gives

$$S_{17}(0) = 0.0243 \pm 0.0018 \text{ keV-barn}$$

with an energy derivative of

$$S'_{17}(0) = -3 \times 10^{-5} \text{ barn}$$

determined from the direct-capture model.

5. Discussion and Conclusions

On the basis of the analysis in Sections 3 and 4 the status of each of the various components of the CN cycle and the p–p chain and their roles in the solar interior would seem to be well determined. *However, this cannot be considered a closed, solved problem until we understand the results of solar neutrino experiments which provide the most direct and crucial tests of our knowledge of the solar interior as interpreted from solar models.* The results of the ^{37}Cl solar neutrino detector (e.g. Davis, 1978; Bahcall, 1979) are particularly sensitive to the high-energy neutrinos from the ^{7}Be and ^{8}B terminations of the p–p chain and thus depend most importantly on S_{34} and S_{17} as well as on S_{11} which determines the temperature in the solar interior. This dependence on the various nuclear reaction rates can be parametrized as follows (e.g. Bahcall et al., 1969):

$$\left[\sum_{\text{all}} (\sigma \Phi) \right]_{^{37}\text{Cl}} = 1.39 \times 10^{-36} \left(\frac{S_{11}}{S_{11}^*} \right)^{-2.5} \left(\frac{S_{33}}{S_{33}^*} \right)^{-0.37} \left(\frac{S_{34}}{S_{34}^*} \right)^{+0.8} \times$$

$$\times \left[1 + 3.47 \left(\frac{S_{17}}{S_{17}^*} \right)^{+1.0} \left(\frac{\lambda_{e7}}{\lambda_{e7}^*} \right)^{-1.0} \right] \text{ s}^{-1} \text{ per } ^{37}\text{Cl atom,}$$

where the S_{ij}^* are the current 'best' values for these parameters, as given in the previous section of this chapter. Because of recent changes in the values for S_{11}, S_{34}, and S_{17} this value, 6.2 SNU, is nearly 20% smaller than the value calculated by Bahcall *et al.* (1982). However, there is still more than a 3 σ discrepancy between the predicted and observed results from the ^{37}Cl neutrino detector. A systematic program to unravel nuclear physics aspects of the current solar neutrino problem must include not only careful remeasurements of $S_{34}(E)$ and $S_{17}(E)$ such as those currently underway for the ^{3}He(α, γ) ^{7}Be reaction, but also a careful remeasurement of the neutron half-life. (A number of alternative ways to destroy ^{7}Be and ^{8}Be in the solar interior have been examined (e.g. Parker, 1972), but they all are at least 10^{10} times less important than ^{7}Be + p and the beta decay of ^{8}B.)

A number of other questions have also been raised concerning the solar neutrino predictions — questions ranging from the determination of the relevant optical opacities, to the distribution of heavy elements and angular momentum in the solar interior, to the possibility of neutrino oscillations, etc. At present, no real solution has been found. The next logical step would seem to be the design and construction of alternative detectors which (with different sensitivities) could answer some of these questions.

The ^{71}Ga detector (using the ^{71}Ga(ν, e$^-$) ^{71}Ge reaction $Q = -233$ keV) (Dostrovsky, 1978; Bahcall *et al.*, 1978) is sensitive to the p–p neutrinos; 63% of ^{71}Ge would be produced by the p–p neutrinos compared to $< 2\%$ produced by the ^{8}B neutrinos. (In comparison, for the ^{37}Cl detector nearly 80% of the ^{37}Ar is produced by the ^{8}B neutrinos, none by the p–p neutrinos, and only 3% by the related pep neutrinos.) Since the flux of p–p neutrinos is determined only by the average solar luminosity and is quite insensitive to the detailed conditions in the solar interior, the ^{71}Ga detector would provide (independent of the details of a solar model) tests of some fundamental questions regarding neutrino oscillations (sensitive to $\Delta m^2 \sim 10^{-11}$ eV^2) or even whether or not the Sun is currently generating energy via nuclear fusion.

At the present time a pilot project, using 1.8 ton of gallium, has been tested and shown to have an efficiency of $\geqslant 95\%$ for the recovery of ^{71}Ge. The full-scale ^{71}Ga detector will require between 15 and 50 ton of gallium, and at the moment its fate is uncertain because of the high cost of gallium.

Studies are also currently underway to examine the practical feasibility of three other solar neutrino detectors involving the separation and counting of the radionuclides produced by the (ν, e$^-$) reaction on ^{7}Li, ^{81}Br, and ^{98}Mo. In all three of these cases there are difficulties associated with the detection and measurement of the resulting radionuclides. For the ^{7}Li detector the problem is related to the low energy of the Auger electron emitted following the ^{7}Be decay (50 eV for ^{7}Be compared to 2.82 keV for ^{37}Ar); for the ^{81}Br and ^{98}Mo detectors the problem is associated with the long lifetimes of the resulting ^{81}Kr and ^{98}Te (2×10^5 yr and 4.2×10^6 yr, respectively). In all three cases it has been suggested that the way around this problem is to make use of Resonance Ionization Spectroscopy as a single atom counting technique (Hurst *et al.*, 1975; 1979; 1983).

None of these three detectors is sensitive to the p–p neutrinos, but they would have different sensitivities to the higher energy parts of the solar neutrino spectrum. While ^{7}Li and ^{81}Br could be utilized in a contemporary detector similar to the ^{37}Cl and ^{71}Ga detectors, it has also been suggested that suitably shielded ores containing ^{41}Ca (Haxton and Cowan, 1980), ^{81}Br (Scott, 1976, 1978; Bennett et al., 1980), ^{98}Mo (Cowan and Haxton, 1982) or ^{205}Tl (Freedman, 1978; see also Rowley et al., 1980) could be used (because of the long half-lives of the resulting radionuclides) to study the time-integrated solar neutrino flux over periods $\sim 10^6$ yr. A meaningful interpretation of any such measurements will clearly require not only an accurate understanding of the nuclear physics involved — in the particular the (ν, e^-) capture reaction — but also a reliable knowledge of the geologic history of the ore sample and its shielding from cosmic rays.

The next generation of solar neutrino detectors beyond the radiochemical ^{37}Cl and ^{71}Ga experiments should involve real-time detectors such as the proposed ^{115}In devices (Raghavan, 1976, 1978; Pfeiffer et al., 1978) which will measure directly the energy spectrum of the interacting neutrinos as well as providing prompt ($\sim$ few microseconds) information of when the neutrino interaction takes place.

Other aspects of the solar neutrino flux that will be of interest to measure include the direction of the neutrinos (to determine their 'solar' (?) origin) and the possible long-term secular variations in this flux. In attempts to relate any observed neutrino flux to the conditions in the solar core, a determination of the direction of origin of this flux is clearly important in order to know what fraction of this flux may be from the Sun and what fraction may be of more general galactic or cosmic origin. One possible way to determine this directionality involves the measurement of the expected 7% annual variation in the *solar* component of the neutrino flux due to the eccentricity of the Earth's orbit; a more direct approach would be to involve a directional detector utilizing, for example, $\nu_e + e^- \rightarrow \nu_e + e^-$ elastic scattering or $\nu + d \rightarrow p + p + e^-$ (e.g. Chen, 1978; Lande, 1978).

Once the thermonuclear reactions in the solar interior and the related solar neutrinos become a solved problem, the solar neutrinos will then become a 'well-known' source of neutrinos which could be used to measure neutrino properties and to carry out neutrino experiments over distance scales unavailable with terrestrial sources.

References

Alexander, T. K., Ball, G. C., Lennard, W. N., Geissel, H., and Mak, H.-B.: 1984, *Nucl. Phys.* A427, 526.

Atkinson, R. d'E. and Houtermans, F. G.: 1929, *Z. Phys.* 54, 656.

Bacher, A. D. and Tombrello, T. A.: 1965, *Rev. Mod. Phys.* 37, 433.

Bahcall, J. N.: 1966, *Astrophys. J.* 143, 259.

Bahcall, J. N.: 1979, *Space Sci. Rev.* 24, 227.

Bahcall, J. N., Bahcall, N. A., and Shaviv, G.: 1968, *Phys. Rev. Lett.* 20, 1209.

Bahcall, J. N.: Bahcall, N. A., and Ulrich, R. K.: 1969, *Astrophys. J.* 156, 559.

Bahcall, J. N., Cleveland, B. T., Davis, R., Jr, Dostrovsky, I., Evans, J. C., Jr, Frati, W., Friedlander, G., Lande, K., Rowley, J. K., Stoenner, R. W., and Weneser, J.: 1978, *Phys. Rev. Lett.* 40, 1351.

Bahcall, J. N. and Davis, R., Jr: 1981, in C. A. Barnes, D. D. Clayton, and D. N. Schramm (eds), *Essays in Nuclear Astrophysics*, Cambridge Univ. Press.

Bahcall, J. N., Huebner, W. F., Lubow, S. H., Parker, P. D., and Ulrich, R. K.: 1982, *Rev. Mod. Phys.* 54, 767.

Bahcall, J. N., Lubow, S. H., Huebner, W. F., Magee, N. H., Jr, Merts, A. L., Argo, M. F., Parker, P. D., Rozsnyai, B., and Ulrich, R. K.: 1980, *Phys. Rev. Lett.* 45, 945.

Bahcall, J. N. and May, R. M.: 1969, *Astrophys. J.* 155, 501.

Bahcall, J. N. and Moeller, C. P.: 1969, *Astrophys. J.* 155, 511.

Bailey, C. L. and Stratton, W. R.: 1950, *Phys. Rev.* 77, 194.

Barnes, C. A.: 1971, in M. Baranger and E. Vogt (eds) *Advances in Nuclear Physics*, Vol. 4, Plenum Press, New York, p. 133.

Bennett, C. L., Lowry, M. M., Naumann, R. A., Loeser, F., and Moore, W. H.: 1980, *Phys. Rev.* **C22**, 2245.

Bethe, H. A.: 1939, *Phys. Rev.* **55**, 434.

Bethe, H. A. and Critchfield, C. L.: 1938, *Phys. Rev.* **54**, 248.

Bondarenko, L. N., Kurguzov, V.V., Prokofev, Yu. A., Rogov, E. V., and Spivak, P. E.: 1978, *Pis'ma Zh, Eksp. Teor. Fiz.* **28**, 328 [1978, *JETP Lett.* **28**, 303].

Bopp, P., Dubbers, D., Klemt, E., Last, J., Schultze, H., Weibler, W., Friedman, S. J., and Scharpf, D.: 1984, *J. Physique* **C3**, 21.

Brown, L., Steiner, E., Arnold, L. G., and Seyler, R.: 1973, *Nucl. Phys.* **A206**, 353.

Burbidge, E. M., Burbidge, G. R., Fowler, W. A., and Hoyle, F.: 1957, *Rev. Mod. Phys.* **29**, 547.

Burchfield, J. D.: 1975, *Lord Kelvin and the Age of the Earth*, Science History Publications.

Byrne, J., Morse, J., Smith, K. F., Shaikh, F., Green, K., and Greene, F. L.: 1980, *Phys. Lett.* **92B**, 274.

Cameron, A. G. W.: 1958, *Bull. Am. Phys. Soc.* **3**, 227.

Caughlan, G. R. and Fowler, W. A.: 1962, *Astrophys. J.* **136**, 453.

Chen, H. H.: 1978, in G. Friedlander (ed.), *Solar Neutrino Conference II*, BNL Report 50879, Vol. II, p. 55.

Christensen, C. J., Nielsen, A., Bahnsen, A., Brown, W. K., and Rustad, B. M.: 1972, *Phys. Rev.* **D5**, 1628.

Clayton, D. D.: 1968, *Principles of Stellar Evolution and Nuclear Synthesis*, McGraw-Hill, New York.

Cowan, G. A. and Haxton, W. C.: 1982, *Science* **216** (4541), 51.

Davis, R., Jr: 1955, *Phys. Rev.* **97**, 766.

Davis, R., Jr: 1964, *Phys. Rev. Lett.* **12**, 303.

Davis, R., Jr: 1978, in G. Friedlander (ed.), *Solar Neutrino Conference II*, BNL Report 50879, Vol. I, p. 1.

Davis, R., Jr, Cleveland, B. T., and Rowley, J. K.: 1983, *Science Underground (Los Alamos, 1983)*, Amer. Inst. Physics, p. 2.

Davis, R., Jr, Harmer, D. S., and Hoffman, K. C.: 1968, *Phys. Rev. Lett.* **20**, 1205.

Dostrovsky, I.: 1978, in G. Friedlander (ed.), *Solar Neutrino Conference II*, BNL Report 50879, Vol. I, p. 231.

Dwarakanath, M. R.: 1974, *Phys. Rev.* **C9**, 805.

Dwarakanath, M. R. and Winkler, H.: 1971, *Phys. Rev.* **C4**, 1532.

Eddington, A. S.: 1926, *Internal Constitution of Stars*, Cambridge Univ. Press, Chap XI.

Elwyn, A. J., Holland, R. E., Davids, C. N., and Ray, W.: 1982, *Phys. Rev.* **C25**, 2168.

Erozolimskii, B. G., Frank, A. I., Mostovoi, Yu. A., Arzumanov, S. S., and Voitzik, L. R.: 1979, *Yadernaya Fiz.* **30**, 692 [*Soviet J. Nucl. Phys.* **30**, 356].

Fetisov, V. N. and Kopysov, Yu. S.: 1975, *Nucl. Phys.* **A239**, 511.

Filippone, B. W., Elwyn, A. J., Ray, W., and Koetke, D. D.: 1982, *Phys. Rev.* **C25**, 2164.

Filippone, B. W., Elwyn, A. J., Davids, C. N., and Koetke, D. D.: 1983, *Phys. Rev.* **C28**, 2222.

Fowler, W. A.: 1958, *Astrophys. J.* **127**, 551.

Fowler, W. A. and Vogl, J. L.: 1964, *Lectures in Theoretical Physics* VI, Colorado Assoc. Press, Boulder, p. 379.

Fowler, W. A., Caughlin, G. R., and Zimmerman, B. A.: 1967, *Ann. Rev. Astron. Astrophys.* **5**, 525.

Fowler, W. A., Caughlin, G. R., and Zimmerman, B. A.: 1975, *Ann. Rev. Astron. Astrophy.* **13**, 69.

Freedman, M. S.: 1978, in G. Friedlander (ed.), *Solar Neutrino Conference II*, BNL Report 50879, Vol. I, p. 313.

Gamow, G.: 1928, *Z. Phys.* **52**, 510.

Gari, M.: 1978, in G. Friedlander (ed.), *Solar Neutrino Conference II*, BNL Report 50879, Vol. I, p. 137.

Griffiths, G. M., Lal, M., and Scarfe, C. D.: 1963, *Canadian J. Phys.* **41**, 724.

Halbert, M. L., Hensley, D. C., and Bingham, H. G.: 1973, *Phys. Rev.* **C8**, 1226.

Hall, R. N. and Fowler, W. A.: 1950, *Phys. Rev.* **77**, 197.

Haxton, W. C. and Cowan, G. A.: 1980, *Science* **210**, 897.

Hebbard, D. F. and Bailey, G. M.: 1963, *Nucl. Phys.* **49**, 666.
Hebbard, D. F. and Vogl, J. L.: 1960, *Nucl. Phys.* **21**, 652.
Hensley, D. C.: 1967, *Astrophys. J.* **147**, 818.
Hester, R. E. and Lamb, W. A. S.: 1961, *Phys. Rev.* **121**, 584.
Hurst, G. S., Chen, C. H., Kramer, S. D., Payne, M. G., and Willis, R. D.: 1983, *Science Underground (Los Alamos, 1983)*, Amer. Inst. Physics, p. 96.
Hurst, G. S., Payne, M. G., Kramer, S. D., and Young, J. P.: 1979, *Rev. Mod. Phys.* **51**, 767.
Hurst, G. S., Payne, M. G., Nayfeh, M. H., Judish, J. P., and Wagner, E. B.: 1975, *Phys. Rev. Lett.* **35**, 82.
Iben, I., Kalata, K., and Schwartz, J.: 1967, *Astrophys. J.* **150**, 1001.
Kajino, T. and Arima, A.: 1984, *Phys. Rev. Lett.* **52**, 739.
Kavanagh, R. W.: 1960, *Nucl. Phys.* **15**, 411.
Kavanagh, R. W.: 1972, in F. Reines (ed.), *Cosmology, Fusion, and Other Matters*, Colorado Assoc. Univ. Press, Boulder, p. 169.
Kavanagh, R. W., Tombrello, T. A., Mosher, J. M., and Goosman, D. R.: 1969, *Bull. Amer. Phys. Soc.* **14**, 1209.
Kieser, W. E., Azuma, R. E., and Jackson, K. P.: 1979, *Nucl. Phys.* **A331**, 155.
Kim, B. T., Izumato, T., and Nagatani, K.: 1981, *Phys. Rev.* **C23**, 33.
Kräwinkel, H., Becker, H. W., Buchmann, L., Görres, J., Kettner, K. U., Kieser, W. E., Santo, R., Schmalbrock, P., Trautvetter, H. P., Vlieks, A., Rolks, C., Hammer, J. W., Azuma, R. E., and Rodney, W. S.: 1982, *Z. Phys.* **A304**, 307.
Krohn, V. E. and Ringo, G. R.: 1975, *Phys. Lett.* **55B**, 175.
Lamb, W. A. S. and Hester, R. E.: 1957a, *Phys. Rev.* **107**, 550.
Lamb, W. A. S. and Hester, R. E.: 1957b, *Phys. Rev.* **108**, 1304.
Lande, K.: 1978, in G. Friedlander (ed.), *Solar Neutrino Conference II*, BNL Report 50879, Vol. II, p. 79.
Lerner, G. M. and Marion, J. B.: 1969, *Nucl. Instr. Meth.* **69**, 115.
Liu, Q. K. K., Kanada, H., and Tang, Y. C.: 1981, *Phys. Rev.* **C23**, 645.
Lorenz-Wirzba, H., Schmalbrock, P., Trautvetter, H. P., Wiescher, M., and Rolfs, C.: 1979, *Nucl. Phys.* **A313**, 346.
McClenahan, C. R. and Segel, R. E.: 1975, *Phys. Rev.* **C11**, 370.
Mingay, D. W.: 1979, *S.-Afr. Tydskr. Fis.* **2**, 107.
Nagatani, K., Dwarakanath, M. R., and Ashery, D.: 1969, *Nucl. Phys.* **A128**, 325.
Neng-Ming, W., Novatskii, V. N., Osetinskii, G. M., Nai-Kung, C., and Chepurchenko, I. A.: 1966, *Yadern. Fiz.* **3**, 1064 [1966, *Soviet J. Nucl. Phys.* **3**, 777].
Osborne, J. L., Barnes, C. A., Kavanagh, R. W., Kremer, R. M., Mathews, G. J., Zyskind, J. L., Parker, P. D., and Howard, A. J.: 1982, *Phys. Rev. Lett.* **48**, 1664.
Osborne, J. L., Barnes, C. A., Kavanagh, R. W., Kremer, R. M., Mathews, G. J., Zyskind, J. L., Parker, P. D., and Howard, A. J.: 1984, *Nucl. Phys.* **A419**, 115.
Parker, P. D.: 1966, *Phys. Rev.* **150**, 851.
Parker, P. D.: 1968, *Astrophys. J.* **153**, L85.
Parker, P. D.: 1972, *Astrophys. J.* **175**, 261.
Parker, P. D., Bahcall, J. N., and Fowler, W. A.: 1964, *Astrophys. J.* **139**, 602.
Parker, P. D. and Kavanagh, R. W.: 1963, *Phys. Rev.* **131**, 2578.
Parker, P. D., Pisano, D. J., Cobern, M. E., and Marks, G. H.: 1973, *Nature Phys. Sci.* **241**, 106.
Pfeiffer, L.: Mills, A. P., Jr, and Chandross, E. A.: 1978, in G. Friedlander (ed.), *Solar Neutrino Conference II*, BNL Report 50879, Vol. II, p. 31.
Raghavan, R. S.: 1976, *Phys. Rev. Lett.* **37**, 259.
Raghavan, R. S.: 1978, in G. Friedlander (ed.), *Solar Neutrino Conference II*, BNL Report 50879, Vol. II, p. 1.
Robertson, R. G. H., Dyer, P., Bowles, T. J., Brown, R. E., Jarmie, N., Maggiore, C. J., and Austin, S. M.: 1983, *Phys. Rev.* **C27**, 11.
Rolfs, C.: 1979, *Bethe – 40th Anniversary Symposium* and Private communication.
Rolfs, C. and Azuma, R. E.: 1974, *Nucl. Phys.* **A227**, 291.
Rolfs, C. and Rodney, W. S.: 1974a, *Astrophys. J.* **194**, L63.

Rolfs, C. and Rodney, W. S.: 1974b, *Nucl. Phys.* **A235**, 450.

Rolfs, C. and Rodney, W. S.: 1975, *Nucl. Phys.* **A250**, 295.

Rolfs, C. and Trautvetter, H. P.: 1978, *Ann. Rev. Nucl. Part. Sci.* **28**, 115.

Rowley, J. K., Cleveland, B. T., Davis, R., Jr, Hampel, W., and Kirsten, T.: 1980, in R. O. Pepen, J. A. Eddy, and R. B. Merrill (eds), *Proc. Conf. Ancient Sun* () p. 45.

Salpeter, E. E.: 1954, *Austral. J. Phys.* 7, 373.

Schardt, A., Fowler, W. A., and Lauritsen, C. C.: 1952, *Phys. Rev.* **86**, 527.

Schilling, A. E., Mangelson, N. F., Nielson, K. K., Dixon, D. R., Hill, M. W., Jensen, G. L., and Rogers, V. C.: 1976, *Nucl. Phys.* **A263**, 389.

Scott, R. D.: 1976, *Nature* **264**, 729.

Scott, R. D.: 1978, in G. Friedlander (ed.), *Solar Neutrino Conference II*, BNL Report 50879, Vol. I, p. 293.

Segre, E.: 1977, *Nuclei and Particles*, Benjamin/Cummings.

Sosnovskii, A. N., Spivak, P. E., Prokofev, Yu. A., Kutikov, I. E., and Dobrynin, Yu. P.: 1959, *Zh. Eksp. Teor. Fiz.* **35**, 1059 [1959, *JETP* 8, 739]; 1959, *Nucl. Phys.* **10**, 395.

Stratawa, Chr., Dobrozemsky, R., and Weinzierl, P.: 1978, *Phys. Rev. D.* 18, 3970.

Tombrello, T. A.: 1965, *Nucl. Phys.* 71, 459.

Tombrello, T. A.: 1967, in J. B. Marion and D. M. Van Patter (eds), *Nuclear Research with Low Energy Accelerators*, Academic Press, p. 195.

Tombrello, T. A. and Parker, P. D.: 1963, *Phys. Rev.* **131**, 2582.

Vaughn, F. J., Chalmers, R. A., Kohler, D., and Chase, L. F., Jr: 1970, *Phys. Rev.* **C2**, 1657.

Vogl, J. L.: 1963, Ph.D. Thesis, Calif. Inst. of Tech., Pasadena, California.

Volk, H., Kräwinkel, H., Santo, R., and Walleck, L.: 1983, *Z. Phys.* **A310**, 91.

Walliser, H., Liu, Q. K. K., Kanada, H., and Tang, Y. C.: 1983, *Phys. Rev.* **28**, 57.

Walliser, H., Kanada, H., and Tang, Y. C.: 1984, *Nucl. Phys.* **A419**, 133; and private communication.

Warters, W. D., Fowler, W. A., and Lauritsen, C. C.: 1953, *Phys. Rev.* **91**, 917.

Weizsäcker, C. F. v.: 1937, *Physik. Z.* **38**, 176.

Weizsäcker, C. F. v.: 1938, *Physik. Z.* **39**, 663.

Wiescher, M., Becker, H. W., Görres, J., Kettner, K.-U., Trautvetter, H. P., Kieser, W. E., Rolfs, C., Azuma, R. E., Jackson, K. P., and Hammer, J. W.: 1980, *Nucl. Phys.* **A349**, 165.

Wiezorek, C., Krawinkel, H., Santo, R., and Wallek, L.: 1977, *Z. Phys.* **A282**, 121.

Williams, R. D. and Koonin, S. E.: 1981, *Phys. Rev.* **23**, 2773.

Woodbury, E. J. and Fowler, W. A.: 1952, *Phys. Rev.* **85**, 51.

Zyskind, J. L. and Parker, P. D.: 1979, *Nucl. Phys.* **A320**, 404.

Wright Nuclear Struchere Laborabory,
Yale University,
New Haven, CT 06520,
U.S.A.

CHAPTER 3

ATOMIC AND RADIATIVE PROCESSES IN THE SOLAR INTERIOR

W. F. HUEBNER

1. Introduction

The Sun is a large sphere of luminous plasma — that is, ionized matter. High densities in its interior are produced by the gravitational pull of its central mass on the surrounding solar matter. A high interior temperature and the consequent thermonuclear reactions keep the entire solar matter ionized. The high central density (about 1.5×10^2 g cm^{-3}) and temperature (about 1.5×10^7 K) define its central pressure (about 3×10^{17} dyne cm^{-2}), which prevents the Sun from collapsing.

The apparent 'surface' of the Sun is only an optical effect, that is, a layer — looking inward — at which the solar plasma becomes opaque to visible light. The temperature of that region, the photosphere (thickness ~100 km), is about 6000 K. Optically thin regions of the Sun — the solar atmosphere, consisting of the upper photosphere, the chromosphere, and the corona — extend further into space. Here we will be concerned only with the region that lies below the photosphere. The radius of the Sun out to the photosphere is 7×10^5 km. Energy flows through the 'surface' of the Sun into space at a rate of 3.9×10^{33} erg s^{-1} (= 3.9×10^{26} W or 6.3 kW cm^{-2}). It must be assumed that the energy flux is in steady state; i.e. total outflow through the 'surface' must equal production by nuclear reactions in the interior. Only then will the Sun's luminosity and the pressure, density, and temperature profiles (as functions of radius in the solar interior) remain constant.

The relationship between pressure, density, temperature, and internal energy of the plasma is determined by the equation of state (EOS). The relationship of these quantities to the radial distance in the Sun is determined primarily by the opacity (the resistance of the plasma to transport of radiative energy), even in the thermally unstable layers which are caused by the excess of the temperature gradient over its adiabatic value. Photon radiation is the main process of energy transport in the Sun, although conduction by electrons begins to compete in the central region, and convection (fluid dynamic transport) dominates near the surface, down to the region where lithium burns. EOS, opacity, and conduction are material properties that depend on the composition, density, and temperature of the medium. The calculation of the opacity is the subject of this article.

In general, energy transport through a medium depends on the phase state and the temperature of the material. In plasmas, gases, and fluids energy is transported by radiation, conduction, and fluid dynamics, e.g. convection. Radiation dominates at high

Peter A. Sturrock (ed.), Physics of the Sun, Vol. I, pp. 33–75.

temperatures. Conduction is important at high densities and low-to-medium temperatures. Figure 1 indicates qualitatively the various processes on a temperature–density $(T–\rho)$

Energy Transport Through A Medium

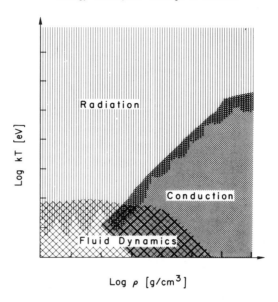

Fig. 1. Schematic presentation of the three energy transport processes in the density–temperature domain. There is considerable overlap where different processes compete. The density and temperature scales are in arbitrary units since the boundaries between regions and the overlap depend on the composition and other parameters of the plasma.

diagram for a general plasma. The boundary region between radiation and conduction is determined by approximate equality between radiative and conductive opacity; the region between radiation and fluid dynamics can be variously approximated depending on the type of fluid flow. For stellar interiors the criterion is the deviation of the temperature gradient from its adiabatic value as first described by Biermann (1938). The main emphasis here will be on opacity, a material property that determines radiation flow. Figure 2 shows where some of the man-made plasmas fall on the $T–\rho$ diagram. Figure 3 shows astrophysical plasmas on the same diagram.

At high density and high temperature electron collisions with plasma constituents are so frequent and of sufficient energy that the plasma is 'collision dominated'. This means that the electronic excitation and de-excitation (including ionization and recombination) are determined by electron impacts (radiation has a negligible effect on the electronic occupation); e.g. the continuum electrons obey a Maxwell (or Fermi–Dirac) velocity distribution and the electronic occupation of the bound states obeys the Boltzmann (or Fermi–Dirac) distribution. If, in addition, the radiation is nearly in equilibrium with the matter – i.e. the angle averaged radiative intensity J_ν is about equal to the Planck intensity distribution (black-body spectrum) $B_\nu(T)$, at the material temperature T – we have total local thermodynamic equilibrium (LTE). These are typically the conditions

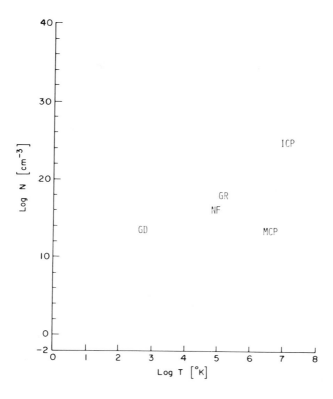

Fig. 2 Schematic presentation of density–temperature regions for man-made plasmas. GD, gas discharge; NF, nuclear fireball; GR, gas-core reactor; MCP, magnetically confined plasma; ICP, inertially confined plasma.

in stellar interiors, but the assumption of LTE is not necessarily valid in an inertially confined fusion (ICF) plasma. The time to establish equilibrium or a steady state may be too short in an ICF plasma. In that case, rate equations must be solved; however, that is beyond the scope of this presentation.

In general, matter at a temperature T emits, absorbs, and scatters radiation. The properties involved in the interaction of radiation with matter are governed by the bound–bound (spectral line absorption and emission), bound–free (photoelectric absorption), free–bound (radiative recombination), free–free (bremsstrahlung and inverse bremsstrahlung), and scattering processes. Figure 4 shows a typical extinction coefficient (the sum of absorption and scattering) as a function of photon energy $h\nu$ for a solar composition at $\rho = 1.88 \times 10^{-7}$ g cm^{-3} and $kT = 1.25$ eV (1 eV is equivalent to a temperature of 11 605 K). Compositions are usually defined by the mass fraction of hydrogen X, the mass fraction of helium Y, and the mass fraction of metals Z (which includes all elements heavier than helium). Opacity is sensitive to the relative abundance among the metals, therefore the abundance for each metal should also be specified in detail.

As Rosseland (1924) has shown, if a plasma is in LTE and the radiation spectrum is close to the black body spectrum at material temperature T, the radiation transport can

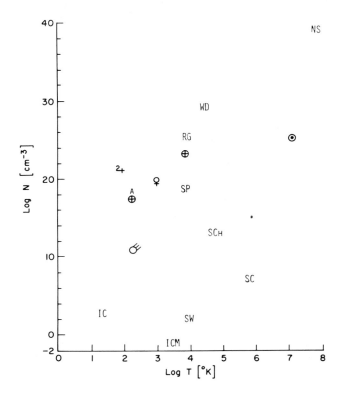

Fig. 3. Schematic presentation of density−temperature regions for astrophysical plasmas. IC, interstellar cloud; ♃, Jupiter atmosphere (near surface), ⊕^{A}, Earth atmosphere (near surface), ☄ comet
coma (~10^{14} cm$^{-3}/r^2$, where r is radius of coma in km, comet at 1 AU heliocentric distance); ♀,
Venus atmosphere (near surface); ICM, intercloud medium; RG, red-giant average density (can range
down to ~10^{19} cm^{-3}); ⊕, center of Earth; WD, white dwarf envelope; ⊙, center of Sun; SP, solar
photosphere; SCh, solar chromosphere (range between photosphere and corona); SC, solar corona;
SW, solar wind; NS, neutron star (average).

be approximated by the use of the diffusion approximation (i.e. radiative energy flux
proportional to the temperature gradient) with the following diffusion cofficient, which
is commonly referred to as the radiative (as opposed to electron conduction) Rosseland
mean opacity.

$$\frac{1}{\kappa_r R} = \int_0^\infty \frac{\partial B_\nu(T)}{\kappa_\nu' \partial T} \, d\nu \bigg/ \int_0^\infty \frac{\partial B_\nu(T)}{\partial T} \, d\nu = \int_0^\infty \frac{\partial B_\nu(T)}{\kappa_\nu' \partial T} \, d\nu \bigg/ \frac{dB(T)}{dT} \, . \tag{1}$$

Here κ_ν' is the extinction coefficient corrected for stimulated emission and ν is the
frequency. The weighting function $\partial B_\nu(T)/\partial T$ peaks at $u \cong 4$ and is very small for $u > 10$.
Here,

$$u \equiv h\nu/kT \tag{2}$$

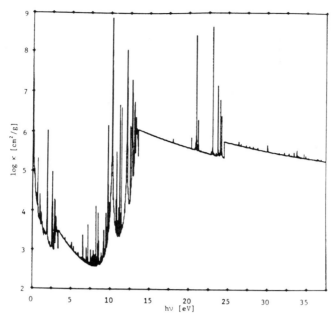

Fig. 4. Extinction coefficients for a mixture with $X = 0.75$, $Y = 0.2321$, and $Z = 0.0179$, at $kT = 1.25$ eV and $\rho = 1.88 \times 10^{-7}$ g cm^{-3}. The dominant line is the hydrogen Lyman α transition; near the center is the Lyman continuum from hydrogen; and at the right is the photoelectric edge of neutral helium. Many of the absorption lines are from the metals.

is the reduced photon energy. For a discussion of the diffusion approximation and the use of Rosseland opacity in conjunction with it see Cox (1965) or Mihalas (1978).

The condition that the angle-averaged intensity $J_\nu \cong B_\nu(T)$ can also be expressed by the requirement that the optical depth

$$\tau_\nu = \int_0^x \rho \kappa_\nu' \, dx \tag{3}$$

is large in the 'windows' between the absorption lines in the region where the Rosseland weighting function is large (typically $u < 10$); this implies that the plasma properties, particularly the temperature, must be nearly constant over the physical depth x. This test is all too often ignored; if this condition is not satisfied, the Rosseland opacity is applied incorrectly. The Rosseland opacity weights the reciprocal of the extinction coefficients (i.e. the extinction in the 'windows' between the absorption lines rather than the strong absorption lines themselves); it is a measure of the least resistance to radiation flow. It also means that the wings of the lines — sometimes the very far wings — and the weak lines must be determined most accurately. In the opposite extreme, where the plasma is optically thin at the peaks of the absorption lines, application of the Planck mean opacity

$$\kappa_P = \int_0^\infty \kappa_\nu B_\nu(T) \, d\nu / B(T) \tag{4}$$

is valid. Optically thin regions occur, for example, in the boundary layer (or emission layer) of a plasma. For this reason the Planck opacity is also sometimes referred to as the emission mean opacity, analogous to Kirchhoff's law for radiation, which relates emittance to absorptance. Unfortunately, there is always an intermediate region near the surface of a plasma where neither the Rosseland nor the Planck opacity is appropriate. There, only the frequency-dependent equation for radiative transfer should be applied. However, a combination of Rosseland and Planck group mean opacities is often used as an approximation.

Astrophysical plasmas contain many different chemical elements. Since the absorption spectrum is different for each element, the absorption lines of constituents can fall in the 'windows' between the absorption lines of other constituents. This increases the Rosseland opacity (because it is a weighted harmonic mean value) nonlinearly with the amount of minor constituents, particularly since they are of a higher atomic number than the main constituents (H and He) of the plasma. A few tenths of one per cent of these and even smaller fractions of heavier elements can double the Rosseland opacity in some temperature regions.

Direct experimental verification of opacities is essentially nonexisting. However, calculations are based on a firm foundation of quantum theory and statistical mechanics, and measured atomic properties (e.g. cross-sections and spectra) do show agreement with theoretical predictions. This should be kept in mind when reading the following sections.

In plasmas for which the refractive index $n_\nu \neq 1$, the radiative Rosseland mean opacity, Equation (1), has to be modified (Cox and Giuli, 1968):

$$\frac{1}{\kappa_r R} = \int_{\nu_p}^{\infty} \frac{n_\nu^2}{\kappa_\nu'} \frac{\partial B_\nu(T)}{\partial T} \, d\nu \bigg/ \frac{\partial B(T)}{\partial T} \tag{1a}$$

(i.e. the weighting is performed on n_ν^2/κ_ν'). Here κ_ν' must include collective effects (i.e. plasma effects). At temperatures and densities where the opacity is determined by continuum processes, the refractive index

$$n_\nu \cong [1 - (\nu_p/\nu)^2]^{1/2}$$

depends only on the plasma frequency

$$\nu_p = \left(\frac{N_e e^2}{\pi m}\right)^{1/2},$$

where N_e is the density of free electrons. In that case the effect on the opacity can be calculated easily; the opacity is

$$\frac{1}{\kappa_r R} = \int_{\nu_p}^{\infty} \frac{[1 - (\nu_p/\nu)^2]}{\kappa_\nu'} \frac{\partial B_\nu(T)}{\partial T} \, d\nu \bigg/ \frac{\partial B(T)}{\partial T} \, . \tag{1b}$$

2. Calculation of Atomic Structure

Now that some of the properties and uses of the Rosseland opacity have been outlined, we shall describe some of the details on calculating the opacity on the atomic scale. As the simplest illustration let us consider a pure hydrogen plasma. At some temperature and density (say $T \cong 6000$ K, $\rho \cong 1 \times 10^{-6}$ g cm^{-3}) hydrogen will be dissociated and most of the atoms will be in their ground state (the additional complexity of molecular states, although important under some conditions, is ignored here because molecules are of no importance at the high temperatures which are our goal for the solar interior). Also H$^-$ (a hydrogen atom with two bound electrons) is momentarily ignored. H$^-$ is an important contributor to astrophysical opacity, but it has significant abundance only if electron donor species are present. (Electron donor atoms, such as Na, Mg, Al, Si, and Fe, have ionization potentials much less than that of hydrogen. They ionize − and thus donate electrons to hydrogen at low temperatures, $T \lesssim 8000$ K − before hydrogen ionizes.) The absorption spectrum for the ground state is determined by the Lyman series $(n_1 = 1)$

$$h\nu_n = hcR(n_1^{-2} - n^{-2}), n > n_1,$$ (7a)

and the Lyman continuum $(n_1 = 1)$

$$h\nu \geq hcR/n_1^2.$$ (7b)

Here R is the Rydberg constant, c is the speed of light, and n is the principal quantum number. At somewhat higher temperature $(T \sim 9000$ K, $\rho \sim 1 \times 10^{-6}$ g cm$^{-3})$ many atoms are in excited states with the $n_1 = 2, 3, 4, \ldots$, levels occupied, and some atoms will be ionized (the degree of excitation and ionization also depends on density). Since the levels with $n_1 \geq 2$ lie close together in energy and are also close to the ionization threshold (but are well separated from the $n_1 = 1$ ground state), differently excited states and ions will coexist. The Balmer $(n_1 = 2)$, Paschen $(n_1 = 3)$, Brackett $(n_1 = 4)$, Pfund $(n_1 = 5)$, etc., series and continua will grow at the cost of the Lyman series. The continua and spectra of these series overlap. Figure 5 illustrates this schematically. At still higher temperatures $(T > 30\,000$ K, $\rho \cong 1 \times 10^{-6}$ g cm$^{-3})$ hydrogen is nearly fully ionized and the spectrum (scattering and free−free) becomes very simple.

Hydrogen and hydrogen-like ions have level degeneracy; i.e. states with different orbital angular momentum quantum number l but the same principal quantum number n have the same energy if relativity or electron spin are not considered. In all other elements the spectrum is much more complex. Not only is level degeneracy removed, but angular momentum coupling between electrons gives rise to many terms. Thus, a transition that appeared as a single line in hydrogen and hydrogen-like ions corresponds to a transition array of many lines in all other elements. This is further complicated by the existence of many multiply excited states in an ion and the coexistence of serveral ions.

The type of information that is needed for opacity calculations includes: energy level (term) values, occupation numbers (abundance of each multiply excited state), ion abundances, and radiative transition probabilities. These data must be known as functions of density and temperature. Except for transition probabilities, this is also the data needed to determine the equation of state. Pioneering work on high-temperature EOS includes papers by Feynman $et\ al.$ (1949), Cowan and Ashkin (1957), Zink (1968, 1970), and Rozsnyai (1972).

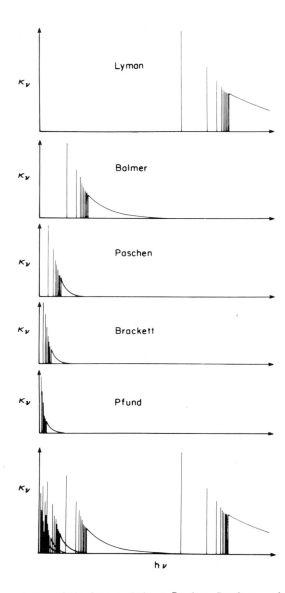

Fig. 5 Schematic presentation of the Lyman, Balmer, Paschen, Brackett, and Pfund spectra of hydrogen. Line profiles and line merging near the continuum edges are not shown. A composite of these spectra showing the overlap of these series is presented at the bottom.

Two approaches are usually considered for opacity calculations. One method, the explicit ion model (also known as the method of detailed configuration accounting) requires the calculation of number densities for all possible ionization stages and all singly and multiply excited states for each ion. This method is the more accurate one, applicable particularly at low temperatures, i.e., in calculations for elements with low or medium atomic number where the important species are relatively few. The right side of Figure 6 illustrates this calculational approach. For temperatures above several

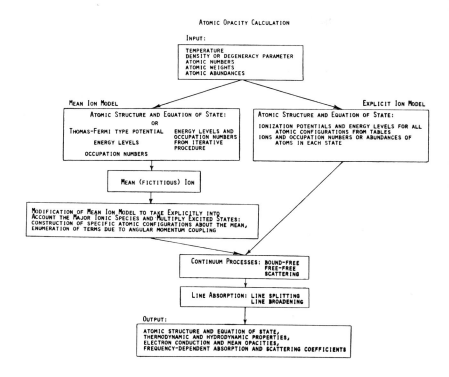

Fig. 6. Calculational procedure for opacity calculations. Left branch: mean ion model; right branch: explicit ion model. Only two types of mean ion models are indicated (See Table I for additional models).

tens of electron volts, data are too sparse and it becomes too cumbersome to apply this model. Here the second approach, the mean ion model, provides the simpler procedure for the calculation of the opacity. In this model an average (i.e. fictitious) atom with fractional electronic occupation numbers is calculated consistent with the temperature and density of the matter under consideration. These fractional occupation numbers immediately give an indication about the real and most probable species of ionization and excitation. The dominant ionic and multiply excited species and their relative abundances can then be approximated from the average atom by perturbation theory. This perturbation procedure in the *mean ion* model replaces the average configuration of the *average atom* model with detailed configurations similar to those of the explicit ion model. Opacities calculated from an average atom model (which omits the perturbation procedure) can be quite inaccurate. The left side of Figure 6 illustrates the calculational approach for the mean ion model.

Basic ideas for an explicit ion model were suggested by Mayer (1948). The method was independently developed and used in opacity calculations by Vitense (1951) and brought to its present state of refinement by Cox (1965). The virtues of this model are that, in principle, detailed information concerning the level structure of the atoms and ions can be extracted from tables based on experimental data and highly sophisticated calculations. The splitting of the level structure in low-Z elements is usually limited to Russell–Saunders (i.e. LS) term values because line broadening in most cases is comparable to, or larger than, the fine structure splitting. For the low-Z elements at low temperature ($kT < 100$ eV) and low density (less than normal solid density) many of these data are available; however, as the atomic number of elements increases, the volume of data required becomes very large while the amount of data available becomes more sparse so that the procedure is limited to elements with $Z \lesssim 30$. To keep calculations tractable, some reduction of level data has to be considered before this value of Z is reached so that not all relevant level information is included even, for example, for iron.

2.1. THE EXPLICIT ION MODEL

In the explicit ion method each ion is considered separately and the interaction with other ion species and free electrons is treated as a perturbation. The calculation starts by considering the ratio of population numbers in adjacent stages of ionization, I and $I + 1$, as given by the Saha equation

$$\frac{N_{I+1}}{N_I} = \frac{Z_{I+1}}{Z_I} \exp\left\{-\eta^* - \left[\chi_I - \frac{e^2 N_f}{2r_0^3}(3r_0^2 - \langle r^2 \rangle_{11}) - E_0\right]\bigg/kT\right\}. \tag{8}$$

Here N_f is the number of free electrons per atom, χ_I is the (positive) ionization energy of ionization stage I for the isolated ion, and r_0 is the ion sphere radius as determined from the volume that the ion occupies, given by

$$V_a = \frac{4\pi}{3} r_0^3 = \frac{M}{\rho N_0}, \tag{9}$$

where ρ is the mass density, M is the atomic weight, and N_0 is Avogadro's number. The partition function is

$$Z_I = \sum_{i=1} g_{Ii} \exp\left\{-\left[\hat{E}_{Ii} - \frac{e^2 N_f}{2r_0^3}(\langle r^2 \rangle_{ii} - \langle r^2 \rangle_{11})\right]\bigg/kT\right\}, \tag{10}$$

where $\hat{E}_{Ii}$ represents the (positive) energy of the excited state i (which may also be the ground state, $i = 1$) above the ground state energy of ionization stage I of the isolated ion:

$$\hat{E}_{Ii} - \frac{e^2 N_f}{2r_0^3}(\langle r^2 \rangle_{ii} - \langle r^2 \rangle_{11}) \le \chi_I - \frac{e^2 N_f}{2r_0^3}(3r_0^2 - \langle r^2 \rangle_{11}) - E_0, \tag{11}$$

and g_{Ii} is the corresponding statistical weight. The correction terms added to χ_I and $\hat{E}_{Ii}$ correspond to the interactions with free electrons. The quadrupole matrix element $\langle r^2 \rangle_{ii}$ is usually approximated from hydrogenic wave functions in the field of a point charge:

$$\langle r^2 \rangle_{ii} = (n_i a_0 / Z_i^*)^2 \, [5n_i^2 - 3l_i(l_i + 1) + 1]/2, \tag{12}$$

where a_0 is the first Bohr radius of hydrogen and Z_i^* is the screened nuclear charge for level i [see also Equation (30)]. The effective degeneracy parameter is

$$\eta^* \equiv \eta + E_0/kT, \tag{13}$$

where the degeneracy parameter η is the chemical potential divided by kT. It is defined such that the nondegenerate limit corresponds to $\eta \ll 0$. For computational reasons the constant energy

$$E_0 = \frac{3}{10} \frac{e^2 N_f}{r_0} \tag{14}$$

is usually introduced. Physically E_0 represents the ion volume-averaged interaction energy of a free electron with the ion core (nucleus and bound electrons localized at the nucleus) and the other (*uniformly* distributed) free electrons. It corresponds to a continuum depression which adjusts the total energy of a free electron to be zero when its kinetic energy is zero. The summation in Equation (10) is over all excited states of an ionization stage. A much used recipe describing cutoffs of the partition function is based on the work by Stewart and Pyatt (1966). In it the interaction of free with bound electrons is replaced by a more general plasma interaction. The model, which is valid for low densities, corresponds in the limit of low temperature to shielding by a uniform distribution of free electrons (similar to the one used above) and at high temperature to the Debye–Hückel screening. These cutoffs are not rigorously consistent, but they provide very simple algorithms that introduce only very small errors in the EOS and opacity if charge neutrality for each ion species is conserved (see, for example, Petschek and Cohen, 1972).

It is convenient to define two polynomials in the ratio N_{I+1}/N_I: P, the ratio of the total number of ions (including neutral atoms) to the number of neutral atoms,

$$P = 1 + \sum_{I=0}^{Z-1} \prod_{I'=0}^{I} \frac{N_{I'+1}}{N_{I'}}, \tag{15}$$

and Q, the number of free electrons per neutral atom,

$$Q = \sum_{I=1}^{Z} I \prod_{I'=1}^{I} \frac{N_{I'}}{N_{I'-1}}. \tag{16}$$

Starting from some initial guess for η^*, the free electron density N_e in terms of these polynomials must be iterated to consistency with the equation

$$N_e = \frac{\rho N_0}{M} N_f = \frac{4\pi(2mkT)^{3/2}}{h^3} I_{1/2}(\eta^*), \tag{17}$$

where $I_{1/2}(x)$ is the Fermi–Dirac integral of order $\frac{1}{2}$

$$I_k(\eta) \equiv \int_0^\infty y^k (e^{y-\eta} + 1)^{-1} \, dy, \tag{18}$$

and

$$N_f = \frac{Q}{P} . \tag{19}$$

Tables for $I_{1/2}(\eta^*)$ are given by McDougall and Stoner (1938); fits to these functions were made by Latter (1955).

Having calculated the ion abundance we can calculate the probability of occupancy of any state i in ion I from the partition function:

$$P_{Ii} \equiv \frac{\hat{N}_{Ii}}{N_I} = \frac{g_{Ii}}{Z_I} \exp\left\{ -\left[\hat{E}_{Ii} - \frac{e^2 N_f}{2r_0^3} (\langle r^2 \rangle_{ii} - \langle r^2 \rangle_{11}) \right] / kT \right\}. \tag{20}$$

Here N_I is the abundance of ion I per unit volume, and $\hat{N}_{Ii}$ is the occupation of state i in ion I.

Knowing the partition functions or the energy levels, statistical weights, and occupation numbers it is a straightforward calculation to obtain the thermodynamic properties of the plasma, but for dilute gases it is frequently easier to calculate the internal energy U per unit volume as

$$U = U_t + U_i + U_x + U_C. \tag{21}$$

Here U_t is the usual translational kinetic energy per unit volume,

$$U_t = (\tfrac{3}{2})NkT + U_f, \tag{22}$$

where N is the number of the atoms per unit volume. The free-electron kinetic energy per unit volume U_f can be written

$$U_f = N_e kT I_{3/2}(\eta^*)/I_{1/2}(\eta^*), \tag{23}$$

where the $I_k(x)$ are the Fermi–Dirac integrals of order k, Equation (18). The ionization energy per unit volume U_i is

$$U_i = - \sum_{z,I} N_{zI} E_{zI}, \tag{24}$$

where N_{zI} is the number density of an ion having atomic number Z and net ionic charge I. The energy, E_{zI}, necessary to remove I electrons is

$$E_{zI} = \sum_{j=0}^{I-1} \chi_{zj} - \frac{e^2 N_f}{2r_0^3} (\langle r^2 \rangle_{ii} - \langle r^2 \rangle_{11}) - E_0, \tag{25a}$$

where the χ_{zj} are the ionization energies for element Z to go from ground state of the isolated ion j to ion $j+1$.

$$E_{z0} = 0. \tag{25b}$$

The energy arising from the excitation of the internal states of the system is given by the term U_x. To calculate this quantity, detailed energy level data are needed along with occupation numbers. The excitation energy per unit volume is

$$U_x = \sum_z N_z \sum_{Im} x_{zI} \bar{E}_{zIm},$$ (26)

where N_z is the number density of species with atomic number Z,

$$\bar{E}_{zIm} = \frac{g_m \, \epsilon'_{zIm} \exp(-\epsilon'_{zIm}/kT)}{\sum_m g_m \exp(-\epsilon'_{zIm}/kT)},$$ (27)

and x_{zI} is the fraction of the particles of type z that are I times ionized. Here ϵ'_{zIm} is the excitation energy of state m above the ground state of ion I and species z, and g_m is the statistical weight of state m. The ϵ'_{zIm} have been corrected for interaction with the free electrons.

The last term in Equation (21), U_C, is the correction to the internal energy per unit volume caused by the averaged Coulomb interactions with the free electrons.

The pressure calculation can be done in a standard manner, but some care must be taken when calculating the corrections caused by interactions. For example, the total pressure P_g for an ionized plasma consisting of ions and uniform charge density of interaction electrons is given by

$$P_g = \left[-\frac{3}{10} \frac{e^2 \overline{N_f^2}}{r_0} + \frac{2}{3} N_f \, kT \, I_{3/2}(\eta^*)/I_{1/2}(\eta^*) + kT \right] \Big/ V_a.$$ (28)

The interaction of a given ion with its surroundings and the resulting perturbation on the energies are major uncertainties of the explicit ion model because not only are the individual levels disturbed and this disturbance is corrected only to first order, but the cutoff of the partition function sum at the depressed continuum leads to even larger uncertainties. It has already been remarked that not all level information can be included practically; e.g. detailed observational data are not available for such heavily ionized ions as Fe XIV. Accurate excited energy levels are necessary for the computation of the partition functions. Figure 7 illustrates the effects of neglecting excited state energy levels in such a calculation. The upper curve shows the fractional abundance calculated using only the ground state level of the potassium-like ion of iron and those which arise from the ground state level by the excitation of a single outer electron [i.e. configurations like $3s^2 3p^6 nl^1$ $(n > 3)$]. The lower curve is the result of adding configurations $3s^2 3p^5 3d^1 nl^1$, $3s^2 3p^4 3d^2 nl^1$, and others (all with $n > 3$) which have excitation energies less than the lowest ionization energy of Fe VIII. These configurations are not listed in Moore's tables of atomic energy levels and therefore are easily overlooked. The curves indicate that the fractional ionic abundance derived from consideration of the ion's ground state levels alone is insufficient. For low-Z elements $(Z \lesssim 30)$ the large amount of information that is available on the energy levels of isolated atoms and ions makes the method generally very practical and accurate.

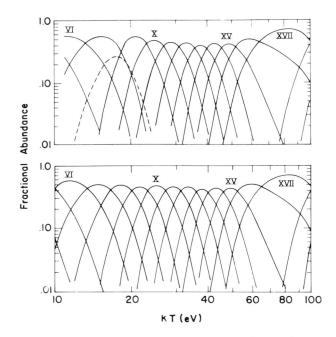

Fig. 7. Fractional abundance of ions of iron at density $\rho = 1 \times 10^{-3}$ g cm^{-3} vs plasma temperature in electron-volts. Upper part is based on extensive but limited data for configurations found in Moore's tables of atomic energy levels. Note that Fe VIII (dashed curve) is never a dominant species. Lower part is the result obtained if additional configurations with excitation energy less than the lowest ionization energy of Fe VIII are added.

2.2. THE MEAN ION MODEL

For heavier elements and for lighter elements at higher density or higher temperature, the mean ion model is used for the calculation of opacity. Only the first four of the seven models enumerated in Table I have been used for opacity calculations. The other three

TABLE I

Mean ion models

1. Screening constant method
2. Thomas–Fermi potential
3. Parametrized Thomas–Fermi potential
4. Relativistic Hartree–Fock–Slater
5. Thomas–Fermi, shells and continuum resonance
6. 'Muffin-Tin'
7. Debye–Hückel–Thomas–Fermi Potential

hold promise, but have not been fully developed. The parametrized Thomas–Fermi (TF) model has been described by Zink (1968, 1970) and the relativistic Hartree–Fock–Slater model by Rozsnyai (1972). Atomic shells and continuum resonances in a TF model were considered by Lee and Thorsos (1978). The 'Muffin-Tin' model (Liberman, 1979) is based on a self-consistent field calculation in which ions surrounding a particular atom are replaced by a positive charge distribution. Electrical neutrality outside and inside a cavity is maintained by an appropriate electron charge distribution. The Debye–Hückel–Thomas–Fermi theory has been described by Cowan and Kirkwood (1958a, b). We shall only describe the screening constant model, which is the one commonly used for astrophysical opacity.

For the screening constant method Mayer (1948) introduced approximate energies for bound electrons:

$$\bar{E}_i = E_i^0 + \frac{e^2 N_f}{2r_0} [3 - \langle (r/r_0)^2 \rangle_{ii}] + E_0, \tag{29}$$

where

$$E_i^0 = -(Z - s_i)^2 n_i^{-2} e^2 (2a_0)^{-1}, \tag{30}$$

where s_i is a screening charge from the bound electrons [compare Equations (31a) or (31b)] and $e^2/(2a_0) \equiv chR$. Since the electron orbit must lie inside the ion's volume, $\langle (r/r_0)^2 \rangle_{ii}$ has to be constrained not to be larger than 1. The second term in Equation (29) is a shielding correction for the bound electron in state i by the N_f free electrons that just neutralize the ion's core charge, while the last term corresponds to the continuum lowering described by Equation (14).

Mayer worked out an approximate 'self-consistent' solution for computing the average occupation numbers N_i for bound electrons. In effect, his method consists of finding the iterative solution of Equations (31) through (36).

$$\bar{E}_i = - \frac{e^2 Z^2}{2a_0 n_i^2} - p_i E_{ii} + \sum_j \bar{N}_j E_{ij} + \frac{e^2 N_f}{2r_0} [3 - \langle (r/r_0)^2 \rangle_{ii}] + E_0, \tag{31a}$$

or, as an alternative, a more accurate form for elements that are only a few times ionized:

$$\bar{E}_i = E_i^{(n)} + \sum_j (\bar{N}_j - N_j^{(n)}) E_{ij} + (1 - p_i) E_{ii} + \frac{e^2 N_f}{2r_0} [3 - \langle (r/r_0)^2 \rangle_{ii}] + E_0, \tag{31b}$$

where the superscript (n) indicates neutral atom,

$$p_i \equiv \frac{\bar{N}_i}{g_i} = [1 + \exp(-\eta^* + \bar{E}_i/kT)]^{-1}, \tag{32}$$

$$\eta^* = \eta + E_0/kT, \tag{33}$$

$$N_f = Z - \sum_i \bar{N}_i, \tag{34}$$

$$N_e = N_f \rho N_0/M, \tag{35}$$

and

$$N_e - \frac{4\pi(2mkT)^{3/2}}{h^3} I_{1/2}(\eta^*) \equiv f(\eta^*) = 0. \tag{36}$$

Here $I_k(x)$ is the Fermi integral of order k (Equation (18)). In (31a), $-e^2 Z^2/(2a_0 n_i^2)$ is the energy of interaction of an electron in state i with the atomic nucleus, E_{ij} is the scaled electron–electron interaction energy based on Slater integrals (for scaling, see Huebner (1970)), and η is the electron degeneracy parameter; the effective degeneracy parameter η^* contains a correction for the continuum depression. The free-electron density per unit volume is N_e and N_f is the number of free electrons per atom needed to neutralize the ion charge. The index i runs over all bound states of the atom.

While Equation (31a) corresponds to Sommerfeld's 'Aufbauprinzip' in which the energy is determined by adding electrons to the bare nucleus, Equation (31b) is based on the inverse process of determining the energy by removing electrons starting with a predetermined value for the neutral atom. Analogously this can be labeled the 'Abbauprinzip'.

If it is assumed that there exists a value of η^* such that all six equations are simultaneously satisfied, a root of $f(\eta^*) = 0$ is a solution. Indeed, at sufficiently high temperature the numerical solution of this set of equations poses no particular difficulty, but at low temperature the well-known nearly discontinuous behavior for the level occupation leads to difficulties in solving Equation (36) and the one for the energy simultaneously. Carson and Hollingsworth (1968) discussed this difficulty and gave a calculational prescription which removed the level degeneracy and thus circumvents the numerical instability.

In another prescription, which is used by us, the width of a level is calculated from collision broadening, and the electronic occupancy of that level is distributed over that width. Then, as the level crosses the continuum boundary — with changing density and temperature — part of the electrons will be bound and part will be free, thus providing for a smooth transition from bound states to free states.

The free electron density is actually a function of atomic radius:

$$N_e(r) = \frac{4\pi(2m)^{3/2}}{h^3} \int_{-V(r)}^{\infty} \left\{ \exp\left[\frac{\epsilon + V(r)}{kT} - \eta \right] + 1 \right\}^{-1} \epsilon^{1/2} \, d\epsilon, \tag{37}$$

where ϵ is the energy of the free electrons. Here $V(r)$ is the (negative) potential energy, and free electrons are those whose kinetic energy exceeds the negative of the local potential energy. The number of free electrons per atom is

$$N_f = 4\pi \int_0^{r_0} N_e(r) r^2 \, dr. \tag{38}$$

If $|V(r) + E_0|$ is small compared to kT, Equation (37) can be expanded. The leading term in the expansion

$$N_e = \frac{4\pi(2mkT)^{3/2}}{h^3} I_{1/2}(\eta^*) \tag{39}$$

is independent of the atomic radius. This approximation is widely used in opacity calculations.

The total mean energy of an ion and neutralizing sphere of electrons in a plasma at temperature T and density ρ, consistent with the equations of this section, is

$$\bar{E} = \sum_i \bar{N}_i \left(\bar{E}_i - \frac{1}{2} \sum_j \bar{N}_j E_{ij} + \frac{1}{2} p_i E_{ii} - \frac{18}{10} \frac{e^2 N_f}{r_0} \right) - \frac{9}{10} \frac{e^2 \overline{N_f^2}}{r_0} +$$

$$+ N_f k T I_{3/2}(\eta^*)/I_{1/2}(\eta^*) + \tfrac{3}{2} kT. \tag{40}$$

The term containing the summation corresponds to the internal energy of the ion core, the next term is the averaged Coulomb interaction of the free electrons, and the last two terms are the kinetic energies of the electrons and the ion. In the nondegenerate limit,

$$\lim_{\eta^* \to \infty} I_{3/2}(\eta^*)/I_{1/2}(\eta^*) = \tfrac{3}{2}. \tag{41}$$

The virial theorem in the nonrelativistic limit gives the gas pressure in the plasma [see Equation (28)].

3. Calculation of Opacity

Having established the structure of the ions in the plasma, one can now proceed with the calculation of the extinction coefficients. The following outlines the calculation of the basic cross-sections and the necessary density–temperature modifications for the bound–bound, bound–free, free–free, and scattering processes.

3.1. BOUND–BOUND PROCESSES

The bound–bound (bb) cross-section for induced emission or absorption per atom is given by

$$\sigma_{if}^{(bb)}(\nu) = \frac{\pi e^2}{mc} |f_{if}| L_\nu = 0.0265400 |f_{if}| L_\nu, \tag{42}$$

where L_ν is the line shape function per unit photon frequency normalized to unity when integrated over all frequencies. The spontaneous process involves the same matrix elements; from the Einstein arguments of detailed balance, the cross-section is:

$$\sigma_{ul}^{(bb,\mathrm{sp})}(\nu) = \frac{\pi e^2}{mc} \frac{g_l}{g_u} f_{lu} L_\nu. \tag{43}$$

The quantum defect methods of Bates and Damaard (1949) and of Burgess and Seaton (1960) are often applied to the calculation of the bb cross-section.

Teller pointed out in 1947 that in a plasma the large number of ions with statistically different occupation configurations will cause a proliferous line spectrum and its contribution to the opacity ought to be significant. The consideration of the line contribution to the extinction coefficient requires consideration of mechanisms for line broadening and line splitting. The line shape function depends on the type of broadening. Table II summarizes types of broadening and approximations that are taken into account in opacity calculations.

TABLE II

Types of line broadening and approximations

Natural broadening
Doppler broadening
Collision broadening
    Impact approximation
        Hydrogen
        Hydrogenic ions
        Nonhydrogenic neutral atoms
        Nonhydrogenic ions
        Resonance broadening
        Van der Waals broadening
    Quasistatic approximation

Doppler broadening is the result of a statistical distribution of line shifts brought about by the thermal motion of an ensemble of radiating atoms. The line shape for Doppler broadening is

$$D_\nu = \gamma_D^{-1} \, \pi^{-1/2} \, \exp(-[(\nu_0 - \nu)/\gamma_D]^2). \tag{44}$$

Here,

$$\gamma_D = (2N_0 kT/M)^{1/2} \, \nu_0/c \tag{45}$$

is the Doppler half-width (half the width of the Gaussian at e^{-1} of the peak value), ν_0 is the frequency at the line center, N_0 is the Avogadro number, and M is the mean atomic weight. Collision broadening, on the other hand, is the effect caused by a statistical distribution of perturbers on each radiating atom in the impact approximation as well as in the quasistatic approximation. Following the prescription of Stewart and Pyatt (1961) and its extension by Armstrong et al. (1966), one obtains, for the half-width (in electron-volts) at half-maximum (HWHM) for collision broadening by electrons,

$$\gamma_c = 0.6374 \times 10^{-22} \, \frac{N_e}{(kT)^{1/2}} \, \frac{1}{(N_f + 1)^2} \, \{n_u^{*2} \, [5n_u^{*2} + 1 - 3l_u(l_u + 1)] +$$

$$+ n_l^{*2} \, [5n_l^{*2} + 1 - 3l_l(l_l + 1)]\}, \tag{46}$$

where kT is in electron volts, the subscripts u and l stand for upper and lower levels,

$$n_u^{*2} = (N_f + 1)^2 \, e^2/(2a_0 E_u), \tag{47}$$

and E_u is the energy of the upper state. A similar equation holds for n_i^{*2}. The total line shape is then a folding of the natural line shape into all of the statistical distribution functions.

If only the impact approximation needs to be considered (which is, for example, the case for a gas at low temperature where only a few ions are present and electron broadening is the most important broadening mechanism), one needs to fold the natural line shape, which is a Lorentz function with HWHM γ, into that of another Lorentz function with HWHM Γ. The result is another Lorentz function centered at ν_0 with a HWHM which is the sum of the natural HWHM and the collision broadened HWHM:

$$\int_{-\infty}^{\infty} \hat{L}(\nu, \nu_0'; \gamma)\, \hat{L}(\nu_0', \nu_0; \Gamma)\, d\nu_0' = \hat{L}(\nu, \nu_0; \Gamma + \gamma). \tag{48}$$

If the quasistatic approximation contributes to the line shape (i.e. perturbations by ions are important), this new Lorentzian will have to be folded with the Stark broadening function. [For a discussion of Stark broadening see, for example, Griem (1964).] If Doppler broadening is of importance, the resulting line shape will have to be folded in with the Gaussian distribution function with Doppler half-width γ_D. In the case where the quasistatic approximation did not contribute significantly to the line shape, a Voigt profile will be obtained from the folding of a Lorentzian and a Gaussian.

Resonance and Van der Waals broadening are only of importance at very low temperatures when the number of free electrons is very small. They are of no concern here.

The line shapes and line widths for the equivalent processes of auto-ionization, Auger transition, and dielectronic recombination have the (usually) asymmetric Fano profile. The line shape may, however, be further modified by Doppler, collision, or Stark broadening, and may therefore require an additional convolution with the auto-ionization profile. Since the effect of the Fano profile on opacity is small for temperatures of interest here, this final folding is usually ignored.

As pointed out earlier, the peaks of the absorption structures are weighted in the calculation of the Planck mean opacity, while the 'windows' between the structures through which radiation can readily flow are weighted in the Rosseland mean opacity. The far wings of the absorption lines can therefore affect the Rosseland opacity much more than the Planck opacity. Indeed, the extreme far wings on the low-energy side of the Lyman α line of hydrogen is a dominant contribution to the Rosseland opacity. The lack of measurement and theory of the extinction properties of these far wings are a major source of uncertainty to the opacity at $kT \sim 0.5$ eV. A similar situation exists for helium, and for hydrogen-like and helium-like ions at higher temperatures. A large effect will also be noticed when a strong absorption line is split into many weak components.

Electrostatic interaction splits the spectroscopic terms, while coupline of magnetic moments associated with angular momentum gives the fine structure splitting. The collection of spectral lines arising from transitions between two terms (taking into account the fine structure) is called a multiplet. The collection of multiplets arising from transitions between two configurations is called a transition array. The following example illustrates the large number of lines that can be contained in one transition array:

Initial configuration   ⟶   Final configuration

Short notation: $s\ p^2\ d\ \longrightarrow$ $p\ \ s\ p^2\ d\ d$

Detailed notation: $1s^2\ 2s^2\ 2p^6\ 3s\ 3p^2\ 3d\ \longrightarrow\ 1s^2\ 2s^2\ 2p^5\ 3s\ 3p^2\ 3d\ 4d$

↓

Multiply excited   Equivalent to

(many multiplets) $2p$

This results in about 6378 lines if fine structure cannot be resolved, or 32 456 lines if fine structure can be resolved. (In a simple average atom model this would correspond to one line.)

At very high temperatures (where the energy between split lines is less than $kT/10$) and for heavier elements (where the number of lines is very large), the exact spacing of the split lines is unimportant; it is then sufficient just to count the split lines. Although details of the splitting depend on the coupling scheme, the number of lines obtained from dipole selection rules (disregarding coincidences) is insensitive to it. Based on its popularity and historical significance Russell–Saunders (LS) coupling is usually chosen for enumerating the split lines. The number of spectroscopic terms may be obtained by various means, e.g. by the original method of Hund, the method of fractional parentage, or the method of coupling of conjugate Young Tableaux. Spectroscopic terms for equiv-✐ alent s, p, d, and f electrons and for some nonequivalent s, p, d, and f electrons have been tabulated by Condon and Shortley (1953), Slater (1960), and Shore and Menzel (1968). The dipole selection rules determine the number of lines in a multiplet. At higher temperatures and particularly for higher Z elements (for example, Fe) the number of split lines becomes very large and can be treated statistically. Moszkowski (1962) finds under the conditions of LS coupling that the statistically weighted distribution of line positions in a transition array has a nonzero second moment. In the case of an array containing many lines he notes that the distribution frequently 'resembles' a Gaussian. Stewart (1965) applies Moszkowski's results to obtain an integrated absorptivity for a Gaussian distribution of Lorentz lines. The result is a Voigt profile similar to the one obtained when a Lorentz profile is folded into a Doppler distribution, but the width of the distribution is now determined from the multiplet splitting.

At elevated temperatures statistical fluctuations of electronic occupation about the mean value of each level cause differing amounts of screening and, in an ensemble of radiating atoms, split each generically one-electron transition into many lines. This splitting is called configuration splitting; i.e. the fictitious, mean ion configuration is split (resolved) into its components. The resulting spectrum corresponding to a one-electron transition from the lower level $(n,\ l,\ j)_l$ to the upper level $(n,\ l,\ j)_u$ is called a transition cluster. The total oscillator strength of a cluster is conserved; i.e. it is equal to that of the corresponding one-electron transition in the mean ion. Clearly, a transition cluster can contain many transition arrays.

For practical reasons only the configurations contributing most heavily to the mean ion are considered in detail in opacity calculation. Partially occupied levels, i, with mean occupation, $\bar{N}_i$, significantly different from zero or from the statistical weight, g_i (i.e. $\delta N_i \leq \bar{N}_i \leq g_i - \delta N_i$), will have the largest fluctuations. Thus, substitution of various integer occupation numbers for the mean values of these levels will give the major configurations contributing to the mean ion. Table III illustrates this for a calculation for

TABLE III

Energies without continuum depression and occupation numbers for the first ten subshells of Fe at $kT = 200$ eV and $\rho = 7.85 \times 10^{-2}$ g cm^{-3}[*]

i	Level	Thomas–Fermi mean ion		Configuration 1		Configuration 51		Configuration 87	
		$-\bar{E}_i$	$\bar{N}_i$	$-E_i^{(1)}$	$N_i^{(1)}$	$-E_i^{(51)}$	$N_i^{(51)}$	$-E_i^{(87)}$	$N_i^{(87)}$
1	$1s_{1/2}$	562	2.000	612	2	563	2	536	2
2	$2s_{1/2}$	111	1.416	138	0	112	2	94.5	2
3	$2p_{1/2}$	105	1.216	134	0	102	1	85.9	2
4	$2p_{3/2}$	104	2.338	132	0	101	2	83.7	3
5	$3s_{1/2}$	43.2	0.032	59.0	0.032	41.0	0.032	30.6	0.032
6	$3p_{1/2}$	41.2	0.028	58.2	0.028	38.8	0.028	30.0	0.028
7	$3p_{3/2}$	40.9	0.055	58.0	0	38.7	0	27.9	0
8	$3d_{3/2}$	38.9	0.048	57.6	0.048	36.0	0.048	24.0	0.048
9	$3d_{5/2}$	38.9	0.071	57.0	0	36.0	0	24.0	1
10	$4s_{1/2}$	21.2	0.007	30.7	0.007	20.1	0.007	13.8	0.007
$P^{(c)}$					0.0003		0.0745		0.0039

iron at a temperature of 200 eV and 10^{-2} times normal solid density. With the arbitrary but typical value $\delta N_i = 0.05$, the partially occupied levels are: $2s_{1/2}, 2p_{1/2}, 2p_{3/2}, 3p_{3/2}$, and $3d_{5/2}$. Eighty-seven configurations (different combinations of integer occupation numbers for these levels) were considered. The *a priori* probability, $P^{(c)}$, for a configuration c is

$$P^{(c)} = \prod_i \binom{g_i}{N_i^{(c)}} \left(\frac{\bar{N}_i}{g_i}\right)^{N_i^{(c)}} \left(\frac{g_i - \bar{N}_i}{g_i}\right)^{g_i - N_i^{(c)}}.$$

(49)

The most probable configurations are those with integer occupation numbers $N_i^{(c)}$ close to the mean values, $\bar{N}_i$. In the example of Table III, $c = 51$ is the most probable configuration. The total possible number of configurations is $(g_i + 1)^r \cdot (g_j + 1)^s \cdot (g_k + 1)^t \ldots$, where $r, s, t \ldots$ are the number of partially occupied levels with statistical weight g_i, g_j, $g_k, \ldots$. In the cited example 1575 configurations are possible. For practical purposes only configurations with probability larger than some (from experience) predetermined value are kept. In the example this value is $P_c = 3 \times 10^{-4}$.

The probability for an initial state configuration strongly modifies the strength of a line or an absorption edge. Binding energies can be computed for each configuration from perturbation theory. The number of bound electrons in each configuration is constrained not to be larger than the nuclear charge, $N_b^{(c)} \leq Z$.

Although the description of the configuration splitting given so far has been in terms of the mean ion model, the splitting is also taken into account in the explicit ionic model.

[*]Energies in units of $e^2/2a_0$.

The important advantage of the mean ion model is that all of the dominant configurations are retained and no effort is wasted on unimportant configurations.

If all permitted configurations are ordered approximately in increasing total occupation, $N_b^{(c)}$, a very good estimate of the spread of the configuration splitting,

$$2\Delta_{lu} = \sum_j (N_j^{(L)} - N_j^{(1)})(E_{lj} - E_{uj}), \tag{50}$$

may be obtained from the difference of the first and last configurations [in Table III the energies of the first and of the last (the 87th) configurations are near the extremes of the distribution]. The superscripts in Equation (50) on the occupation numbers stand for first (1) and last (L) configurations. Assuming that the lines in the transition cluster are randomly distributed they will be resolved if the statistical spacing between lines is less than the average line width:

$$\Delta_{lu} \Big/ \Big(\sum_c N_{LS}^{(c)} \Big)^{1/2} > \max(\gamma_{lu} + \Gamma_{lu}, D_{lu}, \Gamma_{lu}^{(auto)}, \ldots). \tag{51}$$

If lines overlap, transition clusters can be smeared statistically by methods described, for example, by Stewart (1965) or Mayer and Jacobsohn (Mayer, 1948). Mayer and Jacobsohn on the urging by Teller were the first to investigate the effects of configuration splitting on opacity.

3.2. BOUND–FREE PROCESSES

In the single-electron approximation the only possibility for absorbing a photon of energy $h\nu$ in a bound–free (bf) process is given by

$$h\nu = I_i + \epsilon. \tag{52}$$

Here I_i is the ionization energy for an electron initially in bound state i and ϵ is the kinetic energy of the ejected electron at a very large separation from the ion. The calculation of the cross-section for the bf process is carried out formally in the same way as for bound–bound processes. The only difference is that for the continuum wave function the normalization is usually chosen on energy:

$$\int \Psi_\epsilon^*(\mathbf{r})\, \Psi_{\epsilon'}(\mathbf{r})\, dV_a = \delta(\epsilon - \epsilon'), \tag{53}$$

where δ is the Dirac delta 'function' and V_a is the volume of the ion. The photoelectric cross-section per electron in shell i can then be written in the dipole length form

$$\sigma_i^{(bf)}(\nu) = \frac{\pi e^2}{mc} \frac{df}{d\nu},$$

$$= 4\pi^2 \alpha \frac{I_i + \epsilon}{3} \, \Big| \int \Psi_\epsilon^*(\mathbf{r})\, \mathbf{r}\, \Psi_i(\mathbf{r})\, dV_a \Big|^2. \tag{54}$$

Here α is the fine structure constant.

Since hydrogen is very important in astrophysics, we will pay special attention to hydrogenic cross-sections. For hydrogen and hydrogenic ions this cross-section is commonly expressed by the semiclassical Kramers formula and the Gaunt factor, $g_{bf}(\nu)$:

$$\sigma_i^{(bf,K)}(\nu) = \frac{64\pi^4 e^{10} m (Z_i^{**})^4 \, N_i}{3^{1/2}\, 3(h\nu)^3 \, ch^3 \, n_i^5} \; g_{bf,i}(\nu) = 2.815 \times 10^{29} \, \frac{(Z_i^{**})^4}{\nu^3 \, n_i^5} \, N_i g_{bf,i}(\nu), \qquad (55)$$

where

$$(Z_i^{**}/n_i)^2 \, e^2 / (2a_0) \cong (Z/n_i)^2 \, e^2 / (2a_0) - \sum_{j=1}^{i-1} N_j E_{ij} - \frac{1}{2} E_{ii}(N_i - 1). \qquad (56)$$

The hydrogenic Gaunt factor, which takes into account the quantum mechanical corrections to the semiclassical formula, has been calculated by Karzas and Latter (1961). An analytic form of the cross-section for one K-shell electron valid near the threshold of light hydrogenic ions with nuclear charge Z was given by Stobbe (1930):

$$\sigma_1^{(bf,St)}(\nu) = \frac{2^9 \pi^2}{3Z^2} \, \alpha a_0^2 \, [-E_1/(h\nu)]^4 \, \frac{\exp(-4n' \operatorname{arccot} n')}{1 - \exp(-2\pi n')}, \qquad (57)$$

where

$$n' \equiv [-E_1/(h\nu + E_1)]^{1/2}, \qquad (58)$$

and E_1 is the negative binding energy of the $1s$ electron.

Nonrelativistic multipole cross-sections have been calculated by Fischer (1931a, b) and by Sauter (1931a, b, 1933), but they differ from the nonrelativistic dipole calculations of Stobbe only by terms of order $(v/c)^2$. Since these correction terms are of the same order as those obtained with the relativistic Dirac formulation, multipole and relativistic effects should not be considered separately for angle-integrated cross-sections.

An important result for a high-energy photon ionizing a K-shall electron in hydrogenic ions with nuclear charge Z was derived by Sauter

$$\sigma_1^{(bf,Sa)}(\nu) = \frac{2^8 \pi}{3\alpha^2 Z^5} \, a_0^2 \, [-E_1/(h\nu)]^5 \, (\beta\gamma)^3 \times$$

$$\times \left[1 + \frac{3}{4} \frac{\gamma(\gamma - 2)}{\gamma + 1} \left(1 - \frac{1}{2\beta\gamma^2} \ln \frac{1+\beta}{1-\beta} \right) \right], \qquad (59)$$

where

$$\gamma = (1 - \beta^2)^{-1/2} = 1 + (h\nu + E_1)/mc^2, \qquad (60)$$

$$\beta = v/c = [(h\nu + E_1)^2 + 2(h\nu + E_1)\, mc^2]^{1/2}/(h\nu + E_1 + mc^2)$$

$$\sim [2(h\nu + E_1)/mc^2]^{1/2} \sim \alpha Z/n', \qquad (61)$$

E_1 is again the negative binding energy of the $1s$ electron, and v is the free electron speed. In the low-energy limit ($|E_1| \ll h\nu \ll mc^2$), Sauter's result agrees with the nonrelativistic Born approximation for a $1s$ electron:

$$\sigma_1^{(bf,B)}(\nu) = \frac{2^8 \pi}{3Z^2} a_0^2 \alpha [-E_1/(h\nu)]^{7/2}. \tag{62}$$

This in turn agrees with the high-energy limit of Stobbe's result. A higher order Born approximation and high-energy limit formulae have been summarized by Pratt *et al.* (1973). Their statement that the 'Sauter formula is not too useful in practice' may be too pessimistic. Bethe and Salpeter (1957) suggest a semi-empirical cross-section through combination of a factor in Stobbe's formula with Sauter's result. Slightly modified, to give better agreement with Stobbe's formula, the cross-section per electron for light elements is given by

$$\sigma_1^{(bf,BS)}(\nu) = \frac{2^9 \pi^2}{3\alpha^2 Z^5} a_0^2 \left[-E_1/(h\nu)\right]^4 (n'\beta\gamma)^3 \left[1 + \frac{3}{4} \frac{\gamma(\gamma-2)}{\gamma+1} \left(1 - \frac{1}{2\beta\gamma^2} \ln \frac{1+\beta}{1-\beta}\right)\right] \times$$

$$\times \frac{\exp(-4n' \operatorname{arccot} n')}{1 - \exp(-2\pi n')}. \tag{63}$$

The quantities n' and γ are given by Equations (58) and (60). For 200 keV photons this gives a cross-section about 20% higher than the Stobbe formula. At the threshold of hydrogen-like iron the increase over the Stobbe result is only about 1%. For heavy elements neither the nonrelativistic Stobbe formula nor the Born approximation is valid near threshold; cross-sections must be calculated numerically as done, for example, by Pratt *et al.*

For nonhydrogenic cross-sections the quantum defect method can be applied as developed by Burgess and Seaton (1960) and by Bates and Damgaard (1949). The energy dependence of a hydrogenic cross-section can deviate markedly from that of the non-hydrogenic cross-section. The cross-section written in a form convenient for use with the Burgess–Seaton g-function is

$$\sigma_{n_l^*}^{(bf,BS)} = 4\pi^2 \alpha \frac{h\nu}{3\pi} \sum_{l'=l\pm 1} C_{l'} |(n_l^*/Z')^2 g(n_l^* l, \epsilon' l')|^2. \tag{64}$$

Here the coefficients $C_{l'}$ depend on the configuration of the initial and final states of the ion. Tables for $C_{l'}$ are given by Burgess and Seaton and by Griem (1964) for the case of LS coupling.

Very little can be said about the accuracy and range of validity of this method. When comparisons with experiment are possible, the Burgess–Seaton method gives results that are about as good as other methods near threshold. However, as Peach (1967) points out, extrapolations to energies well above the edge are doubtful. Perhaps no more of a succinct statement about the applicability can be made than that made by Burgess and Seaton in the conclusions of their 1960 paper where they state, "From comparisons with results derived by other methods it may be concluded that our method will give good results whenever there is not great sensitivity to the details of the atomic wave functions."

Cross-sections are generally calculated for the isolated atom. Models are now being developed to calculate cross-sections for compressed ions. At very high density (particularly in the case of pressure ionization) the final state in the photoelectric absorption process may be occupied with free electrons. The probability that a continuum state is occupied is

$$P_i(v) = \left\{ \left[\exp\left(-\eta^* + \frac{E_i + hv}{kT} \right) \right] + 1 \right\}^{-1},$$

(65)

and therefore its availability to accept an electron is

$$1 - P_i(v) = \left\{ \left[\exp\left(\eta^* - \frac{E_i + hv}{kT} \right) \right] + 1 \right\}^{-1}.$$

(66)

The term and fine structure splitting distribute edge positions about the mean value. The absorption strength for each of these components is proportional to the ratio of electronic occupation to statistical weight. Thus, if kT is large compared to the splitting, the distribution of strengths is nearly constant. Except for the simplest configurations, the number of term values can be very large. Under these conditions a statistical (Gaussian) distribution may be assumed for the edges in the same way as for the multiplet distribution for line transitions, resulting in the photoelectric edge distribution function

$$\hat{P}_v = \frac{1}{2} \left[1 + \mathrm{erf}\left(\frac{hv + E_i + E_0}{\Delta_E} \right) \right],$$

(67)

where

$$\mathrm{erf}(x) = \frac{2}{\sqrt{\pi}} \int_0^x e^{-t^2} \, dt.$$

(68)

Here $\mathrm{erf}(x)$ is the normalized error function, E_0 is a continuum depression Equation (14), and Δ_E is the half-width of the Gaussian energy density distribution of photoelectric edges at $1/e$ of its peak value. The absorption coefficients should be renormalized to ensure conservation of oscillator strength.

Huebner $et\ al.$ (1978) have shown that good approximations for nonhydrogenic partial (for subshell k) photoelectric cross-sections of ions in their ground state, $\sigma_k^{(i)}$, can be obtained through scaling (according to occupation of subshell k) of available partial cross-sections of the neutral atom, $\sigma_k^{(n)}$:

$$\sigma_k^{(i)}(v) \cong \sigma_k^{(n)}(v) \, (N_k^{(i)}/N_k^{(n)}) \times (Z_k^{**(i)}/Z_k^{**(n)})^4.$$

(69)

The superscript i stands for ion, n for neutral atom, and $(Z_k^{**(i)}/Z_k^{**(n)})^4$ is an additional scaling factor depending on the ratio of wave-function-effective Z. Figure 8 illustrates

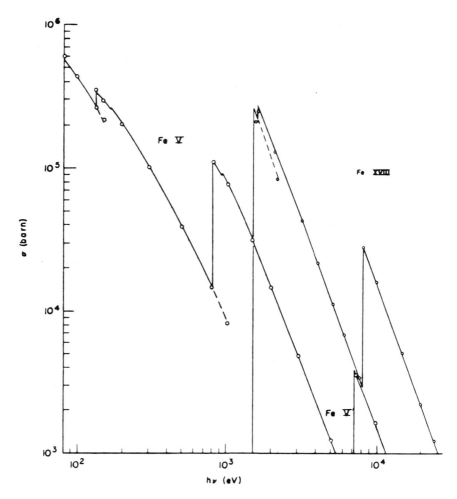

Fig. 8. Some photoelectric cross-sections for iron ions. Circles represent cross-sections scaled from Fe I calculations. Solid lines represent the cross-sections for Fe V and Fe XVIII calculated with the same approximations and to the same numerical accuracy as the corresponding Fe I values. Dashed lines are extensions of partial cross-sections beyond the next higher photoelectric edge. Note the shift of the L-edges at ~0.8 and 0.9 keV in Fe V to ~1.5 and 1.6 keV in Fe XVIII, and the shift of the K-edge from ~7 keV in Fe V to ~8 keV in Fe XVIII.

the scaling of cross-sections for an iron ion from the neutral iron cross-sections. The effective bound—free absorption cross-section with smeared edges for subshell k is

$$\sigma_k^{(i,\mathrm{eff})}(\nu) = \sigma_k^{(i)}(\nu) \, [1 - p_i(\nu)] \, \hat{P}_\nu. \tag{70}$$

The importance of H^- absorption to opacity was pointed out by Wildt (1939). Photodetachment cross-sections for H^- have been computed by Geltman (1962) and by Broad and Reinhardt (1976).

3.3. FREE–FREE PROCESSES

Photoprocesses involving a free electron in both the initial and the final states are commonly called bremsstrahlung (emission) and inverse bremsstrahlung (absorption). To conserve energy and momentum these transitions must occur in the presence of a third body which may be a neutral atom, molecule, or an ion. There is no restriction on the photon energy such as exists for bound–bound and bound–free processes. In the single-particle, fixed potential, dipole acceleration approximation the free–free (ff) cross-section is

$$\sigma^{(ff)}(\nu) = \frac{e^4}{3\pi\nu^3 m^2 c} \sum_{l_i} \sum_{l_f} \max(l_i, l_f) \, \delta\,(l_i - l_f \pm 1) \left|\frac{dV}{dr}\right|^2_{if} P(\epsilon_i)\,q\,(\epsilon_f)\,d\epsilon_i, \quad (71)$$

where dV/dr is the radial derivative of the atomic potential. When both initial and final states are in the continuum it is easier to evaluate the acceleration form of the dipole matrix element than the length or velocity form. In Equation (71) $p(\epsilon_i)$ is the probability of occupancy of the initial state i with energy ϵ_i and $q(\epsilon_f)$ is the availability of the final state f. The integration is to be carried out over all initial states: the energies of final and initial states are related by

$$\epsilon_f = \epsilon_i + h\nu. \quad (72)$$

The form usually encountered in the literature,

$$\sigma_{ff}(\nu) = \sigma^{(ff,K)}(\nu)\,g_{ff}(\nu), \quad (73)$$

involves the Kramers cross-section

$$\sigma^{(ff,K)}(\nu) = \frac{4}{3}\left(\frac{2\pi}{3mkT}\right)^{1/2}\frac{e^6 Z^2 N_e}{hcm\nu^3}, \quad (74)$$

and the Gaunt factor $g_{ff}(\nu)$.

Even though the ff cross-section is commonly written in the form of a Kramers cross-section with a Gaunt factor, this does not have any particular virtue when the cross-section is calculated directly from wave functions. For hydrogenic wave functions the Gaunt factor is

$$g_{ff}(\nu) = \frac{3^{1/2}\,32\pi^3\,h^4}{m^2 Z^2 e^2} \sum_{l_i}\sum_{l_f} \max(l_i, l_f)\,\delta\,(l_i - l_f \pm 1)\left|\frac{dV}{dr}\right|^2_{if}. \quad (75)$$

The normalized, temperature-averaged Gaunt factor for a Fermi–Dirac distribution of electrons at temperature T, including the Pauli exclusion principle for the final state, is

$$\bar{g}_{ff}^{-FDP}(\zeta^2, u, \eta) = \frac{\sqrt{\pi}}{2I_{1/2}(\eta)} \int_0^\infty \frac{g_{ff}(x/\zeta^2, u/\zeta^2)}{[\exp(-\eta + x) + 1]\,[\exp(\eta - x - u) + 1]}\,dx, \quad (76)$$

where

$$\zeta^2 \equiv e^2 \, N_f^2 /(2a_0 kT).\tag{77}$$

The free–free cross-section of H^- (i.e. the photoabsorption of an electron in the field of an H^- ion) has been calculated by Ohmura and Ohmura (1960, 1961).

3.4. SCATTERING PROCESSES

For scattering, the Thomson cross-section per electron,

$$\sigma_s = \frac{8\pi}{3}\left(\frac{e^2}{mc^2}\right)^2 \cong 6.65 \times 10^{-25} \text{ cm}^2,\tag{78}$$

is a sufficient approximation for most applications even for bound electrons. But at the highest temperatures and low densities where scattering is the dominant process more detail must be considered. In that case the most complete form per mean ion is

$$\sigma_{sT} = \sigma_s \, G(u, \, kT) \, N_{ep} \, f(\delta).\tag{79}$$

The subscript T emphasizes that this is the transport scattering cross-section which takes into account the asymmetry of the phase function in the relativistic domain and when collective effects are important. The phase function depends on the angle through which a photon scatters. If the phase function is symmetric (e.g. Thomson or Rayleigh scattering), the transport scattering cross-section is the same as the ordinary scattering cross-section. N_{ep} is the number of electron–positron pairs per mean ion. It equals the number of electrons Z unless the temperature is higher than about 10 keV at low density. It is described by Sampson (1959) together with the function $G(u, \, kT)$ for which he provides a table. At low temperature and low photon energy

$$G(u, T') \cong 1 + 2T' + 5T'^2 + \tfrac{15}{4}T'^3 - \tfrac{1}{5}(16 + 103T' + 408T'^2)uT' +$$

$$+ \left(\frac{21}{2} + \frac{609}{5}T'\right)(uT')^2 - \frac{2203}{70}(uT')^3,\tag{80}$$

where

$$T' = \frac{kT}{mc^2}.\tag{81}$$

The function $G(u, \, kT)$ takes into account the temperature-associated random motion of the electrons in the Klein–Nishina formula. The factor

$$f(\delta) = 1 - \frac{3\delta}{8}\left(\frac{r_D}{r_{De}}\right)^2 \left[(\delta^3 + 2\delta^2 + 2\delta)\ln\left(\frac{\delta}{2+\delta}\right) + 2\delta^2 + 2\delta + \frac{8}{3}\right]\tag{82}$$

takes into account the collective effects which are important when the product of photo wavenumber and Debye radius are about equal to or less than unity. In Equation (82),

$$\delta \equiv \tfrac{1}{2} (2\pi\nu r_D/c)^{-2},$$

(83)

the electron Debye radius is

$$r_{De} = \left[\frac{kT\bar{r}^{-3}}{3\bar{N}_f \, e^2 \, I'_{1/2}(\eta^*)/I_{1/2}(\eta^*)} \right]^{1/2},$$

(84)

and the total Debye radius is

$$r_D = \left\{ \frac{kT\bar{r}^{-3}}{3e^2 \sum_z a_z \, [\bar{N}_{fz}^2 + \bar{N}_{fz} \, I'_{1/2}(\eta^*)/I_{1/2}(\eta^*)]} \right\}^{1/2},$$

(85)

where $\bar{r}$ is the radius of the spherical volume of the mean ion in the mixture, a_z is the abundance of element Z normalized such that

$$\sum_z a_z = 1,$$

(86)

and $I'_{1/2}$ is the first derivative of $I_{1/2}$. The derivation of Equations (84) and (85) is based on a Taylor expansion of the incomplete Fermi–Dirac integrals (Stewart and Pyatt, 1961).

3.5. COMBINING PROCESSES TO OBTAIN THE OPACITY

To determine the extinction coefficient, the cross-sections for each process (*bb*, *bf*, *ff*, and scattering) must be summed at each photon energy. This must be done on a sufficiently fine photon energy mesh so that not too many of the narrow lines are missed. Typically, a mesh with interval $\Delta u = \Delta h\nu/kT = 0.01$ in the interval $u = 0$ to 30 is sufficient.

For a plasma containing several elements, it is important that at a given temperature the electron pressure for each constituent is the same as that of the others. This can also be accomplished by matching temperature T and effective degeneracy parameter η^*. If extinction coefficients are calculated on a $T - \eta^*$ grid, the extinction coefficients weighted by the relative number abundance of each constituent can be added for each photon energy at a given $T - \eta^*$ mesh point to obtain the total extinction coefficients for the mixture. Once the extinction coefficients are known, the Planck or Rosseland opacity can be obtained by multiplying the extinction coefficients (or their reciprocal values) by the appropriate weighting function and integrating over photon energy. The Los Alamos Astrophysical Opacity Library (Huebner *et al.*, 1977) uses this procedure.

Resistance to energy transport by electron conduction can also be expressed in terms of an equivalent opacity. In some circumstances encountered in astrophysical problems (e.g. in the core of red giants) and in laboratory plasmas at high density, the transport of energy by hot electrons may be the dominant mechanism. This process is usually referred to as electron thermal conductivity, or, in connection with Rosseland mean opacities, as 'electron conduction opacity'.

One of the first formulations used in opacity codes was that of Mestel (1950). This method is described by Cox (1965) and by Carson (1971). The Mestel model considers the scattering of electrons by fixed ions but ignores the electron–electron scattering completely. It does take into account Fermi–Dirac statistics.

A second method of computing the electron conductivities popular with plasma physicists is that given by Spitzer (1962). The main shortcoming of this approach is the neglect of Fermi–Dirac statistics for the electrons and at higher densities the ion–ion correlation effects. Under conditions typically achieved in laboratory plasmas, neither of these objections has serious consequences.

At very high densities encountered (e.g. in some astrophysical applications), both the electron degeneracy and the ion–ion correlation may be important. The work of Hubbard (1966), Lampe (1968a, b), and Hubbard and Lampe (1969) has shed new light on the conductivities under these extreme conditions. The result of Lampe's investigations shows that the electron–electron scattering is important over a much larger range of densities than had previously been suspected. The argument that the Fermi–Dirac statistics so severely restrict the volume of phase space that electron–electron collisions may be neglected is shown to be valid only for extremely high densities.

Lampe assumes the ions to be in a Maxwell–Boltzmann distribution. The electron distribution function is solved using a Chapman–Enskog orthogonal polynomial expansion technique. The collision cross-sections are obtained from the Born approximation for shielded Coulomb interactions. In addition he assumed weak coupling for both the electrons and the ions. Weak coupling means physically that the average kinetic energy of the particle is large compared to its average Coulomb interaction energy.

Hubbard and Lampe (1969) have combined the formulation of their previous papers and added interpolation formulae to join various regions. They have also extended the results to include electron thermal conductivity in the 'ion solid state'. These results are presented in their paper for hydrogen, helium, carbon, and mixtures representing red-giant cores and the solar composition. The accuracy stated by Hubbard and Lampe for their theory is approximately 10–15% except in the regions where the ion–ion correlation is large (close to crystallization) where it may range between 50 and 125%. The work of Hubbard and Lampe has been extended by Canuto (1970) to densities where the electrons are relativistic. A further extension was made by Flowers and Itoh (1976) to neutron star densities. Their work is also applicable to high-density white dwarf matter.

The electron conduction opacity κ_c is related to the thermal conductivity Λ_c by

$$\kappa_c = \frac{4acT^3}{3\Lambda_c\rho} , \qquad (87)$$

where a is the radiation constant, c the speed of light, and ρ the mass density. This electron conduction opacity can be combined with the radiative Rosseland mean opacity κ_{rR} to give a total (effective) opacity κ_T:

$$\frac{1}{\kappa_T} = \frac{1}{\kappa_{rR}} + \frac{1}{\kappa_c} . \qquad (88)$$

4. Results

It has already been mentioned that the Rosseland opacity depends strongly and non-linearly on element composition. Element abundances are determined from the intensity of solar spectra in combination with oscillator strengths determined from laboratory measurements or theoretical calculations. The composition of the solar interior cannot be determined through direct observations. Therefore the opacity of the interior has a composition-correlated uncertainty.

The most accepted recent relative abundance of metals (all elements with $Z > 2$) is that of Ross and Aller (1976). An abbreviated list of 18 elements with their abundances is given in Table IV. An order of magnitude revision in the abundance of iron was required

TABLE IV

Abbreviated Ross–Aller metal Abundance (ZRA)

	Number fraction a_Z
C	0.30279
N	0.06326
O	0.50249
Ne	0.02699
Na	0.00138
Mg	0.02892
Al	0.00241
Si	0.03244
P	0.00023
S	0.01151
Cl	0.00023
Ar	0.00073
Ca	0.00163
Ti	0.00008
Cr	0.00037
Mn	0.00019
Fe	0.02297
Ni	0.00138
	1.00000

because of new determinations of the oscillator strength of some solar iron lines about a decade ago. A similar revision was made for neon. Although uncertainties in the metal abundances seem to be much smaller now, it is well to keep in mind the comment made by Ross and Aller (1976): "The uncertainty ±0.08 [in the iron abundance] refers to the consistency of individual measurements; the actual error may turn out to be larger than this when the f-value problem is finally solved and departures from local thermodynamic equilibrium are fully investigated."

Extensive tables based on Ross–Aller relative metal abundances have been presented by Huebner (1978) for solar mixtures with mass fraction for hydrogen $X = 0.75$, mass fraction for helium $Y = 0.2321$, and mass fraction for the total of all heavier elements $Z = 0.0179$, and for $X = 0.35$, $Y = 0.6321$, and $Z = 0.0179$. These did not explicitly contain the collective effects for scattering as indicated in Equations (79) and (82); however, this correction was made in the application of these opacities in the solar model calculations by, for example, Bahcall *et al.* (1980). Plots of these opacities which do contain the correction for correlation in scattering are presented in Figures 9 through 12, and for a more limited density-temperature region of interest to the Sun in Figures 13 through 16.

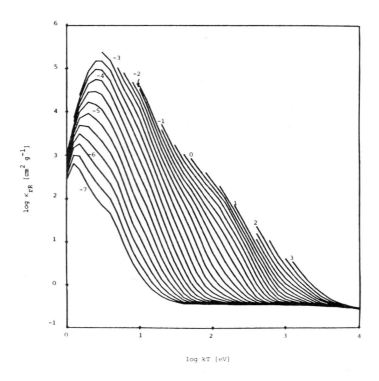

Fig. 9. Rosseland opacity for $X = 0.75$, $Y = 0.2321$, $Z = 0.0179$ vs temperature with density as a parameter. Densities (g cm^{-3}) are logarithmically spaced, three per decade. Relative metal abundances as given in Table III.

Table V contains parameters for fits to the opacities relevant to solar models based on the density-temperature relationship $r \equiv \rho/T_6^3 = 0.035$ and $r = 0.041$, where T_6 is the temperature in units of 10^6 K. The symbol ZRA refers to a mixture with relative metal abundances taken from Ross and Aller as presented in the abbreviated list given in Table IV. (For other mixtures see the next section.) The tabulated cofficients are to be used in the equation

$$\log \kappa = b_1 + b_2 \log T_6 + b_3 (\log T_6)^2 + b_4 (\log T_6)^3, \quad (1 \leq T_6 \leq 15), \qquad (88)$$

where the Rosseland mean opacity κ is in units of square centimetres per gram. These fits reproduce the calculated opacities for any of the mixtures in Table V to about 4%.

TABLE Va

Opacity fit parameters for $X = 0.75$, $Z = 0.0179$

$r = 0.035$					$r = 0.041$			
	b_1	b_2	b_3	b_4	b_1	b_2	b_3	b_4
ZRA	1.721	−0.04829	−3.222	1.806	1.765	−0.08720	−3.206	1.811
ARA, C	1.708	+0.07298	−3.377	1.862	1.752	+0.03845	−3.368	1.872
ZRA, N	1.717	−0.03587	−3.223	1.801	1.761	−0.07208	−3.212	1.809
ZRA, O	1.748	−0.1457	−3.019	1.695	1.791	−0.1871	−2.997	1.698
ZRA, Ne	1.723	−0.07569	−3.178	1.787	1.767	−0.1141	−3.161	1.792
ZRA, Mg	1.723	−0.07496	−3.204	1.809	1.767	−0.1139	−3.186	1.813
ZRA, Fe	1.722	−0.1479	−3.023	1.696	1.766	−0.1830	−3.013	1.704

TABLE Vb

Opacity fit parameters for $X = 0.35$, $Z = 0.0179$

$r = 0.035$					$r = 0.041$			
	b_1	b_2	b_3	b_4	b_1	b_2	b_3	b_4
ZRA	1.698	−0.09306	−3.205	1.794	1.744	−0.1399	−3.173	1.790
ZRA, C	1.686	+0.02527	−3.354	1.847	1.732	−0.01691	−3.331	1.849
ZRA, N	1.694	−0.07879	−3.209	1.790	1.740	−0.1229	−3.182	1.790
ZRA, O	1.723	−0.1717	−3.028	1.693	1.769	−0.2236	−2.986	1.685
ZRA, Ne	1.700	−0.1188	−3.162	1.775	1.746	−0.1649	−3.132	1.773
ZRA, Mg	1.700	−0.1206	−3.185	1.796	1.746	−0.1687	−3.151	1.792
ZRA, Fe	1.699	−0.1999	−2.990	1.674	1.746	−0.2443	−2.962	1.673

(The solar mixtures ZRA are reproduced to within 3.2%.) To obtain opacities as a function of ρ at a given temperature T_6 from opacity values at $r_1 \equiv 0.035$ and $r_2 = 0.041$, the following empirical interpolation is useful.

$$\kappa_r = [\kappa_{r_2} (r^{1/2} - r_1^{1/2}) + \kappa_{r_1} (r_2^{1/2} - r^{1/2})]/(r_2^{1/2} - r_1^{1/2}),$$

$$= [\kappa_{r_2} (r^{1/2} - 0.1871) + \kappa_{r_1} (0.2025 - r^{1/2})]/0.01540, \qquad (89)$$

for $r_1 \leq r \leq r_2$.

Using Equation (1b) instead of Equation (1) − with the lower limit on the integral set to the plasma frequency ν_p − increases the opacity in the center of the Sun by only 0.3%.

5. Uncertainties in the Opacity

The effect of uncertainties in composition on opacity has been discussed by Huebner (1978) where the opacities from different relative metal abundances, but the same hydrogen and helium abundances, were compared. A more systematic series of calculations,

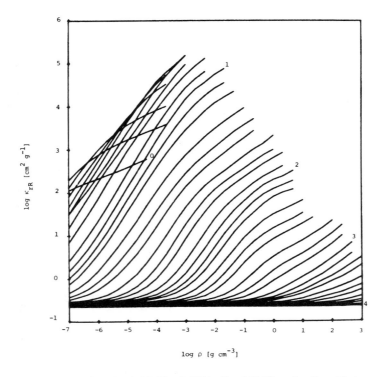

Fig. 10. Rosseland opacity for $X = 0.75$, $Y = 0.2321$, $Z = 0.0179$ vs density with temperature as a parameter. Temperatures (eV) are spaced approximately logarithmically (1.25, 1.5, 2.0, 2.5, 3.0, 4.0, 5.0, 6.0, 8.0, 10.) ten per decade. Relative metal abundances as given in Table III

reducing the number abundance of the metals C, N, O, Ne, Mg, or Fe one at a time by 25% (before renormalization) has been carried out. The results are summarized in Table V. This is useful for checking the sensitivity of solar structure models to the uncertainty in metal abundances.

Comparisons of opacity have been made between various astrophysical opacity calculations including those of Cox (1965), Cox and Stewart (1965), Cox *et al.* (1965), Carson and Hollingsworth (1968), Carson *et al.* (1968), Watson (1969), Cox and Stewart (1970a, b), and Cox and Tabor (1976). Watson's opacities appear to be wrong; possibly the absorption from some lines was counted twice. In general there is good agreement, but it is difficult to pinpoint the cause where opacities disagree because of complexities introduced through the many elements in a mixture. About ten programs have been available at various times to make comparisons for individual elements or simple mixtures in overlapping regions of density, temperature, and atomic number. Four of these opacity programs exist at Los Alamos. Various independent investigators have made comparisons using the Los Alamos programs LEO (light elements opacity) and HEO (heavy element opacity) and the Los Alamos Astrophysical Opacity Library (Huebner *et al.*, 1977) as standards. In all comparisons most disagreements have been traced and explained. LEO, based on a scaled Hartree–Fock model of the atom, and HEO, based on

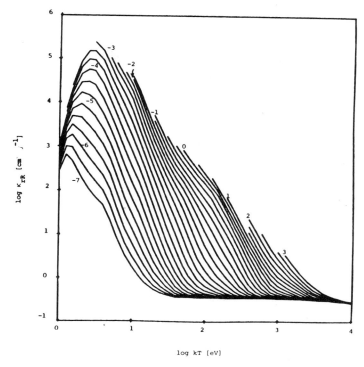

Fig. 11. Same as Figure 9, except $X = 0.35$, $Y = 0.6321$, $Z = 0.0179$.

the Thomas–Fermi potential, have also been compared with each other and the Astro-physical Opacity Library and are in good agreement (better than 30%, see below). From the general nature of these comparisons, and from the many calculations, studies of results, and improvements that have been made (see Magee *et al.*, 1975), some rules have emerged to estimate opacity uncertainties.

An analysis of the accuracy[1] of the Rosseland mean opacity, based on uncertainties of various contributing processes, is very difficult. The most accurate region is the high-temperature, low-density region where scattering dominates. There the accuracy is better than 5%. This is not an important region for the Sun. The region of next highest accuracy is the high-temperature and medium-to-high-density region where inverse bremsstrahlung dominates. Although there are large resonance effects of the frequency-dependent cross-section when an electron is in a quasi-bound state in the continuum, the Rosseland opacity for solar mixtures is not affected by more than a few percent. This has not been investigated extensively for the more general cases. The overall accuracy in this region is better than 10%. In the regions discussed so far, it was assumed that all elements were completely stripped of their electrons. At lower temperatures where a few electrons are bound, the photoelectric effect will dominate and at still lower temperatures (and particularly for heavier elements) spectral line absorption will dominate the Rosseland

[1] Accuracy of opacity is based on the accuracy of the basic atomic data and on agreement of results obtained from various opacity codes.

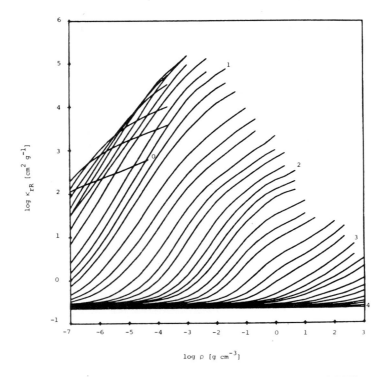

Fig. 12. Same as Figure 10, except X = 0.35, Y = 0.6321, Z = 0.0179.

opacity. In the bound–free dominated region the accuracy is about 15–20% depending somewhat on the perturbing effects of neighboring ions on the photoelectric cross-section near the edges and on the number of spectral lines. Bound–bound transitions can increase the opacity from continuum processes by as much as a factor of ten or more. In this last region, where spectral lines dominate, the accuracy is about 30%. It has already been mentioned that at temperatures less than 1 eV the bound–bound cross-section of the far wings on the red side of the Lyman-α line of hydrogen dominates the Rosseland opacity. This can give large uncertainties at low density where free–free absorption does not dominate over extinction in the far line wing. For further discussions on the uncertainty caused by metals see Magee *et al.* (1984).

Careless application of average atom models[2] to low-Z elements at temperatures of a few eV has shown discrepancies in opacity by one to two factors of ten. This shows that opacity codes must not be applied indiscriminately. Interpolation of opacity between light elements can lead to similar errors, as illustrated in Figure 17; but in this case the cause is the shell structure of the dominant ions. Opacities are highly nonlinear and must be considered carefully.

At temperatures below 1 or 2 eV molecular processes will also contribute to opacity. Although molecular processes are beyond the scope of this presentation, it is well to keep in mind that formation of molecules does not necessarily mean that the opacity

[2] See beginning of Section 2 for distinction from mean ion model.

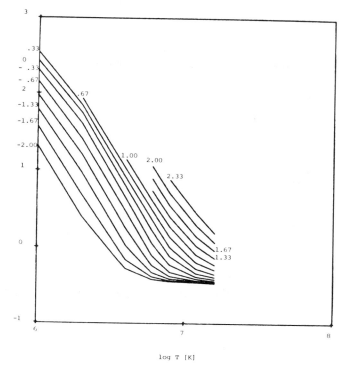

Fig. 13. An expanded region of Figure 9. The logarithm of the density (g cm^{-3}) is given for each curve.

is increased over that from atoms alone, As molecules form, the abundance of some atoms decreases. If these atoms have large extinction cross-sections near the peak of the Rosseland weighting function and the molecules have a low cross-section in that area, the opacity can actually decrease.

Finally it should be mentioned that the pronounced C—N—O bump in opacities as a function of temperature reported by Carson (see, for example, Carson, 1976; Carson and Stothers, 1976; Vemury and Stothers, 1978; Carson *et al.*, 1981) cannot be reproduced by opacity codes at Los Alamos or at Livermore. These codes (which include some models based on the Thomas—Fermi potential, similar to that used by Carson) produce a shoulder in some astrophysical mixtures or at best a very small bump in pure C, N, or O. At the density ($\sim 10^{-7}$ g cm^{-3}) and temperature (~ 50 eV) where this bump is supposed to occur in C, N, and O, the dominant species are hydrogen-like ions; some helium-like ions are of much lesser abundance. It is curious that a pronounced bump should occur in Carson's opacities for these simple ions, particularly since the effects from free electrons at these low densities are very small and hydrogenic theory should be very accurate. Scattering is found to be an important process under the peak of the Rosseland weighting function. This has now been resolved; the C—N—O bump in the Carson opacities does not exist (see Carson *et al.*, 1984).

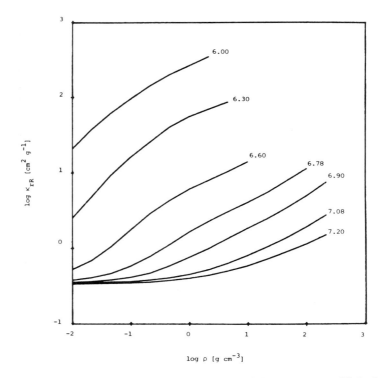

Fig. 14. An expanded region of Figure 10. The logarithm of the temperature (K) is given for each curve.

Acknowledgements

Parts of this presentation are influenced by collaborative efforts on a review with Dr A. L. Merts, to whom I wish to express my gratitude. It is a pleasure to thank Dr A. N. Cox, Prof. L. Biermann, Dr D. Mihalas, Dr E. C. Shoub, and Dr D. H. Sampson for informative discussions and helpful comments, and Mrs S. S. Forrest for careful reading of the manuscript.

This work was performed under the auspices of the U.S. Department of Energy.

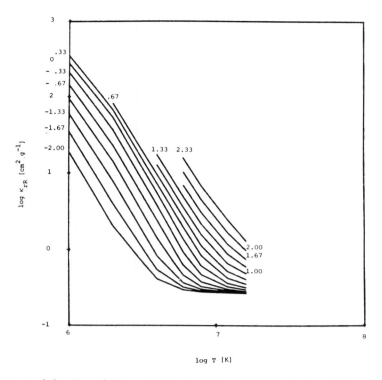

Fig. 15. An expanded region of Figure 11. The logarithm of the density (g cm^{-3}) is given for each curve.

W. F. HUEBNER

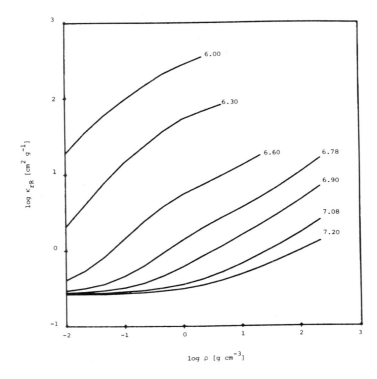

Fig. 16. An expanded region of Figure 12. The logarithm of the temperature (K) is given for each curve.

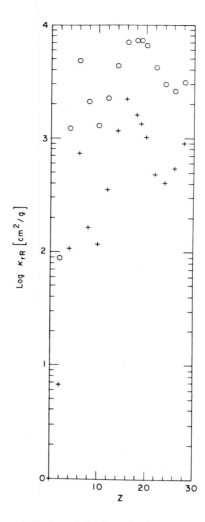

Fig. 17. Example of the nonlinear behavior of the Rosseland opacity with respect to atomic number. Circles correspond to $kT = 100$ eV, $\rho = 4.6$ g cm^{-3}. Crosses correspond to $kT = 100$ eV, $\rho = 0.1$ g cm^{-3}.

References

Armstrong, B. H., Johnston, R. R., Kelly, R. S., DeWitt, H. E., and Brush, S. G.: 1966. 'Opacity of High Temperature Air' in C. A. Rouse (ed.), *Progress in High Temperature Physics and Chemistry*, Vol. 1, Pergamon Press, Oxford, New York.

Bahcall, J. N., Lubow, S. H., Huebner, W. F., Magee, N. H., Jr, Merts, A. L., Argo, M. F., Parker, P.D., Rozsnyai, B., and Ulrich, R. K.: 1980, *Phys. Rev. Lett.* **45**, 945.

Bates, D. R. and Damgaard, A.: 1949, *Phil. Trans. Roy. Soc. London* **A242**, 101.

Bethe, H. A. and Salpeter, E. E.: 1957, *Quantum Mechanics of One- and Two-Electron Atoms*, Springer-Verlag, Berlin, Göttingen, Heidelberg.

Biermann, L.: 1938, *Astron. Nach.* **264**, 361.

Broad, J. T. and Reinhardt, W. P.: 1976, *Phys. Rev.* **A14**, 2159.

Burgess, A. and Seaton, M. J.: 1960, *Monthly Notices Roy. Astron. Soc.* **120**, 121.

Canuto, V.: 1970, *Astrophys. J.* **159**, 641.

Carson, T. R.: 1971, 'Stellar Opacities', in C. A. Rouse, (ed.), *Progress in High Temperature Physics and Chemistry*, Vol. 4, Pergamon Press, Oxford, New York.

Carson, T. R.: 1976, *Ann. Rev. Astron. Astrophys.* **14**, 95.

Carson, T. R. and Hollingsworth, H. M.: 1968, *Monthly Notices Roy. Astron. Soc.* **141**, 77.

Carson, T. R. and Stothers, R.: 1976, *Astrophys. J.* **204**, 461.

Carson, T. R., Mayers, D. F., and Stibbs, D. W. N.: 1968, *Monthly Notices Roy. Astron. Soc.* **140**, 483.

Carson, T. R., Stothers, R., and Vemury, S. K.: 1981, *Astrophys. J.* **244**, 230.

Carson, T. R., Huebner, W. F., Magee, N. H., Jr, and Merts, A. L.: 1984, *Astrophys. J.* **283**, 466.

Condon, E. U. and Shortley, G. H.: 1953, *The Theory of Atomic Spectra*, Cambridge Univ. Press, 1935 (reprinted with corrections 1953).

Cowan, R. D. and Ashkin, J.: 1957, *Phys. Rev.* **105**, 144.

Cowan, R. D. and Kirkwood, J. G.: 1958a, *Chem. Phys.* **29**, 264.

Cowan, R. D. and Kirkwood, J. G.: 1958b, *Phys. Rev.* **111**, 1460.

Cox, A. N.: 1965, 'Stellar Absorption Coefficients and Opacities' in L. H. Aller and D. B. McLaughlin (eds), *Stars and Stellar Systems*, Vol. 8: *Stellar Structure*, Univ. of Chicago Press, Chicago.

Cox, A. N. and Stewart, J. N.: 1965, *Astrophys. J. Suppl.* **11**, 22.

Cox, A. N. and Stewart, J. N.: 1970a, *Astrophys. J. Suppl.* **19**, 243.

Cox, A. N. and Stewart, J. N.: 1970b, *Astrophys. J. Suppl.* **19**, 261.

Cox, A. N., Stewart, J. N. and Eilers, D. D.: 1965, *Astrophys. J. Suppl.* **11**, 1.

Cox, A. N. and Tabor, J. E.: 1976, *Astrophys. J. Suppl.* **31**, 271.

Cox, P. J. and Giuli, R. T.: 1968, *Principles of Stellar Structure*, Vol. 1, Gordon and Breach, New York, London, Paris, Toronto.

Feynman, R. P., Metropolis, N., and Teller, E.: 1949, *Phys. Rev.* **75**, 1561.

Flowers, E. and Itoh, N.: 1976, *Astrophys. J.* **206**, 218.

Fischer, J.: 1931a, *Ann. Phys.* **8**, 821.

Fischer, J.: 1931b, *Ann. Phys.* **11**, 489.

Geltman, S.: 1962, *Astrophys. J.* **136**, 935.

Griem, H.: 1964, *Plasma Spectroscopy*, McGraw-Hill, New York, San Francisco, Toronto, London.

Hubbard, W. B.: 1966, *Astrophys. J.* **146**, 858.

Hubbard, W. B. and Lampe, M.: 1969, *Astrophys. J. Suppl.* **118**, 297.

Huebner, W. F.: 1970, *J. Quant. Spect. Rad. Transfer* **10**, 949; **11**, 142.

Huebner, W. F.: 1978, 'Solar Opacity and Equation of State', in G. Friedlander (ed.), *Proc. Informal Conf. on Status and Future of Solar Neutrino Research*, BNL Rept 50879, Vol. I, p. 107.

Huebner, W. F., Argo, M. F. and Olsen, L. D.: 1978, *J. Quant. Spect. Rad. Transfer* **19**, 93.

Huebner, W. F., Merts, A. L., Magee, N. H., Jr, and Argo, M. F.: 1977, 'Astrophysical Opacity Library', Los Alamos Scientific Laboratory Rept LA-6760-M.

Karzas, W. J. and Latter, R.: 1961, *Astrophys. J. Suppl.* **6**, 167.

Lampe, M.: 1968a, *Phys. Rev.* **170**, 306.

Lampe, M.: 1968b, *Phys. Rev.* **174**, 276.

Latter, R.: 1955, *Phys. Rev.* **99**, 1854.

Lee, C. M. and Thorsos, E. I.: 1978, *Phys. Rev.* **A17**, 2073.

Liberman, D. A.: 1979, *Phys. Rev.* **B20**, 4981.

Magee, N. H., Jr, Merts, A. L., and Huebner, W. F.: 1975, *Astrophys. J.* **196**, 617.

Magee, N. H., Jr, Merts, A. L., and Huebner, W. F.: 1984, *Astrophys. J.* **283**, 264.

Mayer, H.: 1948, 'Methods of Opacity Calculations', Los Alamos Scientific Laboratory Rept LA-647.

McDougall, J. and Stoner, E. C.: 1938, *Phil. Trans. Roy. Soc. London* **A237**, 67.

Mestel, L.: 1950, *Proc. Camb. Phil. Soc.* **46**, 331.

Mihalas, D.: 1978, *Stellar Atmospheres* (2nd ed), Freeman, San Francisco.

Moszkowski, S. A.: 1962, *Prog. Theor. Phys.* **28**, 1.

Ohmura, T. and Ohmura, H.: 1960, *Astrophys. J.* **131**, 8.

Ohmura, T. and Ohmura, H.: 1961, *Phys. Rev.* **121**, 513.

Peach, G.: 1967, *Mem. Roy. Astron. Soc.* **71**, 13.

Petschek, A. G. and Cohen, H. D.: 1972, *Phys. Rev.* **A5**, 383.

Pratt, R. H., Ron, A., and Tseng, H. K.: 1973, *Rev. Mod. Phys.* **45**, 273 and 663.

Ross, J. E. and Aller, L. H.: 1976, *Science* **191**, 1223.

Rosseland, S.: 1924, *Monthly Notices Roy. Astron. Soc.* **84**, 525.

Rozsnyai, B. F.: 1972, *Phys. Rev.* **A5**, 1137.

Sampson, D. H.: 1959, *Astrophys. J.* **129**, 734.

Sauter, F.: 1931a, *Ann. Phys.* **9**, 217.

Sauter, F.: 1931b, *Ann. Phys.* **11**, 454.

Sauter, F.: 1933, *Ann. Phys.* **18**, 486.

Shore, B. W. and Menzel, D. H.: 1968, *Principles of Atomic Spectra*, Wiley, New York, London, Sydney.

Slater, J. C.: 1960, *Quantum Theory of Atomic Structure*, Vols. I and II, McGraw-Hill, New York, Toronto, London.

Spitzer, L., Jr: 1962, *Physics of Fully Ionized Gases* (2nd edn), Interscience, New York, London.

Stewart, J. C.: 1965, *J. Quant. Spect. Rad. Transfer* **5**, 489.

Stewart, J. C. and Pyatt, K. D. Jr: 1961, 'Theoretical Study of Optical Properties', Air Force Special Weapons Center Rept SWG-TR-61-71, Vol. 1.

Stewart, J. C. and Pyatt, K. D.: 1966, *Astrophys. J.* **144**, 1203.

Stobbe, M., 1930, *Ann. Phys.* **7**, 661.

Vemury, S. K. and Stothers, R.: 1978, *Astrophys. J.* **225**, 939.

Vitense, E.: 1951, *Z. Astrophys.* **28**, 81.

Watson, W. D.: 1969, *Astrophys. J.* **157**, 375.

Wildt, R.: 1939, *Astrophys. J.* **89**, 295.

Zink, J. W.: 1968, *Phys. Rev.* **176**, 279.

Zink, J. W.: 1970, *Astrophys. J.* **162**, 145.

Los Alamos National Laboratory,
T-4/T-6,
Los Alamos, NM 87545,
U.S.A.

CHAPTER 4

HYDRODYNAMIC AND HYDROMAGNETIC PHENOMENA IN THE DEEP SOLAR INTERIOR

WILLIAM H. PRESS

1. Introduction

It is a matter of taste to decide where the 'deep' solar interior begins. For definiteness let us take the boundary to be that radius where, moving inward from the surface, the radiative temperature gradient

$$\nabla_r \equiv \left(\frac{\mathrm{d}\ln T}{\mathrm{d}\ln P}\right)_{\mathrm{rad}} = \frac{1}{4}\left(\frac{\kappa\mathscr{F}}{cg}\right)\left(\frac{P}{\frac{1}{3}aT^4}\right) \tag{1}$$

first becomes less than the adiabatic gradient

$$\nabla_{\mathrm{ad}} \equiv \left(\frac{\mathrm{d}\ln T}{\mathrm{d}\ln P}\right)_s = \frac{\Gamma_2 - 1}{\Gamma_2} \tag{2}$$

so that a radiatively stable solution with no fluid motions becomes at least theoretically possible. Whether, or to what extent, the *actual* fluid state is described by this static picture is precisely the issue to be addressed in this chapter. [In Equations (1) and (2), the symbols T, P, κ, $\mathscr{F}$, c, g, s, and Γ_2 have their usual meanings: temperature, total pressure, opacity, flux (energy per area per time), velocity of light, local gravitational acceleration, specific entropy, and adiabatic exponent of the second kind, respectively (see, e.g., Cox and Giuli, 1968, Ch. 13).]

Defined in this way, the deep solar interior is, observationally, an extremely inaccessible place, more remote from us in many respects than the most distant galaxies. Consider that it is shielded from us by an optical depth of, in round numbers, $\tau \sim 10^8$, and that the 'shield' is no passive slab of material through which we might hope to delicately peer, but rather a hydrodynamically active convective envelope, full of turbulence, magnetic fields (and probably the dynamo that generates them), and organized fluid motions on a variety of scales. One is surprised that we should expect to know anything at all about the hydrodynamic state of the deep interior. In fact, we know only very little.

If there is an 'establishment' view of the hydrodynamical and magnetohydrodynamical state of the deep interior, it must certainly be that the deep interior is an unmagnetic, quiescent, hot place, evocative of the sleepy, sun-baked setting for some Graham Greene novel. We can summarize briefly the rationale for this picture:

There ought to be no significant magnetic fields below the convection zone because magnetic buoyancy would have advected them outward on a very short time-scale (Parker, 1955; Parker, 1974 and references therein). This leaves open the question of whether some special topology of primordial field might survive. However, the most widely held view is that equilibrium configurations are possible only in cases where a special symmetry is assumed, and that even these are unstable; with any generic perturbation from special symmetry, hydrostatic equilibrium is not generally possible (Parker, 1965; Jokipii and Parker, 1969).

While recognized as crude, the mixing length theory of convection is generally taken as at least a zeroth-order description of fluid motions outside of the deep interior. Among necessary 'first-order' (though still phenomenological) corrections to the mixing-length theory, it is generally appreciated that some account must be given of convective over-shoot, the penetration of downward-moving 'bubbles' or 'eddies' into the outer part of the deep interior as a consequence of their excess density or momentum at the boundary predicted by the mixing-length theory. There are sensible models of this effect (e.g. Shaviv and Salpeter, 1973) which indicate that the overshoot ought to extend no more than a few tenths of a pressure scale height into the deep interior. If we allowed for an upper bound of *one* pressure scale height (which is only 0.1 of the radius of the boundary defined by Equations (1) and (2) for a standard solar model), we would certainly be in accord with almost all previous thought on the subject.

Of course, the boundary of our quiescent deep interior is now hydrodynamically 'noisy', so we should not expect fluid velocities to vanish *exactly* anywhere within. Since the deep interior is radiatively stable, it makes sense to describe its driven fluid motions in terms of the normal modes of stellar perturbation theory, whose elegant formulation traces back to Cowling (1941), Ledoux (1958) and others (see Ledoux, 1974, 1978). Rhodes and Brown, elsewhere in this volume, review the formalism in detail. Here we can briefly summarize: Modes are characterized by spherical harmonic indices l, m and by their number of radial nodes n. In the WKB limit, the three numbers l, m, n are equivalent to the wavevector $\mathbf{k}$ of a travelling wave. In this limit there is a dispersion relation between $\mathbf{k}$ and ω (the mode frequency); in general, there is an eigenvalue problem for $\omega(l, m, n)$. It is characteristic of stably stratified, compressible, fluid systems that this eigenvalue problem (or dispersion relation) has two distinct branches of solutions. One, called p-modes, has pressure as the dominant restoring force; p-modes are, essentially, sound waves. The other, called g-modes, has gravity (i.e. buoyancy) as the restoring force; g-modes are analogous to internal gravity waves in the Earth's ocean of atmosphere.

It is generally believed that, while the details of the modal spectrum have important *diagnostic* applications to the deep interior (if they can be measured), the amplitude of fluid motions associated with these modes is far too small to have any important physical consequences (such as affecting the nuclear and radiative processes that power the Sun).

We have now almost exhausted the list of widely held beliefs about deep hydrody-namics. The reader who skips on to the next chapter in this volume need feel no inordinate guilt; what remain to be discussed are loose ends, both observational and theoretical, which are left untied by the establishment view, or which might possibly cast doubt on that view.

2. Troublesome Observations

There seem to be three kinds of observations which have led to searches for alternatives to the standard picture of a hydrodynamically quiescent deep interior. Unfortunately, the observations span a large range from highly certain to highly controversial. If one is so rash as to take them all at face value, the observations point in very different directions. One is advised to proceed with caution. (Francis Crick is supposed to have remarked that "no theory should explain all of the facts, since some of the facts are wrong".)

2.1. LIGHT ELEMENT DEPLETION

Least controversial observationally are observations of lithium, beryllium, and boron abundances in the Sun and other low-mass, main sequence stars. (For a review of this subject, see Boesgaard, 1976, and Vauclair et al., 1978.) The elements Li, Be, and B are destroyed by nuclear reactions at relatively low temperatures, much less than those of the solar center but still rather higher than the temperature at the bottom of the convective zone of a standard solar model. For example, Li is destroyed at $T \approx 2.5 \times 10^6$ K, Be at 3.0×10^6 K, while the convective zone is not thought to go much deeper than 2.0×10^6 K. The surprising observational fact, then, is that Li is depleted by a factor of 100 in the Sun, while Be and B are normal.

In fact, the Sun is by no means unusual in this regard. Li is depleted in stars within young clusters, with the depletion systematically greater as one goes down in mass (Zappala, 1972). Comparing clusters of different ages shows that the depletion increases with increasing age (thus showing that the effect operates during main sequence burning rather than, say, in the descent to the main sequence).

The obvious interpretation is that turbulent mixing in the Sun extends to a temperature of between 2.5×10^6 and 3.0×10^6 K. If this mixing is interpreted as due to convective overshoot, that overshoot must extend more than two pressure scale heights into the stable region (Straus et al., 1976), and even deeper for stars more massive than the Sun. In the context of the conventional picture this seems highly unlikely. (In Section 3, we shall discuss the hydrodynamics of convective overshoot and argue for a view somewhat different from the conventional picture.)

2.2. LOW ^{37}Cl NEUTRINO COUNTS

The solar neutrino experiment on ^{37}Cl by Davis and his associates (Davis et al., 1968; Davis, 1969; Davis and Evans, 1973; Rowley et al., 1977) is arguably the most important single observation of stellar and solar astronomy in the last 40 years. (As a corollary, the importance of a possible future experiment with ^{71}Ga should also be emphasized; cf. Bahcall, 1979.) Solar neutrinos probe directly the central core of the Sun without regard to properties of the obscuring solar envelope. As is so well known, a substantial discrepancy exists between the observed neutrino count rate in Davis's ^{37}Cl detector and the theoretical predictions of essentially all standard solar models.

There is no particular reason to tie a solution of the solar neutrino puzzle to any specific *hydrodynamical* effect. Newman, in Chapter 17 of work, discusses a whole range of proposed solutions. Along with many other suggestions, however, a number of hydrodynamical effects have been raised, some of which we will mention in Section 3 below. Generally speaking, one invokes fluid motions *either* to mix the radiatively stable solar core on a nuclear burning time-scale or faster — which, for models of fixed luminosity and radius, decreases the central temperature and neutrino flux — see Iben, 1969; Shaviv and Salpeter, 1971; Rood, 1972; Ezer and Cameron, 1972; Ulrich and Rood, 1973; Schatzman and Maeder, 1981) — *or* as the basis for some mechanism of energy transport as an alternative to radiative diffusion (which decreases the central temperature by decreasing the 'effective' opacity — see Bahcall *et al.*, 1969; Bahcall and Ulrich, 1971; Newman and Fowler, 1976).

2.3. POSSIBILITY OF RAPIDLY ROTATING OR NONAXISYMMETRIC CORE

It is well established that there is a break in the curve of typical rotation velocity as a function of spectral type for stars on the main sequence. Stars earlier than F5 (more massive than $1.3 M_\odot$) rotate rapidly with typical surface velocities in the range 50–200 km s^{-1}; over a large mass range, the specific angular momentum $\mathcal{J}$ (in cm^2 s^{-1}, assuming uniform rotation) is found to vary with mass M (in $M_\odot$) as

$$\log_{10} \mathcal{J} \approx 16.7 + 0.66 \log_{10} M. \tag{3}$$

For spectral types F5 or later, including the Sun, much smaller rotational velocities are found. A uniformly rotating Sun has $\log_{10} \mathcal{J} \approx 14.9$. There is also some evidence from the Hyades and Pleiades that (at least for the first 10^8–10^9 years) rotational velocities decrease with stellar age. For observational reviews, see Kraft (1967, 1968, 1970) and several papers in Slettebak (1970), especially Abt (1970).

The question is: "Where is the Sun's (and other late stars') 'missing' angular momentum?" Dicke (1964) and others, including Roxburgh (1964) and Plaskett (1965), proposed that the large angular momentum is even now present in the Sun, locked in a rapidly rotating core of rotation period on the order of two days or less. Such a core ought to produce a measurable oblateness in the solar figure. Dicke, Hill and Goldenberg designed and constructed an instrument capable of measuring such an effect. A positive result of oblateness $(5 \pm 0.7) \times 10^{-5}$ was found and reported (Dicke and Goldenberg, 1967, 1974; for reviews, see Dicke, 1970, 1974).

The considerable controversy surrounding the solar oblateness measurement cannot be treated adequately here. Complicated effects of photospheric radiative transfer can be blamed for a possibly spurious oblateness signal. Hill and Stebbins (1975) report an independent measurement that is more than five times smaller than the Dicke–Goldenberg result, and not statistically inconsistent with zero. The controversy has sparked theoretical as well as observational work (references are given in Section 3).

In 1976 Dicke put forth an additional interpretation of the same 1966 data. Analysis of the time dependence of the residual variance in the oblateness data appears to show a 12.2 day periodicity, explainable phenomenologically as a nonaxisymmetric, ellipsoidal, rigid rotator, oriented at almost a right angle to the rotation axis, producing $\sim$ 10 km distortion in the shape of the solar photosphere (Dicke, 1976, 1977). There has to date been no observational corroboration of this effect. If it is correct, it must certainly wreak havoc on the standard view of the solar interior's fluid state, both because of the nonstandard claimed period, and because implausibly large nonfluid stresses must be posited to explain the nonaxisymmetry.

2.4. 160 MINUTE OSCILLATIONS

It now seems established that there are solar oscillations, coherent over a scale comparable to that of the solar disk, with a period of 160.01 min and velocity amplitude on the order of 1 m s^{-1} (Severny *et al.*, 1976, 1979; Brown *et al.*, 1978; Snider *et al.*, 1978; Sherrer *et al.*, 1979; see also Rhodes and Brown in this volume). Interpreted solar-seismologically, the oscillation must be a *g*-mode (buoyancy driven), because the spectrum of *p*-mode (acoustic disturbance) eigenfrequencies lies entirely at periods smaller than 1 h — physically related to the sound-velocity crossing time of the whole solar diameter. Because the oscillation is seen coherently across a good fraction of a solar diameter, it must correspond to a low-order spherical harmonic, $l \approx 2$. Because the 140 min period is not vastly *larger* than the sound crossing time (or, more precisely, the period of the *f*-mode that is shared by the *p*- and *g*-mode branches), the radial node number cannot be too large. A value of $n \approx 10$ seems indicated (Christensen-Dalsgaard and Gough, 1976). Buoyancy-driven oscillations are rooted, by their very nature, in a stably stratified region; in an unstably stratified convective zone, their radial eigenfunctions decrease exponentially as an evanescent wave. The conclusion, therefore, is that the 160 min oscillations must reflect a disturbance that extends into the deep solar interior. Since its radial node number is not too large, its evanescent decay in the convection zone is not too fast, and the disturbance extends out to the solar surface, where it is observed. It seems likely, at first inspection, that the driving force of the oscillation is also located in the evanescent-wave region, associated in some manner with the convective motion. What is striking, however, is that only this one mode (or possibly just two others, at 134 min and 148 min; Severny *et al.*, 1979) should be excited. If there is nothing special about 160 min in convection, then one is motivated to search for new sources of excitation in the deep interior itself, or for alternative interpretations of the data as other than a deep *g*-mode.

3. Theoretical Topics

There is at present no generally accepted, unified theoretical treatment of hydrodynamical effects in the deep interior, not least because there are no generally accepted observations (except possibly the 160 min oscillations) which require deep hydrodynamics for their explanation.

Rather than finding in the literature a unified framework, one finds an assortment of specific theoretical proposals, representing a large variety of viewpoints and styles. In this section, only two topics will be considered in any detail: the possibility of nucleothermal destabilization of *g*-modes (first proposed by Dilke and Gough, 1972), and the state of knowledge on convective overshoot. References are listed, without further discussion, for two additional topics: the redistribution of angular momentum in a sun with a rapidly rotating core; and the possible mechanical effects of *g*-modes on the Sun's energy transport.

3.1. NUCLEO-THERMAL DESTABILIZATION OF g-MODES

Dilke and Gough (1972) suggested that the Sun has been unstable to low-order (in l, m, and n) g-modes at certain times in its history of main sequence evolution. If so, then the instability might grow to large enough amplitude to give significant mixing of the core, providing a reduced solar neutrino flux and (more speculatively) an explanation for terrestrial ice ages. The idea of destabilizing g-modes by nuclear-thermal effects has subsequently been considered by Christensen-Dalsgaard *et al.* (1974); Boury *et al.* (1975); Shibahashi, *et al.* (1975); Unno (1975); and others. The related problem for radial modes (which are not g-modes) has been discussed by Defouw *et al.* (1973) and Schwarzschild and Harm (1973); no radial instabilities are found for these modes.

The physical basis of the Dilke–Gough mechanism is not difficult to understand. Locally, g-modes consist of oscillatory motion by undeformed, rod-like fluid elements at some angle α to the vertical, the elements displaced rigidly along their length. Because the fluid is stably stratified, the rods sink back through their equilibrium position and oscillate with a frequency of

$$\omega = N \cos \alpha, \tag{4}$$

where N is the buoyancy or Brunt–Väisälä frequency given by

$$N^2 = g^2 \rho (\nabla - \nabla_{\text{ad}})/P. \tag{5}$$

A small fluid element displaced vertically will oscillate around its mean position at a frequency N. Throughout most of the deep solar interior, the buoyancy frequency has the value 2 to 3×10^{-3} rad s^{-1}, going to zero inside about $r = 1 \times 10^{10}$ cm approximately linearly, as

$$\frac{N(r)}{r} = \frac{\frac{4}{3}\pi G \rho_c^{3/2} (\nabla - \nabla_{\text{ad}})_c^{1/2}}{P_c} = 2.7 \times 10^{-13} \text{ cm}^{-1} \text{ s}^{-1}. \tag{6}$$

(Here subscript c refers to central values of a quantity.) Note that N is an *upper* bound on the frequencies of g-modes, so any more rapid oscillations observed must always be p-modes (acoustic disturbances).

Absent any dissipation, or any variation in the nuclear energy generation rate with wave phase, and the eigenfrequencies of the g-modes are given by a self-adjoint set of equations, so they are real, with amplitudes neither growing nor dying. Of course, this is an idealized state of affairs, not corresponding to the real physics. Dissipation, which damps the modes, occurs principally through radiative viscosity. A fluid element of size λ above (below) its equilibrium position is colder (hotter) than its surroundings, because the medium is convectively stable. It will equilibrate with its surroundings on a thermal time-scale

$$\tau_{\text{eq}} \sim \lambda^2/\sigma, \tag{7}$$

where σ is the heat diffusivity (dimensions $cm^2 \, s^{-1}$) related to the more commonly tabulated opacity by

$$\sigma = \left(\frac{\frac{1}{3}aT^4}{P}\right)\left(\frac{c}{\kappa\rho}\right) \tag{8}$$

(for details, see Press, 1981a, and references therein). The quality factor Q for the oscillation is therefore on the order of

$$Q \sim \omega\tau_{eq} \sim \omega/\sigma k^2, \tag{9}$$

where k is the wavenumber of the mode $\sim 2\pi/\lambda$ and ω is its frequency.

Now consider the effect of a wave-induced displacement of an element in the nuclear energy generating region. The nuclear energy generation rate ϵ ($erg \, gm^{-1} \, s^{-1}$) is an essentially instantaneous function of the local temperature and pressure. (Although the reaction rate per nucleus may be long, the number of nuclei in any macroscopic volume is immense.) An element displaced downward will be hotter than its surroundings, and it may also be richer in internal nuclear fuel. Therefore, it generates excess energy, and excess buoyancy, and therefore will tend to overshoot farther on the rise. This is an antidamping effect whose contribution to the quality factor Q is *negative* and can be estimated as follows: The excess fractional buoyancy is the excess fractional energy generation multiplied by a small number, the fraction of internal energy generated by nuclear processes in one cycle of the wave:

$$\left(\frac{\delta P}{P}\right)_{excess} \sim \left(\frac{\delta\epsilon}{\epsilon}\right)\frac{1}{\omega\tau_{nuc}}. \tag{10}$$

Since the energy generating region of size R_{nuc} is in thermal equilibrium, its energy generation time τ_{nuc} must equal its thermal diffusion time-scale,

$$\tau_{nuc} \sim R_{nuc}^2/\sigma. \tag{11}$$

Since ϵ varies for small changes as some power law in temperature and density, we have

$$\frac{\delta\epsilon}{\epsilon} \sim \frac{\delta P}{P}. \tag{12}$$

(Equation (12) also describes the effect of composition gradients of order unity in one pressure scale height.) So, Equations (10)–(12) give

$$Q = \frac{-(\delta P/P)}{(\delta P/P)_{excess}} = \frac{-\omega R_{nuc}^2}{\sigma}. \tag{13}$$

Note that this antidamping is on the same order as the radiative damping for $k \sim 1/R_{nuc}$, i.e. for low-order modes whose size is comparable to the size of the nuclear burning region. Since the coefficient on the right-hand side of Equation (12) is generally rather large, > 10, one has the definite possibility of net destabilization.

Several calculations (e.g. Christensen-Dalsgaard *et al.*, 1974; Unno, 1975; Noels *et al.*, 1975) have identified modes which are in fact destabilized, apparently by a combination of the strong temperature sensitivity of ^{3}He burning and the gradient of ^{3}He abundance established in the energy-generating core. If the gradient effect is in fact prerequisite, then one can imagine, following Dilke and Gough (1972), that an episode of instability leads to mixing which turns off the instability for perhaps 2.5×10^8 yr until the abundance gradient has had a chance to re-establish itself.

The literature on nucleo-thermal instabilities contains intriguing and elegant results. Nevertheless, one ought probably to adopt a very cautious attitude towards these *g*-mode instabilities. The essential problem is that Q^{-1} from Equation (13), the fractional amplifications per radian, is a very small number. At the solar center one has the numerical values $\sigma \sim 10^5$ cm^2 s^{-1}, $\omega \sim 10^{-3}$ s^{-1}, $R_{nuc} \sim 10^{10}$ cm, so $Q^{-1} \sim 10^{-12}$. Therefore, any neglected form of dissipation at a level $\gtrsim 10^{-12}$ fractional energy loss per radian would tilt the balance to net damping rather than antidamping.

One example of such a neglected effect is likely to be nonlinear wave—wave (or mode—mode) interactions (see Press, 1981a, §VIb). Such interactions can sap energy from a mode and distribute it into different mode numbers and frequencies. The effect is, in general, stabilizing. Since the interactions are nonlinear, the fractional energy loss depends on the amplitude of the wave, which is conveniently parametrized by the Richardson number Ri of the internal flow (the ratio of stabilizing stratification to destabilizing, wave-induced shear). Small amplitude waves correspond to large values of Ri. Nonlinear fractional energy losses per cycle can be estimated to be of order $Ri^{-1/2}$ (cf. Phillips, 1977, Equation 5.5.16). Therefore, any instability of the Dilke—Gough sort probably becomes self-limited due to wave—wave interactions at $Ri > 10^6$. Mixing of the fluid is implausible for this large a value of Ri, corresponding to a tiny mode amplitude.

Another neglected, or at least poorly understood, effect is the interaction of the unstable mode with the overlying convective zone and with the stellar atmosphere. While the mode eigenfunction may be mainly confined to the central region, it does have an evanescent-wave extension which extends (in principle) all the way to the surface of the star. It is not hard to imagine that unmodelled sources of loss at the $> 10^{-12}$ level are in fact present. Sources of *antidamping* are few and far between, requiring as they do a source of free energy; but sources of *damping* are likely to show themselves when any of the model's idealized assumptions are examined in detail.

3.2. DEPTH OF THE CONVECTIVE ZONE

Most calculations of solar and stellar evolution take the boundary between the radiatively stable interior and the convection zone to be at the point of loss of radiative stability (cf. Equations (1) and (2) above). The choice of this prescription is one of convenience only, since it is generally appreciated that, once initiated, the fluid motions of convection are capable of modifying their own boundary. The treatment of this modification falls under two rubrics:

Semiconvection describes effects associated with abundance gradients. These may (i) destabilize, or (ii) stabilize the convection in a region which — if it is convectively mixed — will be (i) stabilized or (ii) destabilized. The two cases can yield, respectively, (i) an intermittent region of convection, or (ii) a metastable region which can be either convective or radiative depending on the past history of its mixing. Semiconvection is not important for the solar convection zone, since the abundance gradients immediately beneath it are thought negligible.

Convective overshoot, on the other hand, describes effects associated with mechanical properties of the convective motion, in particular the fact that downward-moving convective eddies can penetrate the radiative interface by some finite distance, and can mix and entrain material across that interface.

The convective overshoot problem must be classed as a major unsolved problem in understanding the solar convection zone. There are observations suggestive of a solar convection zone extending rather deeper than the value $\approx 1.7 \times 10^{10}$ cm that obtains in standard models. The depletion of light elements has already been mentioned (Section 2.1), and apparently requires a thickness of at least 2.5×10^{10} cm. The spectrum of 5 min solar oscillations of high degree are claimed to require a convection zone at least as thick as 2×10^{10} cm (Lubow *et al.*, 1980; Berthomieu *et al.*, 1980; see also discussion in Christensen-Dalsgaard and Gough, 1980, and Gough, 1979), while Beckers and Gilman (1978) find that the absence of a polar vortex requires a convection zone whose depth is apparently on the order of 3×10^{10} cm or larger.

Various attempts to estimate the magnitude of the convective overshoot, and to express it in a form suitable for stellar evolutionary calculations, are found in the literature. Two rather distinct methodologies have emerged, which we can call the 'modal' and 'model' points of view.

In the modal case, one thinks about modes. One attempts to take the actual fluid equations (or at least an idealized set of equations which closely resembles the fluid equations), and then to solve for actual eigenfunctions or nonlinear solutions of the fluid flow. Pioneered by Veronis (1963), and Spiegel (1965), this approach also includes the more recent work by Gough *et al.* (1976), Latour *et al.* (1976), Toomre *et al.* (1976), Marcus (1979, 1980a, b), and others. Except in the most recent work of Marcus, it has been necessary to make the simplifying 'single mode approximation' to find solutions in the nonlinear (and therefore realistic) regime. In this approximation the planform (or horizontal variation of the convection) is fixed *a priori*, and nonlinear ordinary differential equations are integrated to find the 'lowest eigenvalue' radial dependence. One can summarize to say that the single-mode, modal approach finds *large* convective overshoots, with mixing fluid flows extending of order a pressure scale height λ_p or more into the radiatively stable region.

Unfortunately, the single-mode methodology probably gives the 'right' answer for the 'wrong' reason. A fluid constrained to single-mode motion can be viewed as having a high effective viscosity in some crucial respects; The Reynolds number Re of an unconstrained fluid flow measures the ratio of Reynolds stresses (dynamic stresses induced by the velocity) to viscous stresses. Single-mode theory explicitly removes Reynolds stresses (which would couple to other modes and ultimatley to a turbulent flow) from the modal equations. Therefore, in this sense, the effective Re of single-mode theory is small, corresponding to a large effective viscosity. Even though this effective viscosity does not appear as a dissipation term in the equations, it does constrain the solutions to be 'stiff' on large length scales. The normal tendency of the fluid to 'shear off' into a sharp turbulent boundary layer at the boundary of the radiatively stable region is not allowed by the single-mode equations. It is therefore not surprising that the eigensolutions should penetrate deeply into the stable region, but this result probably cannot be viewed as anything but artifactual. Marcus has a sufficiently large number of fully nonlinear, coupled models in three dimensions to see the beginnings of a turbulent Kolmogorov spectrum; however, all his models to date are *everywhere* convective, so his method has not yet shed light on the overshoot problem.

The 'model' approach stays closer to the local mixing-length theory in its spirit, and constructs a physical picture based on the idea of rising (or falling) 'bubbles' or 'eddies' within the fluid. Model equations are derived not from the Navier—Stokes equations *ab initio*, but from energy and momentum arguments as applied to the bubbles. Although this approach is mathematically less precise than the modal viewpoint, it has the advantage that it can try to include the physical effects of processes of turbulence at high Reynolds number, which are lost to the single-mode theory. In this category, one might include the work of Weyman and Sears (1965), Saslaw and Schwarzchild (1965), Shaviv and Salpeter (1973), Straus *et al.* (1976), Ulrich (1976), and others.

The results of Shaviv and Salpeter (1973) are now taken to exemplify the sort of results that come out of a model analysis. Previous to their work, Roxburgh (1965) and Saslaw and Schwarzschild (1965) had argued that the extent of overshoot was extremely small for realistic stellar conditions (as small as 10^{-2} pressure scale heights) due to the very high efficiency of the convection. High efficiency implies a very small superadiabaticity, and a small convective velocity. These small velocities seemed to imply only small penetrations, going to zero, in fact, as the superadiabaticity goes to zero.

Shaviv and Salpeter showed that the penetration distance remains *finite*, of order one-tenth of a pressure scale height in their model, even as the superadiabaticity goes to zero. This result seems paradoxical at first acquaintance, so it is worth understanding how it comes about. We can obtain it most simply by equating the mechanical energy flux generated by the convection to the power per area needed to maintain mixing in the overshoot layer against the re-establishment of a radiative thermal gradient:

To mix a stably stratified column of volume V and height L requires an energy input of $(\delta\rho V)gL$ to work against gravity. This energy must be renewed once per diffusion time-scale L^2/σ. So the power per volume W/V needed to maintain mixing is

$$\frac{W}{V} \approx \frac{\delta\rho gL}{L^2/\sigma} \approx \frac{\delta\rho\sigma g}{L} . \tag{14}$$

But $\delta\rho$ is given in terms of the actual and adiabatic gradients by

$$\delta\rho \approx \rho(\nabla - \nabla_{\mathrm{ad}})\frac{L}{\lambda_p}, \tag{15}$$

where $\lambda_p = P/g\rho$ is the pressure scale height. So,

$$\frac{W}{V} \approx \frac{\sigma P}{\lambda_p^2} |\nabla - \nabla_{\mathrm{ad}}|. \tag{16}$$

Note that this is independent of L and goes to zero linearly with σ, the thermal diffusivity (in which limit the convective efficiency becomes large).

The mechanical flux of the convection is

$$\mathscr{F}_{\text{mech}} \approx \rho v^3, \tag{17}$$

where v is the typical velocity of the convective eddies. We can relate v to the thermal flux $\mathscr{F}$ carried by the convection (Cox and Giuli, 1968, Equations 14.18 and 14.30),

$$v \approx \left[\frac{\alpha}{4} \nabla_{\text{ad}} \frac{\mathscr{F}}{\rho}\right]^{1/3}. \tag{18}$$

where α is the ratio of the mixing length to the pressure scale height, about unity. We can, in turn, relate $\mathscr{F}$ to the radiative gradient ∇_r, since the convective and radiative regions must carry the same flux,

$$\mathscr{F} \approx \frac{\sigma P}{\lambda_p} \nabla_r \tag{19}$$

(e.g. from Cox and Giuli, Equation 14.11 and Equation 8 above). Now Equations (17)–(19) combine to give

$$\mathscr{F}_{\text{mech}} \approx \frac{\alpha}{4} \frac{\sigma P}{\lambda_p} \nabla_{\text{ad}} \nabla_r. \tag{20}$$

Note that this expression also is linear in σ. Therefore σ, and also P, cancel out when we estimate the depth l of overshoot as

$$l = \frac{\mathscr{F}_{\text{mech}}}{W/V} = \lambda_p \frac{\alpha}{4} \left|\frac{\nabla_{\text{ad}} \nabla_r}{\nabla - \nabla_{\text{ad}}}\right|. \tag{21}$$

The combination $\nabla_{\text{ad}} \nabla_r/(\nabla - \nabla_{\text{ad}})$ should be evaluated self-consistently at a depth of, say, $l/2$ below its zero crossing. A typical case (e.g. the Sun) has $|\nabla - \nabla_{\text{ad}}|$ in the range of 0.1 to 0.2, $\nabla_{\text{ad}} = 0.4$, $\nabla_r \sim 0.3$, so the solution is $l \sim \lambda_p/4$. In any case, apart from the numerics, the *dimensional* scale for l is set by λ_p, not by any other dimensional quantity in the problem.

It is a useful exercise, now, to characterize the convective interface in terms of non-dimensional quantities. There are six relevant dimensional quantities at the interface: the total flux $\mathscr{F}$, gravitational acceleration g, pressure P, density ρ, thermal diffusivity σ, and microscopic viscosity ν. The temperature T enters the equations only through the equation of state involving P and ρ, so it can be viewed as a derived quantity.

We can form some other auxiliary derived dimensional quantities from the basic six, namely the pressure scale height,

$$\lambda_p \sim P/g\rho, \tag{22}$$

the sound speed,

$$c_s^2 \sim P/\rho, \tag{23}$$

the characteristic convective velocity (cf. Equation (18))

$$v \sim (\mathcal{F}/\rho)^{1/3}. \tag{24}$$

From these basic six quantities, we can construct exactly three dimensionless combinations. We take these to be the convective Mach number

$$Ma \equiv v/c_s, \tag{25}$$

the convective Reynolds number,

$$Re \equiv v\lambda_p/\nu, \tag{26}$$

and the Prantl number

$$Pr \equiv \nu/\sigma. \tag{27}$$

(Since radiative viscosity usually dominates, Pr is typically a small number, on the order of $P/\rho c^2$, where c is the velocity of light.) As an alternative to Re, one can use the Peclet number

$$Pe \equiv \frac{v\lambda_p}{\sigma} = Re\ Pr. \tag{28}$$

All the above quantities are local, i.e. defined at a point in depth. We need one nonlocal quantity which will define the spatial 'abruptness of onset' of the convection. Convective instability becomes possible where $\nabla - \nabla_{ad}$ changes sign. We need to know how abrupt (in pressure scale heights) is this sign change.

We can define a dimensionless quantity Ab to parametrize this,

$$Ab \equiv \left[\frac{d(\nabla_r - \nabla_{ad})}{d \ln P}\right]_{\nabla_r = \nabla_{ad}} \tag{29}$$

Finally, we should note that the relation $\nabla_r = \nabla_{ad}$ at the point of onset gives a relation between Ma and Pe. Use of the above definitions (25) and (28), and Equation (19), gives

$$0.4 \sim \nabla_{ad} = \nabla_r = \frac{\lambda_p \mathcal{F}}{\sigma P} = Ma^2\ Pe \tag{30}$$

The nature of the convective boundary in a fluid of uniform composition is now completely determined by three dimensionless quantities, which we can take to be Pe, Re, and Ab. For most models of the convective overshoot, Re is so large that its precise value is irrelevant. In this case Pe parametrizes, roughly, the efficiency of the convection (ratio of thermal time-scale to overturn time), while Ab parametrizes the 'stiffness' of the convective interface. For highly efficient convection, as in the deep Sun, we might expect that the numerical value of Pe also becomes irrelevant. In that case, the overshoot in pressure scale heights can depend only on Ab, presumably as some inverse power law.

It is easy to see that this expectation is in fact realized in the Shaviv–Salpeter model. Equation (21), with $\nabla - \nabla_{ad}$ evaluated self-consistently at a depth l as $(l/\lambda_p)Ab$, can be cast into nondimensional quantities in the form

$$\frac{l}{\lambda_p} \sim \frac{1}{2}\left(\frac{Ma^2\, Pe}{Ab}\right)^{1/2} \sim 0.3\, Ab^{-1/2}. \tag{31}$$

However, the Shaviv–Salpeter estimate is not the only possible estimate of the extent of the convective overshoot. A slightly different (but equally compelling) physical picture gives a different power law dependence on Ab, as well as presumably a different unknown constant of order unity:

The Shaviv–Salpeter model is, we saw, based on an estimate of the mechanical energy flux available from 'free running' mixing-length theory convection, i.e. convection whose 'mechanical load' is just its own internal dissipation into the turbulent cascade. An analogy is with a heat engine with no external mechanical load, but a lot of internal friction. But suppose we can argue that the convective 'heat engine' *is* coupled to a rather larger mechanical load by virtue of being strongly coupled to, and forced to mix, a much deeper overshoot region than that computed energetically in Equation (21). Because of its insufficient mechanical energy flux does the convection then 'stall', i.e. does it come to a halt?

Clearly, the answer must be 'no': if the convection is slowed below that velocity necessary to carry the convective flux $\mathscr{F}$, then the unstable gradient $\nabla - \nabla_{ad}$ will increase until sufficient buoyancy is available to drive the required velocity. Without a detailed (and nonlocal) convective modal, we cannot say what is the velocity versus load behavior of the convective 'engine' (the analog of the r.p.m. versus power curve of an electric motor or internal combustion engine at fixed throttle). Some power law of modest exponent (cf. Equation (18)) seems a likely possibility. In that case, the convection will supply more mechanical power as it is loaded more strongly, so it is not the energetics of free-running convection that determine the depth of overshoot. Rather the extent of overshoot is determined by the depth at which the mechanical load of forced mixing can self-consistently terminate on a *stable* interface to an unmixed region (cf. Press, 1981b). We shall now estimate that depth and find that it has a different functional dependence on Ab from Equation (31).

Figure 1 shows, schematically, the variation of $\ln T$ with $\ln P$ in the fluid for some radiative and adiabatic gradients. The curve AOA′ (here shown as a straight line) is an adiabat. The curve ROR′ is the radiative gradient needed to carry to total flux $\mathscr{F}$. The radiative curve is shallower than the adiabat to the left of O, steeper to the right. (Its radius of curvature is what Ab, Equation (29), parametrizes in nondimensional form.) In local mixing length theory, with efficient convection, the actual gradient would be ROA′, with stability along RO, convection along OA′.

Convective overshoot will mix the fluid to the left of O, restoring it very nearly to

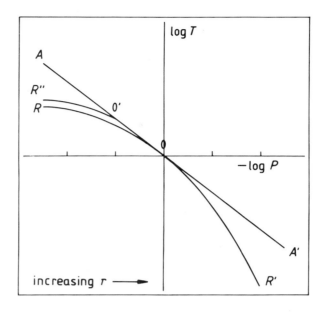

Fig. 1. Variations of temperature with pressure near the base of a convection zone (schematic). The curve AOA′ represents the adiabatic variation, while ROR′ integrates the radiative gradient. According to mixing-length theory, the actual variation (for efficient convection) would be ROA′; convective overshoot to some point O′ should instead yield R″ O′ OA′.

the adiabatic curve. In the Figure, this overshoot is shown as extending from O to O′. To the left of O′, we must return to a radiative gradient sufficient to carry the full flux. The variation in T must also be continuous. We therefore match to a curve R″ O′ which is parallel to the curve RO (so that the two curves have identical derivatives). The actual gradient with overshoot is the R″O′OA′. (We should note that one is also allowed to shift this curve up or down in $\ln T$ by a constant; one cannot determine this constant without knowing more about the boundary conditions to the left and right.)

The key pount, now, is that there is inevitably a discontinuity of slope in R″ O′ OA at the point O′. This means that the Brunt–Väisälä frequency N jumps from (essentially) zero to the right of O′ to some finite value to the left of O′. This discontinuity in N gives the interface a finite stability against continued erosion by the turbulent overshoot. By equating this stability to the strength of the erosion we shall be able to find the location of O′ (depth of the overshoot region).

The buoyancy frequency at some depth l to the left of O′ is (Equations (5) and (29))

$$N^2 \approx \frac{g}{\lambda_p} \, Ab \, \frac{l}{\lambda_p} \, . \tag{32}$$

The local fluid shear Σ is dominated not by the macroscale λ_p and velocity v of the convection, but rather by much smaller eddies in the Kolmogorov cascade. For an eddy of size λ,

$$\Sigma^2(\lambda) = \frac{v^2(\lambda)}{\lambda^2} = \frac{v_0^2(\lambda/\lambda_p)^{2/3}}{\lambda_p^2(\lambda/\lambda_p)^2} = Ma^2\frac{g}{\lambda_p}\left(\frac{\lambda}{\lambda_p}\right)^{-4/3}. \tag{33}$$

There are two possible microscales λ which might at first glance be thought relevant. One is the Kolmogorov microscale $\lambda_p/Re^{3/4}$, where the turbulent cascade terminates. However, the thermal microscale $\lambda_p/Pe^{3/4}$ (where the eddy turnover time equals the thermal diffusion time across the eddy) is much larger (since $Pr \ll 1$). Eddies smaller than the thermal microscale are irrelevant to the thermal structure of the fluid. They do not affect the radiative gradient to the left of O'. In particular they do not reduce the fluid's stability against larger scale flow, so they must die off exponentially in about one eddy length as they propagate to the left of O'. Setting, therefore $\lambda = \lambda_p/Pe^{3/4}$ in Equation (33), we get

$$\Sigma^2 = \frac{g}{\lambda_p} Ma^2 Pe. \tag{34}$$

The Richardson number Ri, which measures the stability of the interface at O is now given by

$$Ri \equiv \frac{N^2}{\Sigma^2} = \frac{1}{\lambda_p}\frac{Ab}{Ma^2 Pe}. \tag{35}$$

Fluid flows are tupically unstable for $Ri < 1/4$. Therefore, using Equation (30),

$$\frac{l}{\lambda_p} \sim \frac{1}{4}\frac{Ma^2 Pe}{Ab} \sim 0.1 Ab^{-1} \tag{36}$$

estimates the depth of convective overshoot.

We might now usefully consider the numerical values of some of the above quantities at the base of the solar convection zone. There, to fair accuracy, one has $c_s = 1.2 \times 10^7$ cm s^{-1}, $v = 6 \times 10^3$ cm s^{-1}, $\lambda_p = 5 \times 10^9$ cm, $v \sim 1$ cm^2 s^{-1}, $\sigma = 3 \times 10^6$ cm^2 s^{-1}. These values imply $Re = 3 \times 10^{13}$. $Pe = 9 \times 10^6$, $Pr = 3 \times 10^{-7}$, $Ma = 5 \times 10^{-4}$. It is not easy to get an accurate value of Ab from published solar models since its value is rapidly changing just at the base of the convection zone. Although Ab is quite large, ~ 3, even half a pressure scale height *above* the interface, its value at and just below the interface seems to be in the range $0.1-0.3$.

With these numerical values, the dimensional estimates of the depth of convection given by Equations (21) and (36) are not very different. However, the constants in front of those equations are not well known. In the case of Equation (21), the constant is substantiated at least by a detailed bubble-mixing *model* (Shaviv and Salpeter, 1973); there is some hope that one might be able to do the analogous model calculation, based on cascade turbulence models, for the dimensional estimate of Equation (36) (Marcus *et al.*, 1983). It seems not at all impossible that overshoot of order two pressure scale heights, maintained by turbulent erosion of the convective-stratified interface, can be found. At present, however, one has no certain information.

3.3. TWO TOPICS NOT DISCUSSED

If the Sun has (or once had) a rapidly rotating core, then one wants to know the nature of processes which bring it to uniform rotation, their time-scales, and their effects on other aspects of stellar evolution. The reader is referred to the work of Goldreich and Schubert (1967, 1968), Colgate (1968), Schwartz and Schubert (1969), Kippenhahn (1969), Fricke (1969), James and Kahn (1970, 1971), Starr (1971), Sakurai (1973, 1975), Schatten (1977), Endal and Sofia (1981).

Gravity waves (*g*-eigenmodes) can have large amplitudes at the solar center. If they were shown to have any significant coupling to the thermal energy content of the core, then they could be a means of mechanical energy transport which effectively reduces the solar opacity. Alternatively, they might be a source of deep abundance mixing, or might mechanically alter the normal radiative diffusivity given by a microscopic calculation. The works of Sakurai (1977), Beaudet *et al.* (1977), Press (1981a, and references therein), Press and Rybicki (1981), are relevant to these considerations.

References

Abt, H. A.: 1970, in A. Slettebak (ed.), *Stellar Rotation*, Gordon and Breach, New York.

Bahcall, J. N.: 1969, *Phys. Rev. Lett.* **23**, 251.

Bahcall, J. N.: 1979, *Space Sci. Rev.* **24**, 227.

Bahcall, J. N., Bahcall, N. A., and Ulrich, R. K.: 1969, *Astrophys. J.* **156**, 559.

Bahcall, J. N. and Ulrich, R. K.: 1971, *Astrophys. J.* **170**, 593.

Beaudet, G., Fontaine, G., Sirois, A., and Tassoul, M.: 1977, *Astron. Astrophys.* **54**, 213.

Berthomieu, G., *et al.*: 1980, in *Nonradial and Nonlinear Stellar Pulsation*, Springer, Heidelberg.

Boesgaard, A. M.: 1976, *Publ. Astron. Soc. Pacific* **88**, 353.

Boury, A., Gabriel, M., Noels, A., Scuflaire, R., and Ledoux, P.: 1975, *Astron. Astrophys.* **41**, 279.

Brown, T. M., Stebbins, R. T., and Hill, H. A.: 1978, *Astrophys. J.* **223**, 324.

Christensen-Dalsgaard, J., Dilke, F. W. W., and Gough, D. O.: 1974, *Monthly Notices Roy. Astron. Soc.* **169**, 429.

Christensen-Dalsgaard, J. and Gough, D. O.: 1976, *Nature* **259**, 89.

Christensen-Dalsgaard, J. and Gough, D. O.: 1980, *Nature* **288**, 544.

Colgate, S. A.: 1968, *Astrophys. J. Lett.* **153**, L81.

Cowling, T. G.: 1941, *Monthly Notices Roy. Astron. Soc.* **101**, 367.

Cox, J. P. and Giuli, R. T.: 1968, *Principles of Stellar Structure*, Vol. 1, Gordon and Breach, New York.

Davis, R. Jr: 1969, *Proc. Int. Conf. on Neutrino Physics Astrophysics, Moscow*, Vol. 2, p. 99.

Davis, R. Jr and Evans, J. M.: 1973, *Proc. 13th Int. Cosmic Ray conf.*, Vol. 3, p. 2001.

Davis, R. Jr, Harmer, D. S., and Hoffman, K. C.: 1968, *Phys. Rev. Lett.* **20**, 1205.

Defouw, R. J., Siquig, R. A., and Hansen, C. J.: 1973, *Astrophys. J.* **184**, 58.

Dicke, R. H.: 1964, *Nature* **202**, 432.

Dicke, R. H.: 1970, *Ann. Rev. Astron. Astrophys.* **8**, 297.

Dicke, R. H.: 1974, *Astrophys. J.,* **190**, 187.

Dicke, R. H.: 1976, *Solar Phys.* **47**, 475.

Dicke, R. H.: 1977, *Astrophys. J.* **218**, 547.

Dicke, R. H. and Goldenberg, H. M.: 1967, *Phys. Rev. Lett.* **18**, 313.

Dicke, R. H. and Goldenberg, H. M.: 1974, *Astrophys. J. Suppl.* **27**, 131.

Dilke, F. W. W. and Gough, D. O.: 1972, *Nature* **240**, 262.
Endal, A. S. and Sofia, S.: 1981, *Astrophys. J.* **243**, 625.
Ezer, D. and Cameron, A. G. W.: 1972, *Nature, Phys. Sci.* **240**, 180.
Fricke, K.: 1969, *Astrophys. J.* **3**, L219.
Goldreich, P. and Schubert, G. 1967, *Astrophys. J.* **150**, 571.
Goldreich, P. and Schubert, G.: 1968, *Astrophys. J.* **154**, 1005.
Gough, D. O.: 1979, *Nature* **278**, 685.
Gough, D. O., Moore, D. R., Spiegel, E. A., and Weiss, N. O.: 1976, *Astrophys. J.* **205**, 536.
Hill, H. A. and Stebbins, R. T.: 1975, *Astrophys. J.* **200**, 471.
Iben, I.: 1969, *Astrophys. J. Lett.* **155**, L101.
James, R. A. and Kahn, F. D.: 1970, *Astron. Astrophys.* **5**, 232.
James, R. A. and Kahn, F. D.: 1971, *Astron. Astrophys.* **12**, 332.
Jokipii, J. R. and Parker, E. N.: 1969, *Astrophys. J.* **155**, 777.
Kraft, R. P.: 1967, *Astrophys. J.* **150**, 551.
Kraft, R. P.: 1968, in H. Y. Chiu, R. Warasila, and J. Remo (eds.), *Stellar Astronomy*, Vol. 2, Gordon and Breach, New York.
Kraft, R. P.: 1970, in G. H. Herbig (ed.), *Spectroscopic Astrophysics*, Univ. of California Press, Berkeley; p. 385.
Kippenhahn, R.: 1969, *Astron. Astrophys.* **2**, 309.
Latour, J., Spiegel, E. A., Toomre, J., and Zahn, J. P.: 1976, *Astrophys. J.* **207**, 233.
Ledoux, P.: 1958, *Handbuch der Physik*, Vol. 51.
Ledoux, P.: 1974, in P. Ledoux, A. Noels, and A. W. Rogers (eds.), *Stellar Instability and Evolution*, Reidel, Dordrecht; p. 135.
Ledoux, P.: 1978, in N. R. Lebowitz *et al.* (eds.), *Theoretical Principles in Astrophysics and Relativity*, Univ. of Chicago Press, Chicago.
Lubow, S. H., Rhodes, E. J. Jr, and Ulrich, R. K.: 1980, in *Nonradial and Nonlinear Stellar Pulsation*, Springer, Heidelberg.
Marcus, P. S.: 1979, *Astrophys. J.* **231**, 176.
Marcus, P. S.: 1980a, *Astrophys. J.* **239**, 622.
Marcus, P. S.: 1980b, *Astrophys. J.* **240**, 203.
Marcus, P. S., Press, W. H., and Teukolsky, S. A.: 1983, *Astrophys. J.* **267**, 795.
Newman, M. J. and Fowler, W. A.: 1976, *Astrophy. J.* **207**, 601.
Noels, A., Gabriel, M., Boury, A., Scuflaire, R., and Ledoux, P.: 1975, in *d'Astrophysique*, Societe Royale des Sciences, Liege.
Parker, E. N.: 1955, *Astrophys. J.* **121**, 491.
Parker, E. N.: 1965, *Astrophys. J.* **142**, 584.
Parker, E. N.: 1974, *Astrophys. Space Sci.* **31**, 261.
Phillips, O. M.: 1977, in *The Dynamics of the Upper Ocean* (2nd ed), Cambridge Univ. Press, Cambridge.
Plaskett, H. H.: 1965, *Observatory* **85**, 178.
Press, W. H.: 1981a, *Astrophys. J.* **245**, 286.
Press, W. H.: 1981b, *J. Fluid Mech.* **107**, 455.
Press, W. H. and Rybicki, G.: 1981, *Astrophys. J.* **248**, 751.
Rood, R. T.: 1972, *Nature, Phys. Sci.* **240**, 178.
Rowley, J. K., Cleveland, B. G., Davis, R. J., and Evans, J. C.: 1977, *Report presented at Neutrino-77 Conference in the USSR*, BNL Rept. 23418.
Roxburgh, I. W.: 1964, *Icarus* **3**, 92.
Roxburgh, I. W.: 1965, *Monthly Notices Roy. Astron. Soc.* **130**, 223.
Sakurai, T.: 1973, *Publ. Astron. Soc. Japan* **25**, 563.
Sakurai, T.: 1975, *Monthly Notices Roy. Astron. Soc.* **171**, 35.
Sakurai, T.: 1977, *Publ. Astron. Soc. Japan* **29**, 543.
Saslaw, W. C. and Schwarzschild, M.: 1965, *Astrophys. J.* **142**, 1468.
Schatten, K. H.: 1977, *Vistas Astron.* **20**, 475.
Schatzman, E. and Maeder, A.: 1981, *Nature* **290**, 683.
Schwartz, K. and Schubert, G.: 1969, *Astrophys. Space Sci.* **5**, 444.

Schwarzschild, M. and Harm, R.: 1973, *Astrophys. J.* **184**, 5.

Severny, A. B., Kotov, V. A., and Tsap, T. T.: 1976, *Nature* **259**, 87.

Severny, A. B., Kotov, V. A., and Tsap, T. T.: 1979, *Astron. Zh.* **56**, 1137; translation in *Sov. Astron.* **23**, 641.

Shaviv, G. and Salpeter, E. E.: 1971, *Astrophys. J.* **165**, 171.

Shaviv, G. and Salpeter, E. E.: 1973, *Astrophys. J.* **184**, 191.

Sherrer, P. H., Wilcox, J. M., Kotov, V. A., Severnyi, A. B., and Tsap, T. T.: 1979, *Nature* **277**, 635.

Shibahashi, H., Osaki, Y., and Unno, W.: 1975, *Publ. Astron. Soc. Japan* **27**, 401.

Slettebak, A. (ed.): 1970, *Stellar Rotation*, Gordon and Breach, New York.

Snider, J. L., Kearns, M. D., and Tinker, P. A.: 1978, *Nature* **275**, 730.

Spiegel, E. A.: 1965, *Astrophys. J.* **141**, 1068.

Starr, V. P.: 1971, *Nature* **233**, 186.

Straus, J. M., Blake, J. B., and Schramm, D. N.: 1976, *Astrophys. J.* **204**, 481.

Toomre, J., Zahn, J.-P., Latour, J., and Spiegel, E. A.: 1976, *Astrophys. J.* **207**, 545.

Ulrich, R. K.: 1976, *Astrophys. J.* **207**, 564.

Ulrich, R. K.: and Rood, R. T.: 1973, *Nature, Phys. Sci.* **241**, 111.

Unno, W.: 1975, *Publ. Astron. Soc. Japan* **27**, 81.

Vauclair, G. Vauclair, S., Schatzman, E., and Michaud, G.: 1978, *Astrophys. J.* **223**, 567.

Veronis, G.: 1963, *Astrophys. J.* **137**, 641.

Weymann, R. and Sears, R. L.: 1965, *Astrophys. J.* **142**, 174.

Zappala, R. R.: 1972, *Astrophys. J.* **172**, 57.

Center for Astrophysics and Department of Physics,
Harvard University,
Cambridge, MA 02138,
U.S.A.

CHAPTER 5

THE SOLAR DYNAMO: OBSERVATIONS AND THEORIES OF SOLAR CONVECTION, GLOBAL CIRCULATION, AND MAGNETIC FIELDS

PETER A. GILMAN

1. Introduction

The magnetic field of the Sun is probably maintained by hydromagnetic dynamo action. Most solar physicists are convinced the seat of the dynamo is the solar convection zone, as well as perhaps the layers immediately above and below it. While the solar corona is profoundly influenced by the magnetic field reaching up from below, its direct contribution to the origin and maintenance of those fields is probably negligible.

Since dynamo action arises from flow of a conducting fluid attempting to cross magneic field lines, solar dynamo theory is concerned with all the major motion fields which occur in the solar convection zone. In terms of spatial scale, these are known to range from the so-called granulation, with typical horizontal scale of 10^3 km, up to the differential rotation, which is global (10^6 km). In between, in ascending order of spatial scale, are the newly discovered mesogranulation, followed by supergranulation, and then possibly giant cells and meridional circulation, the observational knowledge for which is weaker and less direct.

The magnetic fields these motions produce are themselves arranged in complex patterns on a similarly wide range of spatial scales, although the basic element of magnetic field at the photosphere appears to be a magnetic flux tube of diameter no more than a few hundred kilometers (less than the current resolution limit of solar telescopes) but with a peak magnetic field strength inside approaching 2000 gauss. In total, these flux tubes typically occupy no more than a few thousandths of the solar surface, so they are highly intermittent. Nevertheless, flux tubes do often aggregate into larger patterns called active regions, in which are often found 'flux tubes' of much larger scale, namely sunspots. Sunspots, in turn, are the clearest indictors of the most basic observed feature of the solar dynamo, namely the solar cycle. Patterns of flux tubes do result in global patterns of net flux, but these are difficult to see in the photosphere without considerable averaging of magnetograph data. Although not measured there, they are more evident in the corona, as evidenced by coronal structures.

Not only is the solar dynamo complex in spatial structure, it is time dependent on a variety of time-scales. Within a given solar cycle, both the rise to maximum and the subsequent decline to minimum are irregular, punctuated by many 'events' or changes in magnetic structure and activity. Successive cycles vary in strength, and may even nearly disappear for several cycles, as in the Maunder minimum of the late seventeeth century.

Naturally, developing fundamental understanding of such a complex fluid-magnetic

Peter A. Sturrock (ed.), Physics of the Sun, Vol. I, pp. 95–160.

field system presents a formidable challenge to theoreticians, and, while progress has been substantial, we have not reached anything like a full comprehension of what is happening. The great range of spatial and time-scales of significance puts complete, unified theories beyond our reach for many years, and we must resort to less ambitious approaches.

Solar dynamo theory draws on a broad range of fluid dynamics, including particularly the theory of convection and global circulation, as well as low-speed compressible gas dynamics, interaction of velocity and magnetic fields, and hydrodynamic and MHD turbulence theory. A desire to understand the solar dynamo has motivated much basic work in these areas, some for fluid systems much simpler than the real Sun but nevertheless containing essential physics. Two areas of particular note are convection in a rotating spherical shell, and convection interacting with a magnetic field.

The bulk of this chapter describes solar motions of relevance to the dynamo problem, as well as the theories developed to explain them, followed by the major features of the Sun's magnetic field, and the theories for it, including the theory of individual features such as flux tubes and sunspots, as well as 'global' dynamo theories. This main body of the chapter is preceded by a discussion of current knowledge of the gross structure of the convection zone — such properties as its depth, shape, rotation, temperature structure, and heat flux, as well as possible fluctuations with time — to provide a setting within which the fluid dynamics and MHD can be discussed.

Before proceeding further, it is useful to set down the fundamental laws of fluid dynamics and MHD in rather general form to which we shall refer many times in the course of the chapter.

The vector equation of momentum is

$$\rho \frac{\partial \mathbf{v}}{\partial t} + \rho (\mathbf{v} \cdot \nabla) \mathbf{v} = -2\mathbf{\Omega} \times \mathbf{v} - \nabla p - \rho \mathbf{g} + \mathbf{j} \times \mathbf{B} + \rho \mathbf{F} \tag{1.1}$$

in which ρ is the fluid density, p its gas pressure, $\mathbf{v}$ the vector velocity measured relative to a uniformly rotating reference frame of rotation $\mathbf{\Omega}$, $\mathbf{g}$ is gravity (including the centrifugal force of rotation), $\mathbf{j}$ the electric current density, and $\mathbf{B}$ the magnetic induction. From (1.1) the forces of particular interest in the solar dynamo problem are inertial, Coriolis, pressure gradient, gravity (buoyancy), electromagnetic body forces, as well as viscous forces, here denoted by F. F may may take several different forms depending on the particular application. In some more idealized theories we shall mention, it will represent the effect of Newtonian viscosity, while in others it will be the divergence of an anisotropic stress tensor, and in still others, a nonlinear prescription for momentum transfer by small-scale turbulence.

Continuity of mass is represented by

$$\frac{\partial \rho}{\partial t} + \nabla \cdot \rho \mathbf{v} = 0 \tag{1.2}$$

and various approximations will also be made to this form.

If we assume the Sun is a perfect gas, the thermodynamic energy equation is written in the rather general form

$$\rho C_v \left[\frac{\partial \theta}{\partial t} + \mathbf{v} \cdot \nabla \theta \right] + p \, \nabla \cdot \mathbf{v} = H \tag{1.3}$$

in which C_v is the specific heat of constant volume and θ is the temperature. The function H represents all sources of heating for the problem, which could include radiative heating under either optically thin or thick conditions, turbulent transport of heat, and heating by dissipation of momentum or electric currents. In a few applications, particularly convection near the surface of the convection zone, partial ionization is also taken into account.

For prediction of magnetic fields, the standard induction equation of MHD is used, which we write in the form

$$\frac{\partial \mathbf{B}}{\partial t} = \nabla \times (\mathbf{v} \times \mathbf{B}) - \eta \nabla \times \nabla \times \mathbf{B} \tag{1.4}$$

in which η is the magnetic diffusivity. In addition, of course, the condition $\nabla \cdot \mathbf{B} = 0$ (no magnetic monopoles) is invoked. In (1.4), we have assumed η is constant for simplicity but in certain applications it may not be. Applications of various MHD turbulence arguments to (1.4) will produce a different induction equation, for the mean field $\bar{B}$, which we discuss later.

2. Convection Zone Characteristics

2.1. BASIC STRUCTURE AND DEPTH

The precise structure of the solar convection zone is, of course, not known because it cannot be observed. Traditionally, it has been inferred from stellar structure calculations applied to the Sun. In such treatments the convection zone itself is represented through mixing-length theory, which essentially is a crude approximation to the fundamental governing equations (Equations (1.1)–(1.3)), for a collection of convective blobs or elements, ignoring such complications as rotation and magnetic fields. The so-called 'local' mixing-length theory applied to stars is explained in Gough and Weiss (1976) and Gough (1977c). In this version, the properties of the convection of a given level such as its heat flux are assumed to depend on the structure, such as the superadiabatic temperature gradient, at the level. Nonlocal theories have also been explored, such as in Ulrich (1970a, b, 1976), and critiqued and compared with local treatments in Nordlund (1974). Detailed application to the solar case has been made, for example, by Spruit (1974, 1977).

The basic concept is that a buoyant fluid element rising in a superadiabatic layer will move some distance l, called the mixing length, before exchanging its heat and momentum with its surroundings and falling back. The mixing length is generally assumed to be related to the local scale height H, so a fixed ratio l/H, somewhere between 0.5 and

3.0, is generally chosen. Different mixing-length theories, such as that of Böhm-Vitense (1958) and Öpik (1950) assume different details concerning the shape and size of convective elements, but Gough and Weiss (1976) show how they may be reconciled by calibrating them using the present age, luminosity, and mass of the Sun. When this is done the temperature structures they predict are virtually the same except very close to the top of the convection zone where departures from the adiabatic gradient become significant. Figure 1, adapted from Gough and Weiss (1976), illustrates several properties

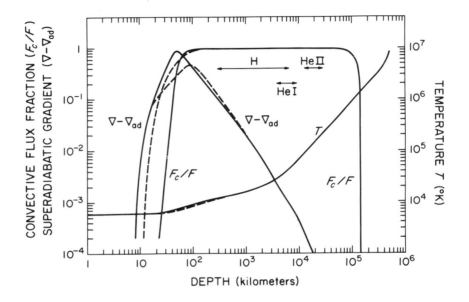

Fig. 1. Structure of two model convection zones with depth. Figure adapted from Gough and Weiss (1976), who computed models based on formulation of Böhm-Vitense (1958), solid lines, and on that of Öpik (1950), dashed lines. Curves from these two formulations coincide except within 10^3 km of outer boundary of convection zone. Arrows labelled H, He I, He II denote 10–9% ionization zones for hydrogen and helium.

of the structure with depth, including temperature and the departure from the adiabatic gradients, as well as the fraction of heat flux carried by convection. We see that superadiabatic gradients in excess of 10% of the adiabatic gradient are confined to the uppermost few hundred kilometers in the convection zone, below which the departure falls to less than 1 part in 10^3 within 10^4 km of the surface. Thus, all the lower part of the convection zone lies on essentially the same adiabat. In this particular calculation, the convection zone extends to about 1.5×10^5 km, or slightly more than 20% of the solar radius. In this case, Gough and Weiss (1976) assumed a metal abundance (fraction by mass) $Z = 0.02$. For a larger metal abundance, the zone is deeper: 1.86×10^5 km for $Z = 0.04$. The increase in depth is simply due to the increased opacity of deeper layers due to more metals, which force the radiative gradient required to carry the solar luminosity to be superadiabatic to a deeper level. The depth depends more weakly on other parameters of the theory, except the ratio l/H, which is fixed in each calculation by the requirement of matching an effective surface temperature T_e = 5800 K when the luminosity $L = L_\odot$.

We note from Figure 1 that the mixing-length theory predicts that convection in the Sun is very efficient, transporting virtually all the required heat throughout the convection zone except in boundary layers less than one scale height in thickness at top and bottom.

Calculations, such as Gough and Weiss (1976) and Spruit (1974, 1977a), yield what are generally regarded as 'standard' solar convection zones with depths between about 20 and 30% of the radius. Recently, debate about the real depth of the convection zone has intensified as a result of interpretation of new measurements of solar oscillations (see Chapter 7 by Brown, Mihalas, and Rhodes) as well as implications of the fact that the element lithium, which burns at a temperature of about 2.5×10^6 K, is depleted on the Sun by a factor of 10^2 or so compared to cosmic values while beryllium, which burns at 3.0×10^6 K, is normal (Vauclair et al., 1978). The arguments are as follows.

As discussed in Brown, Mihalas, and Rhodes (Chapter 7, this volume), the Sun is observed to oscillate with bands of power in frequency—wavenumber space with periods in the neighborhood of 5 min. All solar envelope models containing a convection zone, photosphere, and chromosphere when perturbed predict well-defined bands of oscillations in the same range. The precise positions of the power bands depend on the detailed structure of the models, as well as on model physics and boundary conditions. Ulrich and Rhodes (1977) showed that for a given pressure or p-mode, the frequency dropped for all horizontal wavenumbers as the effective convection zone depth was increased, as measured by increases in the mixing-length ratio l/H. For the solar envelope model they used, Rhodes et al. (1977) found that the best fit to the observed bands required a convection zone depth of at least 25% of the solar radius and perhaps as great as 38%. This upper limit was set not by the oscillations but rather by the requirement that beryllium must not be mixed into the interior. Subsequently, Lubow et al. (1980) have carried out more elaborate calculations which show that inclusion of corrections to the equation of state for Coulomb forces via Debye—Hückel theory, as well as other improvements, slightly reduces the need for deep zones. They now find that a complete standard solar model with mixing-length ratio l/H = 1.65 fits the data best, yielding a convection zone depth of about 30%

Other evidence of oscillations have been interpreted to mean the Sun should have an extremely shallow convection zone, perhaps only 2 or 3% of the radius. These are gravity oscillations of period 66 and 45 min, described by Hill and Caudell (1979). If these oscillations are excited in the stable region underneath the convection zone, then to be seen at the surface with the reported properties a shallow zone would be required. Christensen-Dalsgaard et al. (1980) discuss the possible implications of this claim at length but note that such a model is impossible to reconcile with the positions of power bands in the 5 min oscillations, which are subject to much less uncertainty. Such a shallow zone also requires an extremely low metal abundance for the Sun. Brown and Harrison (1980) argue that the oscillations Hill and Caudell invoke are instead probably confined to the solar atmosphere.

The lithium—beryllium abundance problem for the Sun also tends to favor a fairly deep convection zone, since mixing must occur at least down to the level where lithium burning takes place. But estimates of the amount of mixing in the overshoot region just below the convection zone are quite uncertain (see, e.g., Shaviv and Salpeter, 1973; Straus *et al.*, 1976, Vauclair *et al.*, 1978). Clearer answers will hopefully be forthcoming as theories of the convection zone—radiative interior interface are advanced. We shall see later in this review that this region may also be important for the solar dynamo problem.

We conclude that the present weight of evidence favors a rather deep solar convection zone, most likely in the neighborhood of 30% of the solar radius. When we discuss current theories of solar differential rotation, we shall find there is another argument favoring a deep convection zone.

2.2. OBLATENESS AND LATITUDINAL TEMPERATURE DIFFERENCES

Dicke and Goldenberg (1967) published new observations of radiation from the limb of the Sun which they interpreted as showing that the Sun was more oblate by about a factor of 5 than would be expected from its surface rotation rate. They found an equatorial diameter 87 millarcsec larger than the polar diameter. This result provoked a great flurry of both theoretical and interpretive work, for the implications of a solar interior rotating perhaps 20 times faster than the solar surface are profound. We introduce this topic in the present review because such a fast rotating interior would have important implications for the solar dynamo problem. In particular, the strong radial shears in rotation rate present at the bottom of the convection zone would imply essentially that no magnetic field lines could cross from the low rotation levels of the convection zone to those of high rotation beneath. If they did, they would become stretched out in longitude to very large values, creating large $j \times B$ body forces which would react back on the rotational shear and destroy it in a time much shorter than the age of the Sun.

Most of the arguments about the interpretation of the oblateness observations centered around whether a true visual, geometric oblateness had been measured, or whether instead the larger equator diameter was due to an excess equatorial brightness of some kind, such as due to bright faculae on the surface. Controversy raged until new observations were made by Hill and colleagues (Hill *et al.*, 1974, Hill and Stebbins, 1975) by a completely different technique. The crucial difference in their approach was the use of a very precise definition of the edge of the Sun, which is very sensitive to the shape of the limb-darkening function (and therefore to differences in this function between the equator and the pole), while being quite insensitive to atmospheric seeing. Using a telescope originally designed to measure the gravitational defraction of starlight passing near the Sun, Hill *et al.* (1974) could clearly detect a time varying excess equatorial brightness large enough to explain the Dicke—Goldenberg result. But by picking a time period when this fluctuating brightness was very low, Hill and Stebbins (1975) were able to measure a true oblateness of 18 ± 12 millarcsec, which is consistent with an interior rotation rate similar to that of the solar surface.

During this same time period, a number of efforts were made to measure a temperature difference between equator and pole on the surface, motivated partly by the controversy over oblateness, and partly by its interest for global dynamics of the convection zone. An upper limit of 3 K was given by Altrock and Canfield (1972) for the solar photosphere, largely negating a number of earlier measurements reporting large differences. The interpretation of all such measurements for convection zone dynamics is quite uncertain, because a nonzero temperature difference in the photosphere may not reach far into the convection zone if, for example it is from faculae.

At least one 'loose end' is left over from the oblateness debate of the early 1970s. Dicke (1976a, b, 1977) has found, a roughly 12.2 day period in the fluctuations in his daily solar shape measurements, which he argues is not due to faculae but instead is evidence of a rigidly rotating geometrical distortion in the Sun. If so, this might be evidence of an interior rotating perhaps twice as fast as the surface. Unfortunately, no one else has attempted to verify this effect.

2.3. LUMINOSITY AND RADIUS CHANGES

Changes in the luminosity and radius of the Sun could have important implications for understanding solar convection and the solar dynamo since both could be connected to changes in convective efficiency and/or interactions between convection and the Sun's magnetic field. In addition, of course, luminosity changes could be important for understanding the Earth's climate (see Chapter 20 this work, by Dickinson). Both changes of luminosity and radius are very active subjects now and solar radius measurements are particularly controversial.

Until the advent of rockets and satellites, solar luminosity could only be measured from the ground, and was therefore subject to large uncertainties (several percent) due to variations in atmospheric transmission. Abbott made a lifetime career of such measurememts, and his records are still being analyzed. In particular, Foukal and Vernazza (1979) have found there is a small variation in luminosity, of fractional amplitude $\sim 7 \times 10^{-4}$, which varies with a period of approximately 28 days, indicating persistent solar features rotating at a typical solar rate. They find part of this variation is due to a rise in total flux when the area of white light faculae on the visible disk increases, and part is due to a decrease in flux when more sunspots are present. The two effects do not appear to compensate for each other on short time-scales (days to weeks). On a day-to-day basis, sunspot area is a poor estimator of faculae area.

More recently, measurements of luminosity (Willson et al., 1981) from the Active Cavity Radiometers on board the Solar Maximum Mission spacecraft have produced rather startling evidence of even larger drops in luminosity associated with passage of large sunspot groups across the disk. They observed two such events, in April and May 1980, during which the luminosity fell by a maximum of between 0.1 and 0.2%, with the total event lasting almost two weeks. Willson et al. showed the duration of these events and their magnitude could be explained assuming all the deficit was due to very little radiation coming from spots compared to their surroundings. They could not see any associated rise in luminosity due to faculae. Thus they concluded that there was little, if any, local compensation for the reduced flux.

Foukal et al. (1981) have analyzed similar events in the Nimbus 7 radiometer data (including some with variations as large as 0.45% peak to peak) a few of which correspond to the SMM events. They find some examples where accounting for the presence of bright faculae improves the agreement, but others for which thd drop in radiation is much larger than predicted, assuming no radiation comes from the spots. These latter events suggest there is a larger area below the solar surface over which blockage of the outward heat flux is occurring. Thermal storage models (Foukal et al., 1983; Spruit,

1981c) suggest the blocked heat will ultimately reach the surface over a much wider area than the fields. Dearborn and Newman (1978) (see also Dearborn and Blake, 1980) find that for efficiency changes on time-scales much shorter than the thermal time-scale for the whole convection zone ($\sim 10^5$ yr), the fractional change in luminosity is about half a given change in l/H (and of the same sign), so that a luminosity change of 1% requires a convective efficiency change of $\sim 0.5\%$ Unfortunately, there is no independent means of estimating how much l/H should vary, especially since it is a somewhat arbitrary parameter of a rather crude theory. It is not possible to make measurements of convection and 'determine' a precise value of l/H, since it represents a considerable oversimplification of the real physical situation.

The slight collapse of the convection zone required to supply the energy for a short-term increase in luminosity does not itself imply a decrease in radius at the surface. The change in radius at depth, where most of the thermal and potential energy is stored, required to supply energy for a small, temporary luminosity increase, is extremely small. But the thin layer at the top of the convection zone, where radiation transports a significant fraction of the total energy, expands rapidly to a new hydrostatic balance in response to an increase in luminosity of all layers in the convection zone, leading to an increase in radius of the star (Dearborn and Newman, 1978). However, the ratio, W, of this increase, relative to the increase in luminosity, appears to be heavily dependent on the details of how the outermost layers are treated, though the ratio is always much less than unity. Using different stellar structure models applied to the Sun, Sofia et al. (1979) estimate $W \sim 7.5 \times 10^{-2}$, Dearborn and Blake (1980 get $W \sim 5 \times 10^{-3}$, while Gilliland (1980, 1982) finds $W \sim 8.5 \times 10^{-4}$. So the variation in theoretical predictions is large. Sofia et al. (1979) have carried the chain of inference further to argue that the ratio W can be used to place tight limits on the variations in luminosity in the past century. In particular, they argue that the various solar radius estimates (mostly cited above) constrain radius changes to no more than 0.25 arcsec in the period 1850–1937, implying for their W of 7.5×10^{-2} an upper limit to variations in luminosity of 0.3%. However, if Gilliland's $W \sim 8.5 \times 10^{-4}$ is closer to the correct value, luminosity variations of up to 30% would be allowed, which is two orders of magnitude larger than current estimates from direct measurements of luminosity cited previously. It seems more reasonable to reverse the argument and place limits on the radius change. Thus, with Gilliland's W, a 1% change in luminosity leads to only a ~ 0.01 arcsec change in radius, which is of a similar magnitude to Hill and Stebbins (1975) measure of the difference between polar and equatorial radii (the oblateness), implying such changes will be extremely difficult to detect.

At least two attempts have recently been made to estimate the effects of magnetic fields on changes in solar radius and luminosity more explicit. Thomas (1979) has suggested that an increase in the number of magnetically buoyant flux tubes in the upper convection zone near maximum in the solar cycle compared to minimum could result in an expansion, and drop in surface temperature, without necessarily changing the luminosity. Gilliland (1982) has tested this concept using his stellar structure code, and found that the amount of expansion predicted by Thomas is reasonable but such expansion is always accompanied by a luminosity change of a few parts in 10^4. Furthermore, the sign of this change, as well as that of the surface temperature change, depends on just where in the top layers of the convection zone the flux tubes reside.

Spiegel and Weiss (1980) have argued that magnetic fields at the very bottom of the convection zone, built up by the solar dynamo and extending a short distance into the radiative region below, could significantly alter the superadiabatic gradient required to drive convection. Thus, the strength of magnetic field varying with the cycle could cyclically alter the temperature gradient near the bottom and therefore the thermal energy stored there. They assume this change in thermal energy would reach the surface as a change in luminosity on the time-scale of the solar cycle. They estimate that the luminosity would decrease in going from solar minimum to maximum by $\sim10^{-3}$, consistent with a radius change of $\sim10^{-4}$. From their calculations the ratio $W \approx -0.1$, larger and opposite in sign to the previous estimates described earlier. As Gilliland (1982) points out, the Speigel and Weiss estimate is not based on even a mixing-length stellar structure calculation, and it is not clear how the perturbation they introduce at the bottom of the convection zone ends up being felt at the top. Gilliland (1982) has tested their predictions with his stellar structure code, and finds that instead of the luminosity change being observed at the surface, the model predicts changes in thermal and potential energy reservoirs in the region of magnetic field and the levels just above which nearly compensate. Gilliland concedes that a key question to be answered in the future is how much the arguments would be changed by a nonlocal, nonmixing-length theory of convection. If such convection were to transmit a luminosity rise produced by the mechanism of Spiegel and Weiss to the surface, then the corresponding radius changes should be of the same sign and determined largely by expansion of the outer layers, as discussed in Dearborn and Blake (1980).

From all the above arguments it seems clear, as Gilliland (1982) has concluded, that accurate prediction of small changes in such quantities as solar radius and luminosity requires development of much more sophisticated theories of convection, including interactions with rotation, differential rotation, and magnetic fields. As we discuss in the following sections, this is the same requirement as that for understanding the origins of differential rotation and the solar cycle.

3. Observations and Theory of Solar Convection Zone Motions

3.1. GRANULATION

3.1.1. *Observations*

Granulation represents the smallest clearly defined motion pattern on the Sun. A typical granulation pattern as seen in intensity is shown in Figure 2. We see this pattern is composed of bright irregular polygons separated by thin dark lanes. It is now well established, e.g. Bray *et al.* (1976), that the bright regions contain upward motion, and the dark lanes downward motions. Their estimate is 0.7 km s^{-1} for the upward flow, 1.1 km s^{-1} for the downward flow, taken from filtergram observations of the iron line at 6569.2 Å.

Many facts about granulation are conveniently summarized in the review by Wittmann (1979). For example, the average distance between granules appears to be about 2 arcsec, or about 1400 km at the center of the solar disk. This dimension implies that considerably better than average seeing is required to clearly resolve individual granules. Measurements

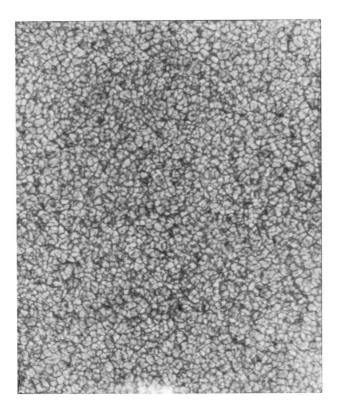

Fig. 2. Photograph of granulation taken at Pic du Midi with excellent seeing conditions, obtained courtesy of R. Muller, Sacramento Peak Observatory. Note the pattern of bright irregular polygons imbedded in a lattice of dark lanes. Resolution roughly 0.25 arcsec.

of both intensity differences between bright and dark regions, as well as Doppler velocities, are significantly influenced by seeing conditions. Estimates of the r.m.s. brightness fluctuations are about 10% when normalized to a standard wavelength of 5000 Å, corresponding to a brightness temperature fluctuation of ~125 K.

Both intensity fluctuations and Doppler velocities change with height. Deductions of velocity changes are influenced by seeing the presence of other velocity fields, particularly 5 min oscillations (see Chapter 7 by Brown, Mihalas, and Rhodes, this volume) and the effects of velocity gradients on line formation. Keil (1980) has recently analyzed this problem in detail, including his own new observations as well as previous studies, and concludes that r.m.s. vertical granule velocities drop approximately linearly from roughly 1.5 km s^{-1} at 100 km above the unit optical depth level, to near zero in a distance of roughly 250 km. This means that granulation penetrates a few pressure scale heights into the solar atmosphere above the top of the convection zone (defined as the level where the radial temperature gradient becomes subadiabatic). However, not all observers agree with this rapid fall-off of velocity with height. For example, Mattig (1980) deduces a vertical granule velocity component of 1 km s^{-1} at the same depths as

Keil, with very little decrease with height. Clearly, further work is needed to resolve these differences.

The horizontal distribution of granulation is often described in terms of its statistical characteristics. For example, Ricort and Aime (1979) find from a two-dimensional power spectrum of the granulation intensity pattern that there is a peak in the power at a spatial period of 2.5 arcsec, which is close to the value of 2 arcsec, mentioned above, that was found by simply measuring distances between adjacent granules. The power also falls off exponentially at higher spatial frequencies, down to the observation limit of perhaps 0.3 arcsec for very good seeing conditions. Aime et al. (1979) have also shown from power spectra that the pattern is neither completely ordered and regular, such as for a spacefilling pattern of hexagons, nor is it completely random and disordered. It can be described as having essentially gaussian departures from a regular hexagonal lattice, with a standard deviation of one-third of the distance between hexagons.

The time evolution of granule patterns is important for understanding their origins. This is particularly difficult to obtain because it is hard to get good enough seeing which lasts several granule lifetimes from ground-based telescopes. Mehltretter (1978) has analyzed photographs taken on the flight of the 'Spektro-Stratoskop' telescope in 1975. These were spaced at approximately 2 min intervals over a period of a little less than an hour. He finds that all granules are formed by fragmentation of earlier granules. About half the granules are destroyed by further fragmentation, while the other half either fade away or merge with other granules. Granules which fragment last about 8 min, those which fade away last 12 min, while those which merge last less than 5 min. When the expansion and fragmentation process occurs in a more or less symmetric manner about a central point, the granule is said to 'explode' (Beckers and Morrison, 1970: Musman, 1972).

Although granulation itself on the Sun is a very small scale pattern, its presence can produce observable global effects. In particular, the correlation between granulation intensity and velocity fluctuations (hot elements rising, cool ones sinking) produces a net blue-shift in spectral lines when measured over an area large enough to include both light and dark elements. This effect has long been a subject of study, and concern for those attempting velocity measurements on much larger scales. It results in the center of the solar disk being blue-shifted relative to the limbs, by as much as 400 m s^{-1} in Doppler velocity measure. Beckers and Nelson (1978) have recently built a model to explain this phenomenon, which is the most likely cause of the limb effect in the Fraunhofer lines used in so many solar velocity measurements. The theory requires both horizontal and vertical motions in granules. One particularly worrisome point is that local fluctuations in granule intensity and size could produce temporary apparent velocities on supergranule or larger scales which could be thought to be real. The way round this difficulty is clearly to observe often enough that such spurious patterns average out.

Aside from such questions as how fast granulation velocities decay with height, it is not clear how much more we really need to know about granulation per se. Proposed observations from the space shuttle (the Solar Optical Telescope—SOT) will undoubtedly give us better resolution and shift the spatial scale for which granulation becomes un-resolved 'microturbulence'. But such measurements will be more interesting for what they can tell us about the relations between small-scale velocity fields, such as granulation, and small-scale magnetic fields. We return to this point in Section 4.

3.1.2. *Granulation Theory*

By the beginning of the 1970s, there was near concensus that granulation patterns represent thermal convection penetrating into the photosphere from the convection zone below. Physical theories for this phenomenon have progressed considerably since then. The correlation between hot granules and upward motion, cold intergranular lanes and downward motion is very good, and estimates of the upward heat flux due to these motions can be made. Such estimates currently vary a great deal, however. For example, Altrock and Musman (1976) estimate that at optical depth 3, about 1% of the flux is carried by convection. By contrast, Edmonds (1974) obtains a convective flux fraction of more than 50% at optical depth unity! As Edmonds illustrates, different mixing-length theories can also give quite different fluxes near the surface. A large part of such differences may depend on how much of the buoyancy work done by a rising warm element goes into vertical acceleration of that element, and how much is immediately converted instead into kinetic energy on still smaller scales by turbulent cascades ('microturbulence'). The resultant granule scale motion, and therefore the upward heat flux, would be reduced. Altrock and Musman (1976) estimate that perhaps five-sixths of the buoyancy work is diverted in this way.

In addition to the convective flux, theories of granulation also should predict the dominant observed spatial scales, the magnitude of velocities and intensity variations with depth as well as limb shift and certain other properties of observed spectral lines. In addition, the characteristic time-dependent behavior of granules described above should be captured, as well as the fact that in granules the rising motion is in the cell center and the sinking motion is on the outside. A number of scientists have been building a succession of convection models of increasing generality, the latest of which have come rather close to this goal. In so doing, these models have progressed closer and closer to full three-dimensional solutions of the governing equations (Equations (1.1)–(1.3)) (with Coriolis and electromagnetic body forces ignored). For example, Nelson and Musman (1977) constructed a two-dimensional steady-state model driven by buoyancy against a nonlinear viscous drag force which generates plausible temperature and velocity variations with height. However, the horizontal wavelength of this convection pattern must be assumed, as well as the temperature perturbation at the bottom of the layer. These are both matched to observations. But using essentially the same model, Nelson and Musman (1978) are able to illustrate qualitatively how the dominant horizontal scale of convection is determined. They show that for scales somewhat smaller than granulation, horizontal radiative transport will damp out the temperature fluctuations and therefore buoyancy forces. For scales larger than granulation, the dynamical pressure (which must be largest in the center of a granule to drive the fluid horizontally outward) varies enough in the horizontal to change the opacity in such a way as to reduce the apparent contrast. In addition, for a given granule depth, there is a limit to how much vertical velocity can be achieved in a rising element due to buoyancy, and therefore in the magnitude of the dynamic pressure peak in the center of the granule. But the larger the horizontal scale, the larger the pressure difference needed to drive the fluid outward against turbulent dissipation. These arguments, however, do not give real sharp limits to granule size, which can come only from even more detailed physical models. But as Nelson (1978) has shown, even this approximate model predicts substantial changes in atmospheric structure due to granulation compared to mixing-length models. Using it, Nelson (1978) has obtained reasonable agreement with the observed limb darkening, center-to-limb variations in intensity fluctuations, as well as r.m.s. horizontal and vertical velocities.

In a parallel effort, Nordlund (1978, 1980) has built a three-dimensional time-dependent nonlinear convection model for granulation. Dravins *et al.* (1981) have computed many characteristics of spectral lines as well as velocities, temperatures, and limb shifts predicted from this model and obtained quite detailed agreement with observations. In this convection model most of the physics has been made quite realistic. Sound waves are filtered out, but the fluid is compressible. The equation of state includes partial ionization (of hydrogen); the absorption coefficients are of the type used in standard stellar atmosphere codes; the radiative transfer is three-dimensional and non-gray. A nonlinear eddy viscosity parameterizes small, unresolved scales. Calculations are done inside a horizontally periodic unit cell of horizontal dimensions 3600 × 3600 km, and a depth of 1500 km. Thus the convection has some freedom to choose its horizontal scale, but still will be strongly influenced by the imposed vertical dimension. Nordlund does see at work the mechanisms for determination preferred granule size described by Nelson and Musman (1978). His model granule patterns are composed of bright regions imbedded in a network of dark lanes, as seen in Figure 2. The downward motion is larger in these lanes than the upward motion in the centers. He also sees fragmentation and 'exploding granules'. It remains to be determined whether this particular phenomenon is more in the nature of a rising expanding bubble or fragmenting vortex ring as Musman (1972) proposed, or more like a secondary convective instability of locally axisymmetric convection, as Jones and Moore (1979) have suggested. We await from Nordlund more detailed physical explanations for the behavior of his model solutions.

Although, in particular, Nordlund's model has apparently brought us close to a complete theory of granulation, the question of just what determines the dominant horizontal scale remains open, requiring numerical modeling in a fluid domain of much larger horizontal and vertical extent. This may not be completely meaningful without consideration of other larger scale motions, such as supergranulation and the newly discovered mesogranulation, which we discuss next.

Dravins *et al.* (1981) have produced such detailed predictions of the behavior of spectral lines in response to granulation that their model may have great potential for deducing properties of convection in other stars.

3.2. SUPERGRANULATION AND MESOGRANULATION

3.2.1. *Observations*

It has been known since about 1955 that there was another pattern of motions, of much larger scale than granulations, covering the Sun in the photosphere. There were first clearly seen by Hart (1956) and more systematically observed by Leighton *et al.* (1962) and Simon and Leighton (1964). These latter studies established the horizontal flow as a cellular pattern with horizontal divergence from a central point in each cell, and the authors gave this pattern the name 'supergranulation'. Typical horizontal distance between supergranule centers is about 3×10^4 km, or about 20 times that of granulation.

The horizontal flow in the photosphere has magnitude of 300–500 m s^{-1}. Vertical velocities have been more difficult to find. Worden and Simon (1976) saw downflows of ~200 m s^{-1} in supergranule boundaries at the sites of magnetic flux tubes (we return to the relation between supergranule velocity fields and magnetic fields in Sections 4.1. and 5.1) as well as possible weak upflow of 50 m s^{-1} in supergranule centers. These flows can easily be obscured by the 5 min oscillations which have a similar horizontal spatial scale unless careful averaging is done.

Unlike granulation, supergranulation does not have a well-defined pattern of photo-spheric intensity fluctnations associated with it. Beckers (1968) found a very small increase in continuum intensity near the cell boundary compared to cell center, but this appeared to be more associated with the chromospheric network and he argued that any horizontal temperature variations would be less than 1 K, in contrast with values of 100–200 K for granulation. Worden (1975) studied this effect further and found, from infrared measurements, temperature drops of 50–500 K compared to the surrounding photosphere but strictly confined to magnetic field elements, confirming that there is no general temperature variation over supergranules at photospheric levels.

Supergranules live much longer than granules. Simon and Leighton (1964) estimate ~20 h, and some are seen to last for several days (Smithson, 1973; Worden and Simon, 1976). Because few supergranule patterns are observed continuously for more than 8 h from the ground, relatively little is known about the details of growth and decay of supergranules.

While granules appear to extend no further than the upper photosphere, supergranule motions have been observed to extend well into the chromosphere (November *et al.*, 1979), at least 11 density scale heights above the photosphere. Furthermore, the velocities seen at high levels are as large as 3 km s^{-1}, with horizontal and vertical velocities of comparable magnitude.

November *et al.* (1981) have also found new evidence of a cellular motion intermediate between granules and supergranules in scale, which they have called 'mesogranulation'. By comparison with granulation and supergranulation, it is a rather weak motion, about 60 m s^{-1}. It has a spatial scale of 5–10 X 10^3 km, and one mesogranule typically lasts about 2 h. This discovery may have been anticipated in part much earlier by Deubner (1972).

3.2.2. *Theory*

It seems likely that supergranulation, and probably also mesogranulation, are both convective motions. However, there is no evidence that they produce an upward heat flux such as has been estimated for granules.

There are no convincing detailed quantitative theories of supergranulation at present. Calculations of the degree of physical realism carried out by Nordlund for granulation have not yet been tried, but they may not be far off. Simon and Leighton (1964) suggested that granulation and supergranulation were produced respectively by the ionization of hydrogen and helium, and November *et al.* (1981) have gone further to speculate that mesogranules come from He$^+$ ionization and supergranules from He^{++}. Figure 1 indicates the depths at which these partial ionization zones occur on the Sun. In each case, the depth is roughly comparable to the horizontal scale of the respective observed pattern, which is a common characteristic of convection. The basic idea is that in each region of partial ionization, the opacity rises, and latent heat of ionization can be released or absorbed in a rising or sinking fluid element; these effects both enhance convective instability. However, the contribution to the total opacity from helium ionization is quite small compared to that for hydrogen, and no opacity 'bump' generally appears in solar structure models at the levels of helium ionization.

It is more likely that the local change in molecular weight of the H–He mixture, and the release of latent heat of ionization, would produce the enhanced instability. These concepts remain untested by quantitative calculations, but the basic theory of convection in a compressible fluid has been advancing rapidly and such a test seems likely to occur within the next few years. If the hypothesis proves correct, the theory which demonstrates it will also have to show why mesogranules are so weak compared to granules and supergranules.

3.3. RELATED THEORIES OF CONVECTION

There are a number of thermal convection calculations for compressible fluids that, while not applicable to solar granulation and supergranulation in detail, are still relevant to those problems. We review some of these calculations here. Those which include such complicating effects as spherical geometry, rotation, and magnetic fields are covered in appropriate later sections when we discuss global circulation and the solar dynamo. There is a very large literature on strictly Boussinesq (incompressible) convection applicable to laboratory situations, which we also do not attempt to cover here.

Linear studies of convection in a compressible medium, — for example, Gough *et al.* (1976), and Graham and Moore (1978) — show that the mode which first becomes unstable and grows is one which extends from the bottom to the top of the convectively unstable layer, even if that layer includes many scale heights. This argues against one of the fundamental assumptions of mixing-length theory, namely that an appropriate vertical scale for convection is the local scale height. This result seems to be relatively independent of what is assumed for viscosity and thermal conductivity of the fluid with depth. Graham's (1975) nonlinear calculations for two-dimensional convection also support this result. These studies do not, however, take into account partial ionization, so they are not in conflict with the concept described above that the distinct scales of convection on the Sun are due to ionization zones for H^+, He^+, He^{++}.

When rigid boundaries are replaced by convectively stable layers above and/or below the unstable layer, the convection does penetrate by an amount which may be as large as the horizontal dimensions of the cell in the unstable region. Thus, as Graham and Moore (1978) have shown, this penetration can be across several scale heights. However, it is generally much less than the upper limit found by assuming that all the kinetic energy gained by a buoyant fluid particle when accelerating upward in the unstable region is converted into work done against negative buoyancy in the stable layer above, as has been assumed in some extensions of mixing-length theories to account for penetration.

Although penetration can be substantial, it is unclear whether it can by itself explain the extreme effects seen by November *et al.* (1979) for supergranules extending into the chromosphere. For that case, one also needs to consider the role played by magnetic fields.

Recent nonlinear compressible convection models have generally incorporated the so called 'anelastic' approximation (see, e.g., Gough, 1969; Latour *et al.*, 1976) in which sound waves are filtered out of the fluid equations, but other effects of compressibility, particularly the large density variation with height, are retained. The advantage is that usually much larger time steps can be employed in the time integration. The approxima-tion is valid as long as the convective velocities remain small compared to the sound speed, which is adequate for most solar and stellar applications. To cut down the computational expense still further, most recent models have severely truncated the horizontal resolution in the convection by resorting to specified plan forms – hexagons, rolls, or squares. Van der Borght (1974, 1975, 1979), Latour *et al.* (1976), Toomre *et al.* (1976), and Massaguer and Zahn (1980) have used both the anelastic approximation and truncated modal representation.

Toomre *et al.* (1976) and Latour *et al.* (1981) have applied their model mostly to the relatively weak convection zones of A stars, and demonstrated that overshoot will produce much more mixing, and greater heat flux, than previously estimated from mixing-length arguments for those stars. In addition, the two separate convection zones supposedly typical of A stars (an upper zone due to by hydrogen ionization; a lower one due to helium ionization) are in fact strongly dynamically coupled. Massaguer and Zahn (1980) ignored the complicating effects of partial ionization to focus on compres-sibility and found the startling and not yet well understood effect that near the top of the most unstable convecting layer the buoyancy force can actually become negative, resulting in a turning of the convective flow into the horizontal before the top is reached. Horizontal pressure fluctuations play an apparently controlling role in the dynamics, and in some solutions the horizontal flow just below the top boundary is substantially larger than at the boundary, even though the boundary is stress free. Latour *et al.* (1981) also found this effect even when some penetration occurs. Massaguer and Zahn (1980) ascribed this property to the assumption of a constant temperature boundary condition at the top, but Glatzmaier and Gilman (1981) have seen the same effect with constant flux boundaries.

The reality of this effect needs to be examined over a wide range of assumptions about boundary conditions, – penetration, nonlinearity, representation of small-scale turbulent stresses, degree of truncation – to be clearly understood. For example, could the effect be an artifact of models which impose a horizontal scale near the top of the convecting layer that is large compared to a scale height, as suggested by Spruit (private communication)? If smaller scales are allowed, perhaps the horizontal flow will break down into smaller eddies which destroy the negative buoyancy.

If the effect is real, one implication for the Sun may be that the top layers of the convection zone actually prevent larger scale convection below from penetrating to the surface in sufficient amplitude to be observed. Turbulent diffusion due to small-scale convection at the surface is probably not, by itself, able to prevent penetration (Stix, 1981b). More work on this problem is definitely needed.

Another problem in need of work is the representation of those scales of motion not explicitly treated in these convection models. A variety of assumptions have been em-ployed, the simplest and perhaps least accurate being that molecular viscosity is replaced by a specified (and much larger) eddy viscosity. Thus the nominal Prandtl number (ratio of viscosity to thermal diffusivity) in stellar convection zones of $10^{-6}-10^{-10}$ becomes of order $1-10^{-2}$, depending on how small the scale of motions explicitly modelled is. Much better prescriptions for small-scale turbulence are needed, presumably built on extensions of turbulence theory. Unfortunately, turbulence theory has generally not yet reached the point of including such physics as compressibility and partial ionization or even buoyancy. In general, departures from homogeneity and isotropy are very difficult to treat, so important *ad hoc* elements are likely to remain in the theory for a long time. The underlying (and often unstated) hope of most scientists actually attempting to model solar and stellar convection is that the details of such prescriptions for small-scale turbulence do not really matter. But at present we simply do not know. For one recent attack on the convective turbulence problem, see Nakano *et al.* (1979). We return to this point in greater detail below when we discuss global convection and circulation models.

3.4. GLOBAL CIRCULATION

3.4.1. *Observations*

The present state of our knowledge of global circulations on the Sun is that the presence of differential rotation is well established and there is some evidence for meridional circulation (mean latitudinal flow) but other global motions, such as might be associated with global convection, have never been convincingly demonstrated to exist. Efforts to find these 'giant cells' continue.

Howard (1978) has recently reviewed observations of solar rotation in considerable detail, much of which we need not repeat here. There are currently two methods widely in use for measuring rotation: one uses Doppler shifts of radiation of the solar plasma, and the other tracks of solar features, usually associated with the Sun's magnetic field. Historically, sunspots have been used most commonly, but more recently filaments, Ca II emission patterns (faculae), magnetic field patterns, X-ray and radio emission, and coronal brightness have all been employed. Each method has its own drawbacks, as discussed in Howard (1978), and generally gives somewhat different values for solar rotation.

By most measures, solar differential rotation takes the form of an equatorial accelera-tion — equatorial regions rotate faster than high latitudes. The angular velocity Ω from Doppler shifts can be represented rather accurately by a formula of the form

$$\Omega = A - B \sin^2\phi - C \sin^4\phi \qquad (3.4.1.1)$$

in which ϕ is solar latitude, and A, B, and C have positive values. For example, Howard and Harvey (1970) found average values of $A = 13.76$, $B = 1.74$, $C = 2.19$ degrees per day (sidereal) for a four-year period beginning in 1966. Thus, the angular velocity is lower by about 30% near the pole compared to the equator. Even without a fit of the form of Equation (3.4.1.1) the angular velocity profile virtually always shows a fairly smooth, monotonic decrease with latitude.

Sunspot rotation rates can usually be fitted to a formula quite similar to Equation (3.4.1.1) for latitudes equatorward of $40°$, where spots are seen. However, even yearly average values show much more scatter, and spots are an inherently biased sample, since the zone of sunspot occurrence migrates towards the equator as a solar cycle advances.

The long-term average solar rotation from spots (see, e.g., Ward, 1966) is systematically higher than the solar plasma at each latitude by about 4%, or 80 m s^{-1} in linear velocity measure at the solar equator.

There are numerous other differences among differential rotation measures. For example, as reviewed in Howard (1978), large-scale magnetic field patterns generally show less differential rotation with latitude than the plasma, although small, short-lived magnetic elements in these patterns differentially rotate at similar rates to the plasma. Long-lived coronal holes (e.g. Wagner, 1975), appear to rotate nearly rigidly, but short-lived ones do show differential rotation (which probably contributes to their short lives!). Even finer variations among tracers can be seen. Large sunspots rotate more slowly (by up to 2%) than small spots (Ward, 1966). On the other hand, longer lived X-ray emission patterns rotate faster by several percent than small compact features (Golub and Vaiana, 1978). Regions of strong magnetic field (outside sunspots) rotate faster than weak field regions (Foukal, 1976). Similar sorts of differences are seen in calcium faculae (Belvedere *et al.*, 1977). Most recently, Duvall (1980) has shown that the pattern of supergranules rotates about 3% faster than the plasma.

Although many measurements of solar rotation have been made, relatively few of them have been systematically carried out and compared for the same time periods. Drawing conclusions from comparisons of Doppler rotation for one time period with sunspot rotation for another can be risky. For example, Scherrer and Wilcox (1980) and Scherrer *et al.* (1980) maintain that the difference between sunspot and Doppler rotation in low latitudes disappeared during the late 1970s. However, they compared their Doppler measurements with the rates for recurrent spots of a much earlier time period (Newton and Nunn, 1951). Their Doppler rate was a few percent higher than that of Howard and Harvey (1970), but sunspots in the late 1970s could also have been rotating faster. This remains to be determined. Also, measurements by Foukal (1979) of the Doppler velocity inside and outside individual sunspots demonstrates spots are indeed rotating faster than the surrounding plasma by several percent.

The above point leads us to the consideration of time variations in solar rotation. Howard (1976) found that the rotation rate of all latitudes increased between 1968 and 1975 by about 4–5%, although the rise was not monotonic. In subsequent years, this rise has begun to level off (Howard, private communication). The general increase has been confirmed by Livingston and Duvall (1979) using Kitt Peak data. They also see evidence that the polar rotation rate changes with the solar cycle, by about 8%, with the most rapid polar rotation occurring near or just after cycle maximum. Stenflo (1977) finds changes in rotation of magnetic patterns with time over a solar cycle. These take the form of zones of larger than average shear in rotation on the poleward side of sunspot zones. These shear zones migrate towards the equator as the solar cycle progresses. More recently, Howard and La Bonte (1980) have found a very small amplitude (±5 m s^{-1}) but quite systematic pattern of alternating high and low-rotation zones which migrate towards the equator in each hemisphere from extremely high latitudes, taking about 22 yr to complete a cycle. The phase of this 'torsional oscillation' is such that the average latitude of sunspots, and peak magnetic flux, coincide with the boundary between a zone of fast rotation on the equatorward side and slower rotation on the poleward side, similar to that found by Stenflo (but of smaller amplitude). Because the migration of a zone to the equator takes ~22 yr, but is followed by another only 11 yr later, there is a great deal of overlap in time of adjacent zones on the Sun, with one low-latitude and one high-latitude zone virtually always present. This property has led Brown (private communication) to suggest that the high-latitude zones may be artifacts of the reduction procedure used by Howard and La Bonte. Brown has shown that a purely low-latitude velocity signal would, when subjected to the global fitting functions used by Howard and La Bonte, alias onto high latitudes to give an apparent migration all the way from the pole to the equator. If this is what is really happening, then the high-latitude signal is not evidence of the next magnetic cycle as Howard and La Bonte (1980) have claimed, but rather an alias of the current one. La Bonte and Howard (1981) have countered this suggestion by processing an assumed low-latitude signal through their own reduction routine, and do not find a significant high-latitude alias. Clearly further study is needed, but as Howard and La Bonte (1980), and La Bonte and Howard (1981) have pointed out, these torsional oscillations, if real, may carry important information about the workings of the solar dynamo in the interior of the convection zone.

On much longer time-scales, Eddy *et al.* (1978) have shown a secular decrease in rotation of sunspots of a few percent in the first half of the twentieth century, followed by an apparent leveling off. During this time, the envelope of the solar cycle amplitudes was rising. Perhaps this behavior connects in time to the rising rotation rate seen in later Doppler measurements by Howard (1976). Systematic study of Doppler and spot rates are needed for the period since 1966 when both can be measured; the author and R. Howard are currently beginning such a study using for sunspots the white light plate collection from Mt Wilson. Going back still further, Eddy *et al.* (1976, 1977) and Herr (1978) have estimated sunspot rotation rates during the first half of the seventeenth century and concluded that equatorial rotation speeded up significantly just prior to the onset of the Maunder Minimum in 1645 (Eddy, 1976) during which solar activity levels remained extremely low until about 1715. However, Abarbanell and Wöhl (1981) have remeasured spot rotations just before Maunder Minimum onset and failed to confirm the higher equatorial rotation. The reduction techniques used are somewhat different, but it is not clear why the results are different. Eddy *et al.* (unpublished) repeated their measurements and obtained the same result as before.

It is commonly suspected that all the time changes in solar rotation described above are connected in some way to the workings of the solar dynamo. High rotation tends to be correlated with low levels of solar activity, although this has not been clearly demonstrated within a single cycle, and may be more a property of the amplitude envelope of solar cycles.

At the other end of the time spectrum, changes in solar rotation on time-scales of less than the rotation period have been reported but are fundamentally difficult to separate from other east—west motions, since not all longitudes are sampled each day (Gilman and Glatzmaier, 1980). Apparent fluctuations of several percent of amplitude and several days of time duration have been seen, for example, by Howard and Yoshimura (1976) but other observations, for example, Duvall (1979) and Scherrer *et al.* (1980) have failed to confirm such large fluctuations. Whether any of these fluctuations represent real solar motions, rotation or otherwise, or are primarily of instrumental or atmospheric origin remains to be determined.

A number of measurements of the mean north—south or meridional motion have also been made, but the estimates of this motion made from Doppler observations differ greatly from those made from sunspots. Ward (1973) and Balthasar and Wöhl (1981) find no statistically significant mean north–south motions of sunspots, with an upper limit of only a few metres per second. On the other hand, Duvall (1979). Howard (1979), and Beckers (1979) have reported net poleward Doppler velocities in each hemisphere in the range 20–40 m s^{-1}. Furthermore, Beckers and Taylor (1980) have argued that proper correction for limb effects actually raises this value to near 70 m s^{-1}. Clearly, if such values are real, sunspots participate in meridional flow hardly at all. In effect, they appear to be much better tracers of mean east–west than mean north–south motions. Meridional circulation is a particularly important input to theories of solar differential rotation, which we discuss further in the next section.

Evidence for giant 'cells' or eddies, – that is, global flow patterns not symmetric about the axis of rotation – has been even harder to obtain than for meridional circulation, although there are many solar phenomena suggestive of such flows. In the Doppler measurements, Howard (1979) has seen large-scale, probably radial flows near the equator in Mt Wilson data which recur for several rotations. But these are seen only occasionally, and can hardly be said to cover the whole Sun the way differential rotation or super-granules do. Howard and La Bonte (1980) have searched for periodic east–west flows by developing synoptic maps and power spectra of the residual velocities after differential rotation and other global corrections are subtracted out. This effort was refined by La Bonte *et al.* (1981) in response to criticisms of the reduction procedure by Gilman and Glatzmaier (1980), resulting in estimates of root mean square velocities per longitudinal wavenumber of about 12 m s^{-1} for wavenumber 1, down to 3 m s^{-1} for wavenumbers larger than 20. But these must be considered upper limits, and the real power could be considerably lower if all instrumental and other nonsolar effects could be removed. Values as low as this stretch current observing systems to their limits of accuracy and beyond.

Schröter *et al.* (1978) have looked hard at the movement of the calcium network with observations at Locarno, to try to find evidence of giant cells or eddies. They find some, but they do not correlate well with either their own Doppler velocity measurements or Mt Wilson measurements. Nor do the Mt Wilson and Locarno Doppler measurements of rotation changes correlate well; the former are highly correlated in latitude, and the latter are not. Perez Garde *et al.* (1981) also report a global eddy in a small sample of Doppler data, but do not compare it with other velocity measures. Part of the difference between Doppler and Ca^{+} velocities may reflect a very complex interaction between the plasma flow and the magnetic fields, but also raises doubts about the accuracy of the measurements, particularly of the Doppler velocities.

The very existence of large-scale patterns in the solar magnetic field, even though the individual elements of magnetic flux in the pattern are very small in spatial scale, is suggestive of corresponding global velocity patterns which have yet to be measured directly. Similar impressions are gained from coronal holes, particularly ones which last several rotations that are not being passively sheared apart by the differential rotation. The arrangement of the pattern of large filaments on the Sun is sometimes suggestive of underlying global disturbance structure (Wagner and Gilliam, 1976), as are the evolution of Hα neutral lines (McIntosh, 1979). The apparent existence of 'active longitudes', areas on the Sun where new active regions preferentially arise, suggests there are persistent velocity patterns bringing up the new magnetic flux.

There is also the old calculation by Ward (1965) of the correlation of east—west and north—south sunspot motions which has been interpreted as evidence of an equatorward transport of angular momentum by eddies presumably of much larger spatial scale than sunspots. More recently, similar but usually larger correlations have been seen in movement of calcium plages by Belvedere et al. (1976, 1978), but the scatter has been too large to determine the profile of transport with latitude. This is one of the flow properties we need to determine for purposes of comparison with differential rotation theories.

To summarize the status of our observational knowledge of giant cells or global eddies, there are many tantalizing bits and pieces, but they do not corroborate each other well yet, and most are not direct measures of the velocity field itself, but some other pattern or tracer. Various theoretical considerations do favor the existence of giant cells and their existence is presumed in most global circulation models applied to the Sun. We discuss this further below.

It appears there are two key questions for the future concerning observations of global flows in the Sun. One is how differential rotation varies with depth; the other is whether giant cells or global eddies really exist. With respect to the first question, the inference has been made by Foukal (1972) that sunspots are 'anchored' at some depth below the solar photosphere, so that their faster rotation rate is interpreted as evidence that the angular velocity increases with depth, at least for the outermost $1-2 \times 10^4$ km or so. A potentially more powerful technique for measuring rotation with depth has been proposed and tested in a preliminary way by Deubner et al. (1979), Rhodes et al. (1979), and Ulrich et al. (1979). This involves measuring the rotational splitting of frequencies of 5 min oscillations. The method is described in detail in Brown, Mihalas, and Rhodes (Chapter 7, this volume). Deubner et al. (1979) report an increase of angular velocity with depth of a few percent in the first 1.5×10^4 km below the surface, but the error estimates are large and more precise measurements are needed to confirm this result and general approach. Claverlie et al. (1981) infer a rotation of the solar interior of 2—9 times the surface value from observations of the lowest order 5 min oscillations, but it is unclear where the angular velocity gradient implied by these results lies. The implications of this new result need to be pursued. Whether the frequency splitting technique will prove to be so sensitive that other global flows besides rotation can be inferred remains to be demonstrated.

With respect to the existence and properties of giant eddies, it would appear that observing systems used up to the present were really not designed to answer this question. From Howard and La Bonte (1980) and La Bonte et al. (1981), global velocities other than differential rotation are weak and yet no stable wavelength references have been employed against which to measure the Doppler shift. The velocity signal at any instant in time is heavily influenced by large amplitude small-scale solar noise: — granules, 5 min oscillations, and supergranules — and yet the typical measurement program has obtained only one or two full disk scans per day. One exception is the short series of observations at Locarno reported in Schröter et al. (1978). To make more progress, the whole solar disk must be observed often enough to average out the granule and oscillations noise, requiring full disk observations at least every minute, and densely enough in time from hour to hour and day to day to reduce the influence of supergranules as far as possible. It appears that supergranules will be harder to get rid of than 5 min oscillations and granules, but we should be able to do much better than we have done so far. The High

Altitude Observatory and Sacramento Peak Observatory have been working together
on the development of one such new instrument, called a Fourier Tachometer (Beckers
and Brown, 1979) which should allow more accurate, more stable, faster measurements
to be made. Other, parallel instrument development efforts would be highly desirable.

3.4.2. *Global Circulation Theory*

Theories of global circulation of the Sun have focused primarily on explaining the differential rotation, for the reason that it is the most prominent motion of global extent,
and by comparison so little is known about meridional circulation or giant cells. There
is also wide agreement that the origin of differential rotation seen in the photosphere
is in the convection zone. It arises from the influence of rotation upon the convection,
through the action of Coriolis forces. The slow decay of solar rotation with solar evolution, due to solar wind torques, is a separate and distinct problem, which has little to do
with the observed differential rotation profile. It has been occasionally suggested, for
example by Schatten (1973) and Alfvén (1971), that solar wind torques might contribute
significantly to the low rotation rate seen at the poles. However, as Parker (1971) and
Gilman (1974) have argued, these torques are far too weak to compete with turbulent
mixing within the convection zone.

In addition to explaining differential rotation, global circulation models must also
satisfy the constraint of virtually uniform heat flux with latitude at the outer boundary.
Some models do rather well in this regard, others do rather poorly.

Despite the lack of observational evidence for giant cells on the Sun, there are theoretical reasons for believing they should exist (Simon and Weiss, 1968). As mentioned in
Section 3.3 above, even when the fluid is compressible and the stratification encompasses
several density scale heights, convection which extends from the bottom to the top is
favored in both linear and nonlinear models. With a solar convection zone depth of 30%,
and convection cells whose horizontal dimensions are comparable to the vertical, we
should expect a pattern of motions whose horizontal dimensions on the surface are
$\sim 2 \times 10^5$ km — much larger than supergranules. Most theoretical models for differential
rotation involve motions on this scale, either explicit or implicitly. Even though from
observations such motions must be relatively weak, they should be very important for
determining differential rotation because their large scale (and weak amplitude) imply
a long turnover time — a month or more on the Sun — during which Coriolis forces have
plenty of time in which to act.

In all models for differential rotation, only the global scale motions are explicitly
calculated, and all the smaller scales are represented in some parametric fashion. At
present, these parametrizations are relatively crude, and need to be improved greatly.
In one class of models, only differential rotation and axisymmetric meridional circulation
are calculated explicitly, and all others are represented by coefficients of turbulent
diffusion. Another class of models calculates global scale convection, including departures
from axisymmetry. We review both below.

Before discussing particular models, it is useful to consider two mechanisms of angular
momentum transport available to generate and sustain the solar differential rotation.
Since angular momentum can be convected by the fluid itself any. flow which has a
component in either the radial or latitudinal direction can change the rotation profile

in the meridian plane. It is convenient to consider the effects of axisymmetric meridional circulation, and all departures therefrom — global and small-scale eddies or convective cells — separately, with the aid of Figure 3. At the top, two schematic meridional

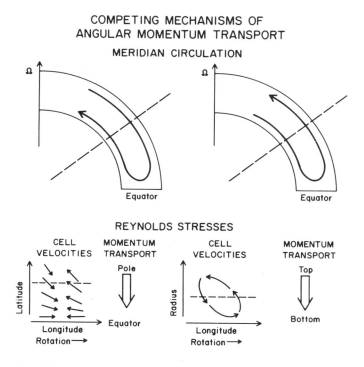

Fig. 3. Schematic of flow patterns responsible for angular momentum transport in various global circulation models applied to the solar differential rotation problem. Adapted from Gilman (1980).

circulations are shown, of opposite sense. If either pattern started up in a fluid originally in solid rotation, and no other transport processes were acting, high latitudes and deeper layers would tend to spin up, because the circulation would conserve angular momentum. But now if sufficient diffusion is present (most likely associated with small-scale motions in the convection zone) to link the different layers of fluid and keep the angular velocity nearly constant, the circulation on the left may produce a net equatorial acceleration (Kippenhahn, 1966). This is because fluid moving towards the equator on the outer branch, (for example, crossing the dashed line) would contain more angular momentum than fluid moving toward the pole underneath, where the moment arm is shorter. The only exception to this argument would be if the differential rotation were exactly constant on cylinders concentric with the axis of rotation. In that case, the momentum is constant on surfaces of constant angular velocity, so a meridional flow across such a surface transports no angular momentum in the net.

The lower schematics of Figure 3 illustrate angular momentum transport by non-axisymmetric motions, through correlations between east—west and either north—south or radial motions, called Reynolds stresses. Both left and right schematics illustrate common motion patterns, plotted relative to a uniformly rotating reference frame, actually seen in models. On the left is a horizontal flow which leads to angular momentum transport towards the equator. Flow towards the west (in the same direction as the rotation of the whole system) also has a component towards the equator, while flow at adjacent longitudes towards the east has a poleward component. If we average in longitude − e.g. along the dashed line, − we get a net correlation which implies momentum flux towards the equator, even if there is no net mass flux. In the right schematic is a typical convective circulation pattern in a local longitude-radius plane that by similar arguments leads to a net flux of angular momentum inwards. In both cases, the necessary tilts in the velocity vectors are induced by Coriolis forces acting on the convective motion. These same forces also determine which convective modes are preferred.

Turbulent diffusion of momentum is usually also present in some form in differential rotation models. Some models explicity compute the convection patterns and, therefore, Reynolds stresses typified in Figure 3, but treat transport by smaller scale motions with ordinary diffusion of the form $\nu \nabla^2 \mathbf{v}$, in which ν is the kinematic (eddy) viscosity. Others use extensions of mixing-length arguments to estimate how Coriolis forces locally generate Reynolds stresses, and then include these estimates in a parametric way in a global model in which only the meridional circulation is explicitly calculated in addition to differential rotation. Therefore, in the former case the Reynolds stresses enter in Equation (1.1) through the component of nonlinear inertial term $\rho(\mathbf{v} \cdot \nabla)\mathbf{v}$ in the east–west direction, while in the latter case the transport effects enter through the term $\rho \mathbf{F}$ with $\mathbf{F}$ treated as the divergence of the Reynolds stress tensor. Still other models invoke inhomogeneities or anisotropies in diffusion of momentum or heat, in combination with meridional circulation, to obtain differential rotation. Some of these ignore the effect of rotation on turbulent transfer altogether, or invoke it only as a weak effect.

Longitudinal pressure forces in Equation (1.1) vanish for an axisymmetric differential rotation, and electromagnetic body forces ($j \times B$) are probably too small on the Sun to exert much influence, except perhaps as a drag in the neighborhood of magnetic flux tubes. (We return to this point when discussing possible theories for the torsional oscillations observed by Howard and La Bonte (1980).) In addition, buoyancy forces do not act in the azimuthal direction. Thus, we are left with the motions themselves as the direct determinant of differential rotation. In the real situation, all the motions compete with each other in determining the resulting differential rotation. Which angular momentum transport by which motion dominates can only be determined by actual model calculation and the result is also bound to be somewhat model dependent. Coriolis forces play a crucial role in determining convective mode size and shape, and therefore their momentum transport properties. It follows that a very important parameter determining the kind of Reynolds stresses that are produced in the convection is the ratio of turnover time for the convection to rotation time for the whole system. For the Sun, this ratio is much less than one for granules, somewhat less than one for supergranules, and probably greater than one for giant cells. Therefore, we can expect these different scales of motion to contribute differently to the observed differential rotation.

Let us first consider axisymmetric models. In one class of models, the eddy viscosity is assumed to be anisotropic, i.e. there are different transport rates in different directions. This approach originated with Biermann (1951) and was exploited by Kippenhahn (1963), Cocke (1967), and in greater detail by Köhler (1970, 1974). The hypothesis is that this anisotropy is introduced by the presence of gravity but Coriolis forces are ignored. When the anisotropy is included, solid rotation is no longer a solution of the equations of fluid motion (1.1) and (1.2). Both meridian circulation and differential rotation result, and with suitable choice of the sign and magnitude of anisotropy, the solar equatorial acceleration can be reproduced. If the viscosity in the model is relatively small, then the angular velocity is also nearly constant on cylinders concentric with the axis of rotation, and so decreases with depth. The mechanism is essentially the one described earlier in connection with Figure 3. The meridian circulation set up has flow towards the equator in the outer branch, flow back towards the pole in the inner branch. With angular velocity nearly constant on cylinders, there is more equatorward transport in the outer branch than poleward transport on the inner branch crossing a radial line. A typical meridional circulation velocity in this model is only 2 m s^{-1}, which is smaller than can be observed. As Köhler (1970) points out, due to boundary conditions, the differential rotation is not exactly constant on cylinders. It is precisely this small departure from cylindrical symmetry which allows a net equatorward transport by the meridional circulation. In the steady state, this is balanced by a weak poleward diffusion of angular momentum due to the turbulent viscosity (Kippenhahn, 1966; Köhler, 1970, 1974).

To get equatorial acceleration requires that the eddy viscosity for horizontal momentum transport be larger than that for radial transport. With the reverse, deceleration results (due to a reversed meridional circulation). But there is no physical argument which clearly favors either sense of anisotropy. Also, no account has been taken of the influence of rotation upon the eddy viscosity. Finally, no thermodynamics are included in the model.

A second class of models which have been carried further have invoked convective heat flux weakly influenced by rotation, such that it becomes a function of latitude. This was first tried by Durney and Roxburgh (1971) and later developed further by Belvedere and Paterno (1976, 1977, 1978) and Belvedere *et al.* (1980a). Here again, the sign and magnitude of the variation in heat transport coefficient with latitude is chosen so as to give the observed equatorial acceleration. Again, the dominant meridian circulation has flow in the outer part towards the equator. Belvedere *et al.* (1980a) have shown that if the eddy diffusion of momentum is assumed to be much less than that for temperature, the required meridian circulation can be very small, and be consistent with extremely small differences with latitude in surface temperature. However, there appears to be little physical justification for this assumption. Later versions of the model also contain a similar density variation with depth as the solar convection zone is thought to have, unlike many previous models. A typical differential rotation with depth produced by the model has angular velocity increasing inward, being nearly constant on surfaces perpendicular to the rotation axis.

One problem with this particular model is that it relies heavily on the assumption of weak influence of rotation upon convection (turnover time short compared to rotation time) which is a poor assumption for the deep parts of the solar convection zone. When the influence of rotation is more accurately taken into account a different answer may result. Schmidt (1981) has calculated solutions for this model when the influence of rotation is assumed not to be small. However, the same functional form for turbulent diffusivity of heat is used as in the low-rotation case, and there are questions concerning the convergence of these high-rotation solutions (Schmidt, private communication).

There is also some question as to whether the answers obtained are reasonable even in the weakly rotating case. In the author's experience with nonaxisymmetric convection in rotating spherical shells, it has been difficult to construct examples in which convection weakly influenced by rotation could sustain an equatorial acceleration. Global convection weakly influenced by rotation normally gives high latitude acceleration, as well as large oscillations in rotation rate, when enough degrees of freedom are included in the calculation to represent finite amplitude effects reasonably well.

Clearly, if this particular approach is to be pursued further, the basic assumption of convective heat flux being a weak function of latitude would need to be put on a much firmer foundation in terms of arguments based on the theory of turbulent convection.

Durney and Spruit (1979) have made a promising start on a much more general form of axisymmetric differential rotation model, in which turbulent transports of momentum and heat from all motions other than differential rotation and meridional circulation are treated as tensors. The tensor components are evaluated using extensions of mixing-length arguments which take into account locally the influence of Coriolis forces on the convection. This is done by assuming the turbulent flow may be reasonably characterized by a single dominant eddy of specified dimension. A closely related formulation has been developed by Gough (1978) in the limit of low rotation, but was not directly applied to the differential rotation problem. Unlike the earlier models, there is some nonlinear coupling between the turbulent flow and the axisymmetric mean state, principally through induced changes in the superadiabatic temperature gradient. First calculations of actual differential rotation (Durney, 1981) have been somewhat disappointing, in that at best a weak equatorial deceleration is obtained. However, heat flux differences at the top of the model are within observational limits even though the turbulent viscosity and conductivity are equal in the limit of zero rotation, unlike in Belvedere *et al.* (1980a). Durney's calculations were done assuming the latitudinal and longitudinal dimensions of the dominant eddy were the same and independent of latitude, but other combinations need to be tried. On the other hand, from nonaxisymmetric global convection calculations we describe below, there is no particular reason to expect there *is* a single dominant eddy. Rather, a broad spectrum of convection is generally produced, in which various sizes and shapes are included with rather similar amplitudes. It remains to be determined just how good the 'single eddy' approximation is for purposes of estimating differential rotation, but such an approximation is worth pursuing, at least for its computational economy.

The theory of convection in a rotating spherical shell has been developed greatly during the 1970s and calculations for a compressible fluid with several scale heights variation in density are now beginning to be done. Earlier work was virtually always for a stratified liquid. In these early liquid models, the shell is heated uniformly at the bottom, cooled at the top, and all small-scale diffusion is assumed to be isotropic and independent of position and time. Usually the diffusion rates for temperature and momentum are assumed to be equal. Analogies are then drawn to the Sun by identifying the model diffusion with small scale eddy diffusion of momentum and heat. These diffusion coefficients are passive, however, in the sense that, in the absence of global motions, solid rotation and uniform heat flux are all that result. If we think of this class of model as representing a classical Newtonian fluid in a spherical shell held together by a central gravity, rather than an approximation to a stellar convection zone, then we have a completely well-defined physical system with no *ad hoc* assumptions or para-metrizations. We could imagine building such a system as a physical experiment, and have some confidence our model calculations would accurately describe the observed dynamics.

Early, mostly linear analyses of convection in a rotating spherical shell of stratified liquid by Busse (1970, 1973), Durney (1970, 1971), Yoshimura and Kato (1971), and Gilman (1972, 1975), demonstrated the preference for convective modes that transport momentum towards the equator, via the Reynolds stress mechanism illustrated in Figure 3. More recent, nonlinear calculations by the author (Gilman, 1976, 1977a, b, 1978, 1979) have exploited this fact to determine in detail when equatorial acceleration occurs, and with what amplitude, relative to the convection which drives it and relative to the basic rotation rate. As with the thermally driven, axisymmetric models described earlier, some of the early calculations referenced above were done in the limit of weak influence of rotation, mostly as a mathematical convenience. But in that limit, the preference for convective motions which transport momentum towards the equator is very weak, and is easily overpowered when finite amplitude effects are taken into account. Therefore, some of the inferences and extrapolations made from these early papers have not been borne out, even as to sign, by later nonlinear calculations. This is partly because the first attempt in these early papers to represent nonlinear effects involved severe truncations of the system, down to essentially the first unstable mode, which naturally gives too much advantage to this mode in competition with others. Some of these early calculations also ignored the role played by radial transports of angular momentum in determining the final differential rotation profile.

The basic model the author has been using is fully nonlinear. It is formulated for a finite difference grid in the meridian plane and is Fourier analyzed in longitude. Typically, 16 to 24 longitudinal wavenumbers are retained, and all the nonlinear interactions among them included. Thus the model is capable of producing a kinetic energy spectrum which can be compared with an observed spectrum, such as estimated by La Bonte *et al.* (1981). Boundary conditions are usually stress free top and bottom, as well as constant heat flux bottom, constant temperature top. Finite amplitude equatorial acceleration, as in the Sun, is produced in this model only when the influence of rotation upon convection is strong, i.e. the rotation time is less than the turnover time for convection. In these circumstances, the angular velocity also decreases inward; when the rotational constraint is very strong, the angular velocity predicted is nearly constant on cylinders concentric with the axis of rotation. The angular velocity is fastest at the equator because the equatorward transport of angular momentum from high latitudes by the Reynolds stresses in the convection is the dominant mechanism determining differential rotation.

When the convection zone is deep, say one-third of the radius or more, the equatorial acceleration profile with latitude is broad, with essentially monotonic decrease in angular velocity to the poles, as seen on the Sun. On the other hand, when the convection zone is shallow, say 20% or less, and the rotational influence is strong, the angular velocity reaches a minimum in mid latitudes, and then increases again towards the poles. Thus, a polar vortex is present in these thinner layers. Since no polar vortex is observed (Beckers, 1978), this is an argument in favor of a relatively deep solar convection zone (Gilman, 1979). The width of the equatorial acceleration is determined by the depth of the layer: the shallower the layer, the narrower the width. The monotonic decrease of angular velocity with latitude all the way to the poles for deep layers arises because the Reynolds stresses which transport angular momentum towards the equator reach to higher latitudes, and the moment of inertia of the polar cap is a smaller fraction of the total for the shell when the shell is thick, so the poles are easier to spin down.

For both deep and shallow convection zones, with weaker rotational influence (increased convective velocities) the profile switches from equatorial acceleration to deceleration. Angular velocity now increases with depth. There is an intermediate stage in which the angular velocity is highest in mid latitudes and lower near the equator and near the pole, while still decreasing with depth. None of these cases corresponds to the Sun, but they could to other stars. The changeover from equatorial acceleration to deceleration comes about because inward radial transports of angular momentum in equatorial regions, due to both convective cells and mean meridional circulation, become more powerful than the equatorward transport from high latitudes by the cells in determining the differential rotation profile.

In virtually all these model calculations, there is also produced a secondary meridional circulation with flow towards the poles near the outer boundary, as has been observed. This is in contrast to most axisymmetric models described above, for which the dominant meridional circulation has flow towards the equator in the outer part of the fluid region.

The model does have certain undesirable thermal characteristics. In particular, even though constant heat flux is assumed at the bottom, the heat flux does vary at the top with time and with latitude by 10–20% (with constant flux boundary conditions, the temperature would vary at the top by a similar amount). This is, of course, much larger than observed on the Sun. Much of this effect is probably due to the assumption of an incompressible stratification and can be improved. Two effects of large stratification are involved. One is that in this case the turbulent thermal conductivity should increase rapidly with depth, resulting in rapid lateral transfer of heat from a temperature perturbation (Spruit, 1976). The other is that the departures from the adiabatic gradient below the supergranule layer are no more than one part in 10^4, strongly limiting the magnitude of temperature perturbations presented to the top layers by global convection below (Gough, 1977a).

By suitable adjustment of the thermal forcing for convection in a relatively deep layer, it is possible to obtain from the model a surface differential rotation which is quite similar in amplitude and profile to the observed one. Furthermore, in the most recent (unpublished) calculations from this model, the kinetic energy per longitudinal wavenumber associated with the convection which drives the differential rotation is sufficiently small to be consistent with the observed upper limits described in La Bonte et al. (1981).

All conclusions drawn from an incompressible model for convection must remain suspect in their application to the Sun at least until the model is generalized to include compressibility and the conclusions tested. Gilman and Glatzmaier (1981) and Glatzmaier and Gilman (1981a, b, c, 1982) have developed a compressible extension of the same approach and examined such properties as the Reynolds stresses associated with the most unstable convective modes, as well as the initial differential rotation to be expected from them. They find that results depend strongly on how the magnitudes of the diffusion coefficient for heat and momentum as well as the superadiabatic temperature gradient vary with depth.

If the kinematic eddy viscosity ν and heat diffusivity κ are nearly constant with depth, as suggested from mixing-length arguments, and the superadiabatic gradient is like that in Figure 1, with the largest superadiabaticity near the outer boundary, then unstable global convection modes are much larger in amplitude near the top of the layer, and are confined to low latitudes close to the equator particularly when the influence of rotation is strong. Because of the small latitudinal extent of the modes, Coriolis forces in latitude are very weak, so Reynolds stresses that transport momentum in latitude are also weak. Inward radial transport dominates, resulting in an equatorial deceleration.

The convective modes used for these calculations are linear. In the nonlinear case, it could happen that the flow will 'penetrate' much more strongly from the most unstable levels into the less unstable ones, resulting in a more even velocity distribution with depth. The horizontal scale of the motion should expand in both longitude and latitude. In these circumstances, latitude transport should compete much better against radial transport in determining differential rotation, since the modes will reach to higher latitudes where horizontal Coriolis forces will produce the larger Reynolds stress needed for latitudinal transport. But all of these arguments must be tested with real nonlinear compressible calculations.

A broad equatorial acceleration is still more likely to occur when the superadiabatic gradient does not change greatly with depth, assuming the influence of rotation upon convection is strong. For these assumptions, the angular velocity also is likely to be constant on cylinders (and therefore decreasing with depth) at least in the deeper part of the convection zone as in the incompressible case. The global convective velocities either peak near the bottom of the convection zone, or are fairly uniform throughout. In this case the dynamical eddy viscosity $\rho\nu$ and conductivity $K = \rho c_p \kappa$ are nearly constant, so the kinematic diffusivities ν and κ decrease with depth. These assumptions are much less realistic from the point of view of mixing-length arguments, but the results that follow from them could be more indicative of what happens in the nonlinear case, as suggested in the previous paragraph.

Nonlinear calculations are also needed to estimate the amplitude of any differentials in radial heat flux reaching the outer boundary. They should be much smaller than in the incompressible case, because a temperature perturbation reaching the surface is bounded by the small departure from the adiabatic gradient needed to produce the required average heat flux. Glatzmaier and Gilman (1982) have demonstrated that the percentage temperature difference in latitude in their initial tendency solutions for differential rotation is usually much less than the percentage rotation difference (the ratio of differences needs to be $<10^{-2}$ to be consistent with solar observations). Gough (1977a) has argued from very simple considerations that such variations in temperature at the surface should be no larger than 10 K or so, which is only slightly above observational limits (see also Spruit, 1977b).

When making application to the Sun, or any star, one of the weakest points of the modeling approach just described is the reliance on scalar eddy diffusivities for momentum and heat for those scales of motion which are not explicitly calculated. The influence of rotation on these transport coefficients should be taken into acount in the future, as well as the nonlinear turbulent breakdown of large eddies into small ones. On the other hand, one of the strengths of this approach is that the nonlinear transport of momentum and heat are explicitly and nonlocally calculated for the large eddies —

and a full spectrum of these eddies is retained. An implicit assumption is that it is these large eddies that are the most important for determining such global features as the mean differential rotation, as well as the solar dynamo. We contend this is a plausible assumption, but is certainly not proven.

The Durney—Spruit approach, on the other hand, has the advantage that the influence of rotation is estimated for all eddies that are included. These eddies are rather restricted in scale and shape, but presumably the approach can be generalized still further. In effect, no real distinction is made in this approach between little eddies and big ones; all enter into the momentum stress tensor and all are convectively driven. The effects of eddies of all scales are estimated essentially locally, and the global character of the large eddies is mainly ignored. It may be possible to improve this aspect in the future, but for the present we feel the full three-dimensional model calculations capture the effects of the large eddies better. As long as calculations using the Durney—Spruit approach remain axisymmetric (with longitudinal averages taken over the eddy transports), these calculations require much less computation time than for the full three-dimensional simulations, so that more extensive explorations of the parameter space can be made — for example, the roles played by different sizes and shapes of eddies can be extensively studied.

We suspect that in the end these two approaches will converge towards a single hybrid, building on the insights gained previously from each individually. Already, they have substantial elements in common, with differing strong points. They are complementary, not contradictory. They both contain elements of convective turbulence, and have roots in mixing-length theory, but neither is complete or conclusive.

In addition to the mean differential rotation profile, we obviously would like to be able to explain many of the differences in rotation rate found by different measures, as described in Section 3.4.1 above. For example, the 4—5% faster rotation of sunspots compared to Doppler values has been interpreted by Foukal (1972) and Foukal and Jokipi (1975) as meaning the angular velocity increases inward down to a level where sunspots are 'anchored' in the plasma. The increase is produced by rising and sinking fluid particles in supergranules conserving their angular momentum. By this argument, sunspots are anchored somewhere near the bottom of supergranules, since a 4—5% rise in angular velocity would correspond to an inward displacement of 2—2.5% of the solar radius (at the equator) for such a fluid particle. Gilman and Foukal (1979) have verified that nonlinear convection in a thin rotating spherical shell will produce a layer whose average angular momentum per unit mass becomes nearly constant with depth as the thermal forcing for the convection is increased. We expect this result to hold also in the compressible case because no new force is introduced which is likely to prevent angular momentum conservation (but angular momentum mixing in a contained fluid due to turbulence is a subtle problem — see, e.g., Bretherton and Turner, 1968). Since virtually all models for spherical shell convection strongly influenced by rotation predict an equatorial acceleration and angular velocity decreasing with depth at least in the deep part of the convection zone, an important remaining question is: "At what depth does the radial angular velocity gradient change sign?" Perhaps the inferences from frequency shifts in solar oscillations will tell us in the near future. As we discuss later, where (and if) this gradient reversal occurs could be very important for the workings of the solar dynamo.

The fact that many magnetic field patterns rotate much more rigidly than the plasma itself is usually taken as evidence of globally coherent waves. These might represent coherent global convection patterns (Simon and Weiss, 1968), global hydromagnetic waves (Suess, 1975), dynamo waves (Stix, 1971, 1977), or some combination. There is no generally accepted explanation. Some of the difficulties are reviewed in Gilman (1977b).

The torsional oscillations seen by Howard and La Bonte (1980) represent an even smaller effect to be explained, since in magnitude they correspond to rotation changes from peak to trough of no more than 0.5% (or 5% of the average differential rotation). Yoshimura (1981) and Schüssler (1981) have suggested such perturbations are a result of $j \times B$ forces from the solar dynamo reacting back on the differential rotation. It is clear that virtually any dynamo which predicts migration of the Sun's toroidal (east–west) magnetic field toward equator at the right rate to explain the migration of the zone of sunspot occurrence will have associated with it a $j \times B$ force, but no model is currently able to convincingly predict the observed magnitude of the oscillation. The density of the convection zone increases very rapidly with depth near the top, and as the depth at which the $j \times B$ force is actually applied is not known, it is difficult to estimate the local inertia of the fluid against which the $j \times B$ force will be working. Furthermore, these early models ignore the much larger forces (from the divergence of the Reynolds stress tensor) also acting in the azimuthal direction whose near balance determines the mean differential rotation profile. Very small perturbations of this balance could easily swamp a small $j \times B$ force, and, in addition, this force could equally produce perturbations in the Reynolds stress balance. The problem is actually rather subtle, and not too much effort should be expended on it until the observational result is more firmly established. La Bonte and Howard (1981), who first observed the oscillation, believe it may be the driver of the solar cycle rather than a reaction to it. This seems quite unlikely, since the velocity shears associated with the oscillations are so weak compared to those found in the mean differential rotation. It is very difficult to see how the effect of one could be shut off while the other is allowed to work. One possibility is that the oscillation has much larger amplitude beneath the surface.

The theory of solar differential rotation is in a state in which a small number of different, complementary approaches are being pursued, primary examples of which have been described above. Efforts to reconcile and unify the different approaches have been made by Rüdiger (1980) using turbulence arguments. Certainly all these approaches are confronted with the same difficult problem: how to represent accurately the role of convective turbulence. We need to make progress beyond simple linear eddy diffusion coefficients, and beyond dependence on mixing-length arguments and their (worthwhile) extensions to include effects of rotation. Basic turbulence theory has gone significantly beyond this point in applications with simpler physics, such as engineering and laboratory fluid dynamics as well as dynamic meteorology and oceanography. These developments need to be examined for their utility in the solar case, with the inclusion of more relevant physics, such as large density variations with height.

Models for differential rotation which have had some measure of success in explaining the solar equatorial acceleration can, in principle, be generalized to make estimates of differential rotation for other stars. Examples of such efforts are Belvedere *et al.* (1980b) and Gilman (1980). In the former case, the slow rotation approximation is used throughout, with the Sun as a calibration point. Belvedere *et al.* (1980b) conclude that equatorial acceleration would be present in all stars of spectral type F5 and a minimum around G5. Gilman (1980) argues from estimates of the ratio of convective turnover time to rotation time that stars later than the Sun should also have broad equatorial accelerations unless their rotation rates are much lower than the Sun's, because they have deep convection zones. For stars earlier than the Sun on the main sequence, there should occur somewhere in F stars a transition to equatorial deceleration, even though the rotation rates of these stars are usually much larger than for the Sun. This is because their convection zones are estimated to be shallow and their turnover times short. But such arguments currently represent only tempting speculations, since there are virtually no observations of differential rotation in any other stars. This is principally because it is an extremely difficult observation to make. However, Bruning (1981) has shown that equatorial acceleration and deceleration can give significantly different spectral broadening. Gilman (1980) has argued that the place to start looking for differential rotation is in the fastest-rotating star one can find that has a convection zone of substantial depth, say at least 20% of the stellar radius. (Related arguments for a different purpose have been made by Durney and Latour (1978).) Results from new observations made at Mt Wilson by Vaughan *et al.* (1981) of rotational modulation of calcium emission in stars may tell us which stars to observe for differential rotation effects in their Doppler broadening. To further test differential rotation theories, it would be extremely useful to build up a sizable statistical sample of differential rotation measures for a variety of stellar types. This may require more sensitive techniques than those used at present, but more effort in this area would be a worthwhile investment.

4. Phenomenology of the Sun's Magnetic Field and Related Features

In Section 3 we described observations and theory for convective and circulatory motions on the Sun in order of ascending horizontal dimensions, from granulation to the differential rotation. Observations and theory for each scale could be considered somewhat independently: supergranules do not appear to be made up of collections of granules, for example. By contrast, the magnetic field of the Sun appears to have just this character. That is, most patterns of magnetic field in the photosphere of the Sun are made up of collections of small magnetic flux tubes containing very high field strength. Relatively little field appears to be present between these flux tubes. Thus we choose to describe our present knowledge of the magnetic field of the Sun on all spatial scales in a unified manner before discussing theories in detail.

The phenomenological hierarchy of magnetic fields is roughly as follows. The smallest element is the individual magnetic flux tube, followed by small bipolar groupings and rings of flux tubes on granule scale, followed by a 'network' of flux tubes which follows the boundaries of supergranule cells. This network is also seen in brightness patterns in many spectral lines, such as Ca K, Hα, as well as photospheric continuum. On still larger scales, network fields are divided into two types: quiet Sun and active regions. The latter regions are generally quite complex in magnetic structure, and are the sites of virtually all sunspots. On still larger scales, there are more nearly global patterns of magnetic flux, with different patterns characteristic of low and high latitudes. Generally speaking, the larger the spatial scale of the particular pattern, the longer it lasts. Most patterns show systematic evolution through the solar cycle, and differences from one cycle to the next. Most magnetic patterns have important manifestations in the chromosphere and, particularly, ones of larger scale are felt in the corona. We leave discussions of those aspects to other chapters in this work, particularly Chapter 10 by Orrall and Pneuman.

The principal technique currently used to determine magnetic fields in the solar photosphere is to measure the polarization induced in magnetically sensitive spectral lines by the Zeeman effect. So far, principally the line of sight component of the field (towards or away from the observer) has been measured in this way, because it is easier to obtain than the transverse component. In its simplest form, the Zeeman effect for a magnetic field oriented along the line of sight produces a splitting in wavelength into left-handed and right-handed circularly polarized light. The amount of the splitting is linearly proportional to the magnetic field strength. For more general field orientations, and fields varying in amplitude and direction on spatial scales smaller than the resolution of the magnetograph, interpretation becomes much more complex and somewhat un-certain, requiring careful radiative transfer calculations. The details of the Zeeman effects expected in the solar case as well as methods for deducing magnetic fields from them are well described in a review by Stenflo (1978). Stokes polarimeters, which measure all of the Zeeman components and from which, in principle, all the magnetic field components can be obtained, have been developed by the High Altitude Observatory and the Institute of Astronomy at the University of Hawaii. However, interpretation of results from these instruments is proving to be extremely difficult, and so far little new information about photospheric fields has been obtained that is considered reliable. Most of the observations which are described below refer to the line of sight (longitudinal) magnetic field obtained from simple Zeeman splitting. Even these measurements are subject to uncertainties of interpretation, and in many cases the field patterns may be more significant than apparent field amplitudes.

4.1. FLUX TUBES AND NETWORK

The magnetograph measures the average magnetic flux over the scanning aperture of the telescope, not the magnetic field strength itself. It generally cannot resolve the individual magnetic flux elements which contain most of the Sun's magnetic field, so that the true local field strength cannot be determined by observing Zeeman splitting of a single spectral line. Stenflo (1973) found a way around this difficulty by measuring the ratio of Zeeman splitting for two neighboring lines (he chose Fe I 5250 and 5247 Å) whose only difference was their sensitivity to magnetic field. With very strong actual magnetic fields, both lines become 'saturated' (so the relation between splitting and field strength is no longer linear) but to different degrees, from which the actual field strength can be estimated. Using this technique, Stenflo found field strengths of ~2000 gauss for quiet network fields. By estimating roughly the number of magnetic elements expected in the scanning aperture element (in Stenflo's case '2.4 X 2.4') from other information such as the number of bright points making up the network pattern (as seen, for example, in Hα or Ca K emission). Stenflo estimates a typical flux tube diameter to be between 100 and 300 km, considerably smaller than granules. Subsequent estimates, — for example, by Wiehr (1979), using different techniques — have yielded slightly larger values. Wiehr (1978) found similar field strengths using different iron lines for a wide variety of small magnetic features. Occasionally atmospheric seeing has been good enough to nearly resolve these magnetic elements, and similar field strengths have been found (see, e.g., Harvey and Hall, 1975, and Tarbell and Title, 1977).

By similar sorts of comparisons of the Zeeman signal in two spectral lines, Howard and Stenflo (1972) and Frazier and Stenflo (1972) conclude that at least 90% of the total magnetic flux emerging from the Sun is contained in these flux tubes. On the other hand, Tarbell *et al.* (1979) estimate an upper limit of 80% of the total solar flux could reside in weak fields in quiet Sun network, and 25–30% in the neighborhood of active regions. But the actual weak fields could be much weaker, still. Tarbell *et al.* believe that much of this weak field is bipolar in a spatial scale of a few arcseconds, and therefore does not reach into the chromosphere and corona with significant amplitude compared to that from strong field flux tubes. Thus, while being weak in amplitude, it is not diffuse in a global sense, only somewhat less concentrated than the strong field flux tubes.

Estimates of lifetimes of magnetic elements vary from $\sim$5 min for the smallest (Mehltretter, 1974) up to something less than 2 h (Howard and Stenflo, 1972). Thus, individual magnetic elements have much shorter lifetimes than the overall network pattern in which they are found, which approaches supergranule lifetimes of a day or so (Simon and Leighton, 1964).

Since only the line-of-sight component of the magnetic field is measured, it is not possible to determine the inclination of a typical flux tube to the vertical. However, by comparing magnetograms from lines formed at different levels, as well as comparing the positions of magnetic field signals with those of Ca brightness above, some inferences can be made. Howard and Stenflo (1972) determined that the field lines diverged rapidly with height above the photosphere. Kömle (1979) found field lines isotropically distributed inside a cone centered on the local vertical with half angle $\sim$20°. On the other hand, Wiehr (1978) deduced a systematic eastward tilt with height (opposite to the sense of solar rotation) of up to 55°. Clearly, these widely differing results need to be reconciled. Some theoreticians have repeatedly invoked twisted magnetic flux tubes, but there seems to be little observational evidence that flux tubes are commonly twisted.

Magnetic flux tubes are observed to be the sites of strong downflow, of up to 0.6 km s^{-1} at photospheric levels (Giovanelli and Slaughter, 1978) and up to 1–2 km s^{-1} in Hα and Hβ 100 km or so above (Simon and Leighton, 1964). The precise relation between these downdrafts and those expected at the outer boundaries of supergranules where most of the magnetic elements are found is not clear, but there is evidence that the region of downflow is wider than the tube itself (Stenflo, 1976; Frazier and Stenflo, 1978). Oscillatory motions are also found inside and around magnetic flux tubes (Giovanelli *et al.*, 1978). They observe oscillations directed along the axis of the flux tube, with amplitude $\sim$± 0.27 km s^{-1} in the high photosphere, increasing to ±0.75 km s^{-1} in Hα. These oscillations are probably closely related to the 5 min oscillations found in the photosphere and low chromosphere generally.

Determining the temperature of magnetic flux tubes at photospheric levels relative to their surroundings has proved particularly difficult, again because of inability to spatially resolve the elements. Stenholm and Stenflo (1977) have shown that inferences of temperature from radiative transfer models for spectral radiation based on the assumption of local thermodynamic equilibrium (LTE) and neglect of lateral transfer effects are questionable. Many previous LTE models deduced temperature enhancements at photospheric levels and above. Stenholm and Stenflo are able to produce the same spectral behavior by assuming the inside of the tube has a Wilson depression (lower pressure and temperature compared to the surrounding plasma at the same level, arising from lateral pressure balance) with extra radiation channeled into the tube by scattering from the hot side walls. While the Stenholm–Stenflo calculation may overestimate the role of hot walls, relatively low resolution measurements in the infrared continuum by Worden (1975) showed a temperature deficit at all levels including the deep photosphere. More recently, Foukal et al. (1981) have found the first good evidence of a continuum brightness deficit in the visible in the region of magnetic flux tubes when observed near the center of the solar disk. As with earlier observations, they also find an excess brightness in these same features (the so-called photospheric faculae) near the solar limb. They conclude that their observations provide evidence of both the 'hot wall' effect and a 'hot cloud' of plasma overlying the cool material below. Previous measurements missed the photospheric temperature deficit because the continuum pass bands they used contained residual opacity from Fraunhofer lines formed in the hot cloud. The work of Foukal et al. should thus help to resolve a number of questions with respect to the thermodynamics of flux tubes, but it needs to be tested further with additional observations and calculations, particularly to establish whether regions of flux tubes always look dark away from the solar limb. Spruit (1976) argues that very thin tubes can look bright at the disk center.

So far we have concentrated on characteristics of individual magnetic flux tubes. A typical distribution of flux tubes on the Sun is illustrated in the high-resolution quiet Sun magnetogram seen in Figure 4. This example from Kitt Peak is provided by J. W. Harvey. Bright regions represent positive magnetic polarity; dark regions negative polarity. The total area is approximately 210 by 250 arcsec (1.6 by 2.0 $\times$ 10^5 km) with a resolution element of roughly 0.4 arcsec on a side. A number of features are quite evident. Much of the magnetic field pattern is 'cellular', in the form of reasonably well-defined rings about a central region containing much less flux. The ring can be identified with supergranule boundaries. Occasional flux observed inside these rings are referred to as inner network fields. There are also some regions where the flux tubes are close together in apparently more irregular patterns. Ramsey et al. (1977) have found that magnetic flux tubes in active regions can form cellular patterns of typical diameter 2–3 arcsec, an order of magnitude smaller than the supergranule patterns. These patterns appear to be related to granule motions.

Figure 4 also illustrates that, even in relatively small regions, the field polarities are well mixed. Several small bipolar pairs can be seen, the opposite polarities separated by no more than a few arcsec. Clearly, then, coarser magnetograms will cancel out contributions to the net flux from neighboring flux elements, resulting in a lower average flux.

In addition to the photospheric continuum facular darkenings and brightenings observed, for example, by Foukal et al. (1981), it is well known that many brightness patterns originating higher in the solar atmosphere correlate well with magnetograms such as seen in Figure 4. Ca K, Hα, Hβ and many other spectral lines all show quiet network and active region enhancement called 'plages'. Some of these patterns are found to have fine structure down to the apparent dimensions of individual magnetic flux tubes although it is not clear that this structure coincides spatially in precise detail with such magnetic elements in precise detail. Dunn and Zirker (1974) have examined such fine

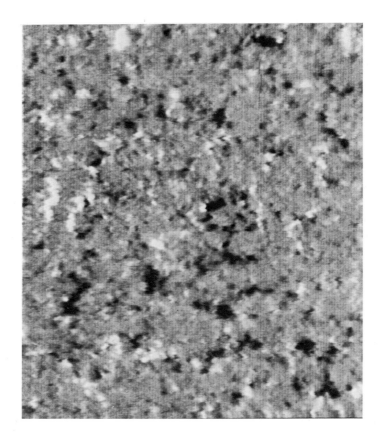

Fig. 4. High resolution (0.4 arcsec) magnetogram taken near quiet disk center, from Kitt Peak National Observatory (courtesy J. W. Harvey). White features represent positive polarity (toward the observer) black features negative polarity. Displayed area is approximately 210 × 250 arcsec. Note existence of (a) several nearly circular areas nearly devoid of field, but ringed by field of mixed polarity; (b) many small bipolar pairs; (c) patches of irregularly distributed field. Magnetograph saturation occurs at ~100 gauss for this resolution.

structure in Hα in the form of lacy chain-like patterns which they call 'filigree'. Granulation is usually abnormal in shape near the filigree (Ramsey *et al.*, 1977) and a rather complex interaction appears to be taking place.

We shall not explore radiative model calculations for these brightness features in detail, but will refer to them in various places as indicators of the pattern of magnetic field.

4.2. SUNSPOTS, ACTIVE REGIONS, AND GLOBAL PATTERNS

Unlike small-scale magnetic flux tubes discussed in the previous section, many basic properties of sunspots have been well observed for a long time. A standard reference for early work is the book *Sunspots* by Bray and Loughhead (1964). Sunspots have the strongest magnetic fields observed on the Sun, 3000–4000 gauss. This field is a maximum near the center of the spot where it is nearly vertical, and tilts away from the local vertical and decreases in amplitude nearer the edge of the spot (e.g. Beckers and Schröter, 1969). Detailed spot structure can vary widely from spot to spot, particularly in sunspot groups. Sunspots represent the largest flux tubes, large spots having an area comparable to a supergranule. Sunspots are also the darkest features on the solar surface, with effective temperatures as much as 2000 K below that of the surrounding photosphere. Sunspots form depressions in the photosphere of isotherms and isobars of typically 650 km (Wittmann and Schröter, 1969).

In large spots particularly, the dark umbra is surrounded by a brighter penumbra (still darker than the photosphere). The penumbra appears striated with roughly alternating bright and dark lanes or filaments extending more or less radially from the umbra to the surrounding photosphere. The penumbra is the site of the so-called Evershed effect, interpreted as a radial outflow of average value $1-2$ km s^{-1}. This flow appears to be highly nonuniform in azimuth about the axis of the spot, being confined largely to the dark penumbral filaments (Mattig and Mehltretter, 1968; Beckers, 1969), reaching magnitudes of up of 6 km s^{-1} there. In the chromosphere above, there is an inflow toward the axis of the spot which is similarly confined to channels (Bones and Maltby, 1978). But the Evershed flow is not the only radial outflow associated with sunspots. Sheeley and Bhatnagar (1971) and Sheeley (1972) have found radial outflows at photospheric levels extending out $10\,000-20\,000$ km beyond the penumbral boundary, of magnitude $0.5-1.0$ km s^{-1}. These flows seem more uniform in azimuth than the Evershed flow in the penumbra, and it is not clear what the connection is between them. This flow is most often present during the decaying phase of the sunspot. Also, this flow is distinctly different from that of neighboring supergranules although of similar spatial scale. Granules near sunspots are also apparently modified. Macris (1979) finds that granule diameters within a few arcsec of the penumbral boundary are smaller for spots with stronger umbral magnetic fields. Spots with fields as high as 3000 gauss are surrounded by granules whose diameters are only 60% that of average granules further away from the spot.

There is a variety of evidence of wave motions in sunspots, seen mostly above the photosphere. Beckers and Schultz (1972) find vertical velocity oscillations in sunspot umbra with periods of $\sim$180 s and amplitude $\sim$1 km s^{-1}. They find oscillations with distinctly longer periods in penumbras. Giovanelli (1972) found similar periods ($\sim$165s) and even larger amplitudes ($\sim$3 km s^{-1}). He also observed waves propagating from the umbra to the penumbra with phase velocities $\sim$20 km s^{-1}. Similar 'running' penumbral waves have been seen by Zirin and Stein (1972) in Hα intensity. There are also very short period, sporadic brightenings in Ca K emission, called umbral flashes (Beckers, 1975). Waves may also be the cause of some microturbulent broadening of certain spectral lines in sunspot umbras. Waves in sunspots are discussed in more detail in Chapter 7 by Brown, Mihalas, and Rhodes.

The typical sunspot umbra seen in white light looks relatively uniform and unstructured compared to the penumbra. But in fact the umbra has a great deal of fine structure in intensity. This is revealed particularly well in high-resolution photographs taken with special precautions to reduce the presence of scattered light entering the telescope from the neighboring penumbra and photosphere. Using different exposure times, Loughhead *et al.* (1979) have demonstrated that the umbra is full of 'dots' of horizontal dimensions

0.4–0.5 arcsec, i.e. at the resolution limit. It is not known whether detailed magnetic field patterns of the umbra would correspond to these dots. Within the bright striations in penumbras there are also many dots.

Sunspots can have a duration between a few hours and several solar rotations. The largest spots generally live longest. The growth and decay of sunspots are generally related to the growth and decay of the active regions in which they are found, but there is one observed phenomenon peculiar to sunspots which is apparently important for their growth and decay. This phenomenon is the systematic inflow and outflow of magnetic flux elements to and from sunspots. Vrabec (1974) has reviewed these effects in detail. Outflow appears to occur more universally. This was inferred from UV spectroheliograms by Sheeley (1969) who observed bright points moving steadily outward from sunspots into the surroundings at speeds of the order of 1 km s^{-1}. Harvey and Harvey (1973) did the first systematic study of this phenomenon with the Kitt Peak Magnetograph. They found that this outflow of flux was present only in decaying spots, and that the rate of net magnetic flux migration agreed with the spot decay rate. Interestingly, flux elements of both polarity were observed to move away from a unipolar spot, usually in bipolar pairs. The total flux of both signs leaving the spot was much greater than expected from the spot decay rate. These features are explained as being due to sunspot flux tubes developing additional loops during the expulsion, with each loop showing up as a small bipolar pair in the photosphere.

Spots showing this decay process always have around them an annular 'moat' through which the flux tubes move. This moat is essentially free of long-lasting magnetic structure, and occupies the same region outside the spot as the annular outflow velocity observed by Sheeley (1972) that we described earlier. As the Doppler and magnetic element velocities are about the same, the presumption is strong that the outflow is carrying the flux tubes away from the spot. The moats, once developed, last an average of six days. A sunspot surrounded by a moat is never observed to outlive it, though the moat may last for one or two days after the spot has disappeared. (Pardon *et al.*, 1979).

Inflow of magnetic flux into a sunspot is confined to its growing phase (Vrabec, 1974) and may be the principal method of growth. This process has been studied less, and appears to happen less often but that may be because the elements moving into the spot are small and difficult to see. It must compete, of course, with the eruption of new flux in the formation of a sunspot. There does not appear to be a 'moat' phenomenon associated with flux tube inflow.

Active regions, in which sunspots are always imbedded, begin with a small region of new emerging flux, rather than by coalescence of previously existing surface flux. The vast majority of such regions never grow to the point that sunspots are formed. These features are called ephemeral active regions, because, typically, they last for only a day or two (Harvey and Martin, 1973). Hundreds per day can be seen. Active region birth can be clearly seen in Hα spectroheliograms in the form of new dark arches (West, 1970; Glackin, 1975); in high resolution magnetograms as small bipolar regions (Harvey and Martin, 1973; Martin and Harvey, 1979) and in X-ray images as bright points (Golub *et al.*, 1977; Golub, 1980). Ephemeral regions have a broad spectrum of sizes. The smallest regions occur most frequently, and have the shortest lives. Those regions which do grow and persist long enough to produce sunspots initially have rather random orientation of arches and magnetic polarities. However, within a few days they align themselves in nearly an east—west orientation, with usually a slight tilt towards the equator at the leading edge (in the sense of the solar rotation) (Weart, 1970; Martin and Harvey, 1979). Sunspots form within the active region by coalescence of smaller flux elements that have recently emerged. Within a given solar cycle, the spots in the leading part of all active regions in a given hemisphere have the same magnetic polarity, with opposite polarities in north and south hemispheres. Following spots have the opposite polarities. In successive cycles, the polarities all reverse. This is known as Hale's polarity law.

Active regions can vary a great deal in size as well as total magnetic flux, with long-itudinal extent frequently reaching several hundred thousand kilometers or a few tens of degrees in longitude. Occasionally, several eruptions of new flux will occur in one region, leading to an even larger 'complex' of activity. Thus, active regions are much larger in scale than individual supergranules. In fact, the regular supergranule pattern of the quiet Sun appears to be substantially altered within active regions, with a much greater variety of cell sizes and shapes present, which accommodate to the strong, irregular magnetic flux patterns present.

In the decay phase of an active region, the magnetic field patterns are dispersed from long-lived sunspots by the outflow process described above, and the flux blends in eventually with the quiet Sun network. Decay of an active region generally takes much longer than its growth phase, which is usually completed in less than one solar rotation. Zwaan (1978, 1981) gives a useful summary of active region morphology and evolution, on which we have relied heavily in the above description.

A typical full disk magnetogram made near solar activity maximum shows several active regions of various ages in each hemisphere. An example from Kitt Peak is shown in Figure 5. The large extent of active regions, as well as their persistence and overall organization (as seen in Figure 5) suggests that whatever process forms them is itself large in spatial extent. Giant cell convection is a likely candidate, but since velocities of this scale never have been found in the velocity data, no clear association is possible. Many other processes involving dynamics of flux tubes, which we review later, probably also contribute. The only flow field clearly associated with active regions is a downflow seen prior to the birth of sunspots (Karvaguchi and Kitai, 1976) and over the active region generally (Howard, 1972), but it is not clear that this flow has anything to do with convection below.

Many aspects of sunspots and active regions have been observed in great detail, but there are important properties we have not been able to observe. For example, the detailed process of destruction of magnetic flux remains mysterious. It is widely suspected that field line reconnection is taking place in the solar convection zone and atmosphere, but such an event has never clearly been observed on a magnetogram. Such impulsive events as flares and X-ray brightenings may well be reconnection events in which magnetic energy is suddenly released, but this has not yet been demonstrated by observations. We also would like to know the depth reached by such coherent features as sunspots. At what depth does the east—west or toroidal field implied by Hale's polarity law reside? These questions, unfortunately, may never be answered from observations alone.

There is evidence for magnetic fields organized and persistent on still larger spatial scales than active regions. For a review of global field patterns, see Howard (1977). Some of the largest scale patterns, as seen for example in Figure 5, may last in identifiable

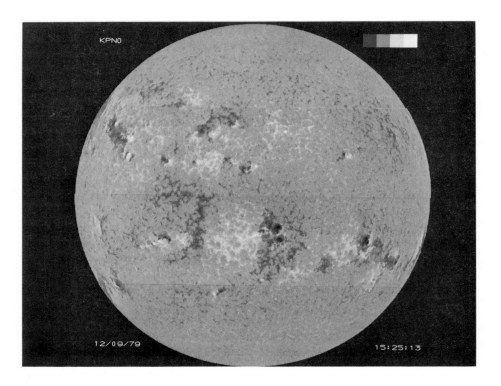

Fig. 5. Full disk magnetogram (1 arcsec resolution) of active Sun from Kitt Peak National Observatory
(courtesy J. W. Harvey). Note particularly (a) network structure of field; (b) bipolar regions in low and
middle latitudes in each hemisphere; (c) more nearly unipolar (but weak) field at high latitudes; (d)
several dense 'active' regions of field within these bipolar regions — the dark and light solid areas are
sunspots. Magnetograph saturation occurs at ±1024 gauss for this spatial resolution.

form for more than a year. Occasionally neighboring fields of predominantly one polarity
can appear to stick together, becoming a single giant unipolar magnetic region. Persistent
very large scale bipolar regions can also be found (Bumba and Howard, 1969; Ambroz
et al., 1971). In many of these patterns, the same magnetic polarity is found at the same
longitude north and south of the equator. This is similar in character to the solar 'sector'
fields inferred from interplanetary magnetic field data by Wilcox and collegues (see,
e.g., Svalgaard and Wilcox, 1978). Active regions can also be seen to recur in the same
longitude interval much longer than the lifetime of a single active region, giving rise to
the term 'active longitude'. On the other hand, the magnetic flux magnitude and pattern
can change abruptly over a very wide range of latitudes and longitudes (Howard, 1974;
Levine, 1977). The rate of emergence of new magnetic flux over the whole Sun, as
measured by X-ray bright points, can also change in just a few rotations (Golub and
Vaiana, 1980). These events are further evidence that the processes controlling the
magnetic field patterns on the Sun are at least partly global in scale.

Global field patterns are most easily seen in magnetograms by heavily averaging the
data into large spatial elements. Levine (1979) has done this with Kitt Peak magnetograms

for the 1973–74 Skylab period, and found evidence of a warped neutral line centered on the equator which crosses it twice in 360° longitude. The simplest magnetic pattern that has this characteristic is a tilted dipole, whose axis is at an angle to the rotation axis. Formally, of course, one can represent the entire solar field in a spherical harmonic expansion, and thereby identify a whole sequence of multipoles, but the physical significance of any of these harmonics considered individually is not clear.

Polar magnetic fields on the Sun typically have a rather different character than low-latitude fields. While still being made up of individual flux elements of high strength, they are largely unipolar through most of the solar cycle. The exception occurs near solar maximum, where the polar fields in each hemisphere change their dominant polarity.

Evidence for global magnetic field patterns on the Sun can also be found in coronal brightness patterns, because the corona is clearly dominated in its structure and dynamics by the magnetic field. The warped neutral sheet seen in averaged magnetograms by Levine (1979) is clearly evident in the characteristic coronal brightness patterns (Hundhausen, 1977; Hundhausen *et al.* 1981). Photospheric regions of unipolar magnetic field are also seen to be the locations of coronal holes (regions that are dark in X-ray and scattered white light) (Levine *et al.*, 1977; Levine, 1977; Harvey and Sheeley, 1979).

4.3. THE SOLAR CYCLE: MANIFESTATIONS IN SUNSPOTS, MAGNETIC FIELDS, AND OTHER PROPERTIES

A new solar cycle is classically defined to begin with the appearance of new sunspots at the highest latitudes they are observed to occur, which is in the neighborhood of 40°. The new spot groups in each hemisphere have magnetic polarities reversed from the previous cycle. The polarities of leading spots are always opposite in north and south hemispheres, the so-called Hale's sunspot polarity law. The subsequent progression of the sunspot cycle is marked by a migration of the zone in latitude where new spots are formed towards the equator in each hemisphere. Individual spots do not appear to participate in the migration. At the end of the cycle, the zone of spot occurrence is generally quite close to the equator, but it is extemely rare for a spot in one hemisphere to cross over to the other. If the latitude of occurrence of each sunspot is plotted against time for each hemisphere, the figure takes the form of a 'butterfly', whose axis lies along the equator, usually called the Maunder butterfly diagram, after its originator.

The amplitude of the sunspot cycle is usually defined by an empirical index, the sunspot number, which is a weighted sum of the number of spots and spot groups observed on the disk at any one time with a telescope of standard aperture. The sunspot number generally peaks three or four years after the beginning of the cycle, and then decays over the remaining seven or eight years. The daily sunspot number fluctuates a great deal, and even monthly average values do not show a monotonic rise to solar maximum, followed by a monotonic decay. Thus, the cycle has a substantial irregular or random component in amplitude — but Hale' sunspot polarity law is virtually never violated. Some of these irregularities may be repeated in more than one cycle. Gnevyshev (1977) has compiled evidence suggesting the average cycle may have two maxima, with a slight dip in average activity in between. The early maximum is associated with total sunspot area peaking near 25° latitude, followed about three years later with a second peak due to sunspots tending to cluster around latitude 10°.

Since sunspots are always contained in active regions, active region occurrence also migrates towards the equator with the progression of a solar cycle. There is also an association of frequency and latitude of occurrence of the short-lived, ephemeral active regions (ERs) with the solar cycle, but different indicators give rather different results. Martin and Harvey (1979) show that ERs, seen as small bipolar flux groupings on the Kitt Peak magnetograms, precede the start of a new sunspot cycle with an early increase in their occurrence at high latitudes. They found new cycle bipolar pairs occurring more than one year before the new spots of Cycle 21 (which began in 1976) were seen. Furthermore, the total number of ERs seen on the disk reached a minimum in 1975, a year before the minimum in sunspot number. Thus, if a true solar cycle should be defined as beginning with the first new ERs in high latitudes, and ending with the last sunspot near the equator, there will be a substantial overlap between cycles, with the average cycle being considerably longer than the usual eleven years.

In contrast with the above picture, evidence of ER changes with the solar cycle, as measured by X-ray bright point occurrence, shows a pattern more nearly out of phase with the cycle (Davis *et al.*, 1977; Golub, 1980). They found that global occurrence of X-ray bright points was much higher in 1976 near sunspot minimum than in 1973. It is important to resolve this conflict. To do so probably requires much more detailed study of simultaneous magnetograms and X-ray images of high resolution, to determine whether all bright points represent emerging magnetic flux and all detectable emerging flux shows up in X-ray emission (Golub, 1980). It is possible that current magnetograms miss many of the smallest ERs, which might still give rise to a detectable bright point (Martin and Harvey, 1979).

The evolution of a solar cycle can be seen clearly in the behavior of the Sun's global magnetic field. This is particularly true of the axisymmetric (mostly radial) part of the field, obtained by taking a running average of longitudinal magnetograms over several rotations, and then displaying the result as a function of time and latitude. The most complete recent study of this type is that of Howard and La Bonte (1981), who used Mt Wilson magnetograms for the period 1967–1980. Earlier studies of a similar nature were carried out by Stenflo (1972) and Yoshimura (1976), among others.

Howard and La Bonte (1981) examined both the average of the signed magnetic flux and the average of the absolute value of the flux, to distinguish between contributions to net flux at a given latitude, and contributions from bipolar flux eruptions to the total magnetic energy seen at the surface. They find that almost all the magnetic flux on the Sun detectable at their resolution ($12''5 \times 12''5$) originates in active region latitudes. These latitudes exhibit a net flux of the sign of preceding sunspots for that hemisphere, because this preceding flux is more concentrated and therefore predominates in the magnetogram signal. They find that polar fields of the Sun are first built and then reversed by the drift of field of following polarity towards the poles from the high-latitude side of the belt of active regions in each hemisphere. This drift is not continuous, but rather episodic, with a few large events dominating the whole cycle. The movement is also quite unlike a diffusion process, in that the following polarity fields move coherently without great spreading. This suggests the fields are being carried by a net poleward mass flow of magnitude $\sim 10 \text{ m s}^{-1}$.

Polar field reversals were seen in the north in 1971 and 1980, and the south pole field reversed in 1969 and 1981. Babcock (1959) observed polar field reversals in the previous cycle. The existence of polar fields and their reversals near sunspot maximum has been inferred from polar faculae variations back to the beginning of the twentieth century by Sheeley (1976).

According to Howard and La Bonte (1981), the absolute value of the magnetic flux peaks strongly at active region latitudes, and is very small by comparison in polar regions at all phases in the cycle. Polar regions appear to contain no more than 1% or so of the total flux measured in this way. The flux peak appears to be closely in phase with sunspot number, with flux at maximum about three times as large as at minimum. Howard and La Bonte argue that flux amplitude measured by X-ray bright points — as discussed in Golub *et al.* (1977) and Golub (1980), where the amplitude appeared to be out of phase with the sunspot cycle — cannot contribute greatly to the total flux seen by the magnetograph (either Mt Wilson or Kitt Peak).

As was described above for X-ray bright points and Kitt Peak magnetogram data, Howard and La Bonte (1981) have seen rapid increases in total flux, but find most of the increase comes in active region latitudes. Sometimes this new flux can spread quite rapidly over the solar surface (at a rate of 50–100 m s^{-1}). The typical flux emergence rate is large enough that the total flux present would be replaced in about 10 days. Since the flux does not increase at anywhere near that rate for extended periods even during the ascending phase of the solar cycle, a destruction process of nearly equal magnitude must also always be at work.

In summary, the axisymmetric radial field of the Sun (sometimes called the poloidal field) originates in active latitudes, followed by migration of leader field polarity toward the equator and follower field towards the poles. The latter cancels out the pre-existing polar field of opposite sign near the maximum of the sunspot cycle, while the former retains its polarity near the equator for several more years until the end of the current sunspot cycle. This equatorial radial field is thus nearly in phase with the subsurface east–west or toroidal field responsible for the sunspots and active regions. These properties are also clearly seen in the earlier studies of Yoshimura (1976) and Stix (1976a).

Since successive solar cycles generally do not have the same amplitude, the envelope of solar cycle peak amplitude varies, by a factor of 4–5 in sunspot number, for well-observed cycles. But the range of envelope behavior may in fact be much greater, as argued by Eddy (1976). He re-established convincing evidence indicating periods of several decades when solar activity is so low that sunspots are rarely seen at all, and many terrestrial manifestations of solar activity are largely absent. The most recent such period, which Eddy called the Maunder Minimum, occurred between about 1645 and 1715. Largely from the record of ^{14}C in tree rings, he argued for the existence of several previous minima, as well as maxima, during which solar activity was at much higher levels than it is now.

The connection between solar activity and ^{14}C in tree rings is as follows. ^{14}C is produced in the high atmosphere by galactic cosmic-ray bombardment, and the cosmic-ray flux seen at the Earth is larger when solar activity is low than when it is high. This ^{14}C is subsequently transported to the lower atmosphere by turbulent mixing, whence it enters the biosphere and is deposited (among other places) in trees. Analysis of the ^{14}C record for solar signals has been carried much further now — for example, by Stuiver and Grootes (1980) — with primarily confirming results. Siscoe (1980) has also examined the auroral record much more completely, confirming the existence of the most recent of these minima and maxima in solar activity. A number of other studies of aurorae have claimed to refute their existence, but these are generally based on much less extensive surveys of the available data.

Several other kinds of evidence concerning solar activity in the past have been discussed in various articles in the new book *The Ancient Sun* (Pepin, Eddy, and Merrill (eds.), 1980). The ^{14}C record can give us information about solar activity for, at most, only the past few thousand years, and because of being recycled through several biological and geophysical reservoirs, time resolution is no better than 10–20 years. A promising alternative which may surmount both these problems is the isotope ^{10}Be found in ice cores (Raisbeck and Yiou, 1980). Its half-life is 1.5×10^6 yr, compared to 5730 yr for ^{14}C, so a record of at least half a million years can potentially be built up. Furthermore, the residence time of ^{10}Be in the atmosphere is quite short (~1 yr), since it is generally fixed on aerosols and precipitated out. Thus it may be possible to establish whether the solar cycle was still operating during envelope minima, for example. ^{10}Be has not been used before because much more sensitive techniques are required to measure it than are required for ^{14}C. Effort is already under way, particularly in France, to exploit Be for solar signals.

The Sun is hardly unique in having a magnetic cycle. For many years, Wilson (1978) observed variations in calcium H, K chromospheric emission in main sequence stars, and found many with cyclic variation of spectral type G2 and later. The magnitude of these variations are typically of the same order as that for the Sun (20–40%), now well observed by White and Livingston (1981). Wilson also finds periods of similar length to the solar cycle. The stellar calcium monitoring has continued at Mt Wilson by Vaughan and colleagues (Vaughan and Preston, 1980; Vaughan, 1980) from which many further important effects are emerging. Of particular importance, it appears that only older stars are cyclic. But these stars have a relatively low calcium emission compared to younger stars, which have higher activity levels but lack cyclic variations. Thus, there may be two quite different types of stellar magnetic activity, and therefore presumably two rather different types of stellar dynamo. Vaughan *et al.* (1981) have also demonstrated that the stars with cyclic variations have rotation periods longer than 20 days, while noncyclic emitters tend to have shorter periods. These results are discussed in detail by Noyes in Chapter 19 of this work.

5. Theories of the Solar Magnetic Field

5.1. FLUX TUBES AND NETWORK

Theoretical work to explain the existence of magnetic flux tubes on the Sun with field strengths as large as 2000 gauss has made great progress during the past six or seven years. Concise reviews of complementary aspects of this problem are given in Spruit (1981a) and Galloway *et al.* (1977). Many of the relevant physical processes and analytical models are discussed at length in the book *Cosmical Magnetic Fields* by Parker (1979a).

It became generally accepted during the 1960s that the horizontally divergent motion in supergranule cells should sweep magnetic fields to the cell boundaries, there to pile up until the reaction back on the flow was strong enough to prevent further amplification. These fields then give rise to the network seen in Ca K and other emissions. It was argued intuitively that the limit of field amplification was determined by local equipartition of energy between kinetic and magnetic forms. But supergranule flows at photospheric levels are in equipartition with magnetic fields no larger than 100 gauss, which we now know is a factor 10–20 too small. So further mechanisms of concentration are required. Parker (1974a, b) argued that turbulent pumping from granules outside an already somewhat concentrated flux tube, as well as the Bernoulli effect of fluid sloshing back and forth vertically inside the tube, could both lead to further compression and amplification of the field, up to an equipartition limit based on granule velocities of ~500 gauss – still a factor of 3 or 4 short in amplitude (and therefore a factor of 10 or more in magnetic energy).

But none of the above arguments explicitly takes into account the compressibility of the medium. It now appears most likely that 2000 gauss fields are achieved in flux tubes by the partial evacuation of the tube, which results in lower pressure and temperature inside. Low temperatures are now supported by the recent observation of Foukal *et al.* (1981), described earlier in Section 4.1. The collapse phenomenon appears to be basically a form of convective instability inside the flux tube, as analyzed principally by Parker (1978a), Webb and Roberts (1978), Spruit (1979), Spruit and Zweibel (1979), and Unno and Ando (1979). In Parker's (1978a) version, the fluid inside the tube sinks down because, due to local inhibition of the convection by the magnetic field, the tube becomes thermally insulated from its surroundings and fluid flowing down in the tube will quickly acquire a virtually adiabatic temperature gradient. Since the surrounding medium is, on average, superadiabatic, the material inside the tube, therefore, will continue to sink due to its negative buoyancy. Fluid attempting to replace that lost from the tube converges on it and compresses it further. The upper limit of the evacuation process is given by the total magnetic pressure approaching the ambient gas pressure, which allows field strengths of several thousand gauss even just below the photosphere.

Webb and Roberts (1978) and Spruit and Zweibel (1979) treated this problem somewhat more formally as a problem in hydrodynamic stability. They both find that for convective instability to be initiated inside the flux tube, the field strength must be less than between about 1000 and 1350 gauss near the solar surface, and further concentration occurs only for downdrafts. Flux tubes with stronger fields are stable to both updrafts and downdrafts, which presumably contributes to their observed lifetime. Spruit (1979) has also argued that the strength of the collapsed flux tube depends on the local field strength before collapse, with weaker initial fields curiously resulting in stronger final fields. But for an initial field in energy equipartition with local granulation, a reasonable final field strength of ~1650 gauss is predicted.

All these models require a downdraft for the collapse to take place, but it is not clear whether this downdraft is present for the whole life of the flux tube, or just its collapse phase. Presumably if it were present only for the collapse phase, it would not be commonly observed over existing flux tubes, as has been seen. This aspect of flux tubes remains to be understood. It seems likely that the thermal structure of flux tubes is rather like that modelled by Spruit (1976), with a dark depression in the photosphere similar to, but less extreme than, that for a sunspot. The brightness of the tube as seen in photospheric faculae near the limb comes largely from the bright walls of the tube.

The above theoretical picture of flux tubes seems plausible, but it has not been tested yet with a fully consistent compressible convection model that allows the several relevant physical processes to compete to determine the net outcome. However, much important work has been done on nonlinear convection in a magnetic field for an incompressible (Boussinesq) fluid, for example by Galloway *et al.* (1978), Peckover and Weiss (1978), Proctor and Galloway (1979), Galloway and Moore (1979), and Weiss (1981a, b, c). It had been established a decade before in many kinematic studies that when induction of magnetic fields by fluid motions is large compared to diffusion of fields in the bulk of the fluid (the magnetic Reynolds number R_m is much larger than unity), the magnetic field will be largely expelled to the boundaries of a pattern of convective eddies, where the field will amplify until locally diffusion balances induction. The amount of this amplification is proportional to R_m. What the newer nonlinear studies have shown is that in three-dimensional convection this amplification need not be limited to an equipartition value by feedbacks due to $\mathbf{j} \times \mathbf{B}$ forces. The field strength may exceed this value by a factor proportional to $(\nu/\eta)^{1/2}$ $(\ln R_m)^{-1/2}$, in which ν is the kinematic viscosity of the convecting fluid and η the magnetic diffusivity as in Equation (1.4). Thus, if the viscosity exceeds the magnetic diffusivity by a large factor, and R_m is not extremely large, field values well above equipartition can be attained. This result was explored in detail for axisymmetric convection in cylindrical geometry by Galloway *et al.* (1978) and Galloway and Moore (1979). Interestingly, in this case the magnetic flux becomes concentrated in a stagnant region along the axis of the convection pattern, rather than at its outer boundaries. This is true for a cell with either central updraft or downdraft. The new upper bound on the amplified fields comes about because ohmic dissipation due to large field gradients in the flux tube cannot greatly exceed the viscous dissipation rate for the flow. If it did, buoyancy could not sustain the flow against the total dissipation and the velocity amplitudes would fall, so that less magnetic field concentration would occur.

Whether this result is important for concentration of solar magnetic fields is not clear. Molecular values for ν in the photosphere are much smaller than for η, so on that basis amplification is not possible. But for a flux tube even as small in diameter as 10^2 km, both diffusion times are large compared to a typical flux tube lifetime, so molecular diffusivities do not seem relevant. If these are replaced by turbulent diffusivities, the ratio ν/η is most likely to be somewhere near unity, making amplification beyond the equipartition limit unlikely. However, in the neighborhood of strong, highly nonuniform magnetic fields, scalar isotropic turbulent diffusivities may be too much of an oversimplification. Clearly, further exploration of the representation of these turbulent processes is needed before much more can be said.

In the above calculations, when the magnetic field is strong enough to feed back significantly on the convective flow, but not so strong as to suppress it, hydromagnetic oscillations can be excited. In the case of a flux tube in a compressible medium, both longitudinal and transverse oscillations are possible (see, e.g., Spruit, 1981a), and such oscillations have been invoked to contribute to cooling of the flux tube (e.g. Parker, 1976, 1979b). For further details on waves in flux tubes, see Chapter 7 by Brown, Mihalas, and Rhodes.

5.2. ACTIVE REGIONS AND SUNSPOTS

It is obvious from the observations that the birth and growth of an active region on the Sun is a complex dynamical process, not very amenable in a detailed sense to mathematical modeling. A major issue is to determine the primary mechanism responsible for the emergence of new magnetic flux. The two principal candidates are advection of flux tubes upward by the convection itself, and magnetic buoyancy. The first mechanism is self-explanatory. As for the second, it was demonstrated by Parker (1955a) that an isolated flux tube in lateral total pressure equilibrium and thermal equilibrium with its surroundings will be less dense and, therefore, buoyant. At a given level in the fluid, the strength of this magnetic buoyancy force is proportional to the square of the field strength in the tube. For a given field strength, a flux tube is much more buoyant near the top of the convection zone than the bottom, because its magnetic pressure is a much larger fraction of the ambient gas pressure near the top. On the other hand, a rising flux tube will expand against the lower ambient pressure, weakening its field strength and therefore its magnetic buoyancy. As the tube moves, it also encounters frictional resistance from the surrounding medium.

As a consequence of all these effects, estimates of rise times for flux tubes from various levels of the convection zone to the surface are uncertain. Estimates vary greatly, according to what assumptions are made. For example, Parker (1975a) estimates a small flux tube of field strength 100 gauss would rise to the surface from a depth of 10^5 km in about two years, on the assumption that the tube motion is resisted by aerodynamic drag. However, if turbulent viscous effects are taken into account, the same flux tube may take as long as 80 years (Unno and Ribes, 1976; Schüssler, 1977; Kuznetsov and Syrovatskii, 1979). It is not known what the true viscous interaction between a small flux tube and its surrounding is in the typical solar case, so it is difficult to narrow this wide difference by these arguments. But Stix (1976b) points out that if the rise time from deep in the convection zone is longer than a small fraction of a solar cycle, then flux tubes from more than one cycle (with opposite polarity fields) would be rising at the same time. Since different tubes having different field strengths would rise at different rates, at least two cycles would become mixed, with both arrangements of polarities in active regions, which is never observed. Parker (1975a) argued that if the rise time is too fast, the solar dynamo will fail to amplify the magnetic fields sufficiently to sustain itself. But Schüssler (1979) shows that with turbulent viscosity resisting the flux tube movement, there is plenty of time for amplification to take place. Fortunately, this result is very insensitive to the value of eddy viscosity assumed, since that parameter is so poorly known.

Schüssler (1980a) has carried his analysis of flux tube motions still further to show that rising tubes should stay nearly horizontal until they get very close to the top of the convection zone where the local scale height is small. There instability to loop formation will result in vertical loops with horizontal scale of perhaps ten local scale heights. He argues that it is this local instability which produces the small active regions seen at the surface.

The formation of an active region obviously involves the rising to the surface of several distinct, but neighboring, flux tubes. Parker (1978b, 1979c) has shown that hydrodynamic forces between two parallel rising tubes generally tend to push them together. These include the simple Bernoulli effect when the tubes are side by side at the same level, as well as the wake of the upper tube drawing in the lower one if they are rising along the same path. In addition, undulations and pulsations of neighboring tubes produce attraction. From all these effects, there will be a tendency for neighboring tubes to coalesce, and these effects may be partly responsible for the building of a sunspot. On the other hand, the theory usually assumes these flux tubes neither expand nor deform, so the surface pressure distribution on the tube can take whatever value the external dynamics dictate. It is not clear how much the results would be modified when the tube's internal dynamics allow adjustment to these pressure forces.

As stated before, convection itself could be largely responsible for bringing new magnetic flux to the solar surface. We can easily imagine a horizontal flux tube being caught up in the updraft in the center of a supergranule, for example, and then spreading out as the foot points are carried by the divergent horizontal flow to the cell boundaries. The interaction between a flux tube and a convection cell is bound to be complex, depending on the relative sizes of tube and cell, as well as field strength and cell amplitude. A flux tube in convective shear flow may experience aerodynamic lift (Parker, 1979d; Tsinganos, 1979). Meyer *et al.* (1979) have done the most extensive study of how a flux tube would be carried around in a supergranule, and found that while small tubes brought up are swept towards the network at the edge of the cell, larger ones tend to remain floating vertically in the cell center.

On scales much larger than supergranules, it seems possible that giant cells could be responsible for bringing-up large arrays of magnetic flux (such as seen in large persistent active regions) and complexes of activity. Giant cells might also be responsible for the occasional globally coherent abrupt rises in the rate of emergence of new flux seen in X-ray bright points and magnetograms (see Section 4.2).

Which of the two mechanisms — magnetic buoyancy or convection — is primarily responsible for new flux emergence presumably depends on the relative rates of rise expected, which may depend on depth in different ways in the two cases. At the bottom of the convection zone, the turnover time for convection is of the order of a month or two, and we should expect flux tubes present there to be brought near to the surface in that time, unless they are sufficiently intense to resist by $j \times \mathbf{B}$ forces or are largely excluded from updraft regions. Below that level of flux tube field strength, the rate at which such tubes are brought up should not depend much on field strength. On the other hand, magnetic buoyancy gets stronger as the square of the field strength. Flux tubes with fields small enough to imply a rise time by magnetic buoyancy much longer than a month will most likely be brought up by convection. Thus the 100 gauss flux tubes considered by Parker (1975a) and Unno and Ribes (1976) should be brought up by convection, not magnetic buoyancy. On the other hand, a 5000 gauss flux tube found at the bottom of the convection zone should rise primarily due to magnetic buoyancy. By contrast, a 100 gauss flux tube at a density of 10^{-5} g cm^{-3} in the supergranule layer should experience magnetic buoyancy and convective lifting of similar amounts. A 10^3 gauss flux tube there should rise much faster due to magnetic buoyancy than supergranule convection.

Theories of sunspots in recent years have focused primarily on dynamical effects. Magnetohydrostatic model studies do continue (see, for example, Low, 1980) but they cannot address such important questions as: what processes lead to development of a spot?, what makes the spot stable?, what cools it?, and what eventually causes it to decay? Only dynamical models can answer such questions as these. For a recent review of dynamical theories of sunspots, see Spruit (1981b).

While many magnetohydrodynamic theories motivated by the sunspot problem have been developed, none of these can be said to provide a complete self-consistent theory in which all the relevant processes are allowed to compete to determine which will dominate. Instead, models have emphasized detailed examination of one or more parts of the problem separately, results from which are coupled together in a heuristic manner. Much insight has been gained in this way, but the time is now becoming ripe for a more direct assault on the complete problem. As with the theory of small flux tubes, this assault will most likely involve development of the nonlinear theory of convection of a compressible fluid in a magnetic field with full feedbacks allowed between the field and the flow.

Until about 1974, it was widely accepted that, at least in a qualitative sense, sunspots are cool because the magnetic field in them largely inhibits the outward convection of heat relative to the surrounding plasma (Biermann, 1941). But Parker (1974c) argued that such inhibition should lead to excess pressure and temperature immediately underneath the spot due to the net flux from below being dammed up there. Some of this heating would lead to an observable bright ring around the spot as a result of turbulent diffusion. The excess pressure should also work to spread out the magnetic flux and thereby destroy the spot. But many spots last for several rotations despite the excess pressure, and no substantial bright rings have been seen. This led Parker (1974c) to postulate an additional mechanism to sustain a cool spot, namely the vigorous production of Alfvén waves, which carry away the excess energy built up at the bottom.

Parker's arguments triggered many studies of heat transfer around obstacles in the solar convection zone, Alfvén waves in sunspots, and thermal models of spots. The net result of all these studies has been to weaken considerably the force of Parker's arguments, but they have not ruled out Alfvén or other MHD waves as an important ingredient in spot dynamics. For example, Spruit (1977c) demonstrated that if the turbulent thermal conductivity increases with depth in the convection zone in proportion to the fluid density, as expected from mixing-length theory, then thermal energy built up under a sunspot will diffuse laterally much faster than towards the surface, spreading the excess energy reaching the surface over such a wide area that no significant bright ring should be expected (and note that the recent SMM Active Cavity Radiometer measurements, discussed in Section 2.3 above, indicate that the extra heat flux is not coming out even over a broad area around the spot). Clark (1979) and Spruit (1977c) showed that only shallow spots are likely to produce a bright ring intense enough to be observed.

Cowling (1976) also criticized Parker's arguments in several respects, doubting that a sunspot could generate Alfvén waves with the required 80% efficiency. Such a question can only really be answered by a nonlinear calculation with a specified excitation mechanism, which has not been done. Parker argued the most likely exciter is the convection itself (see also, e.g., Roberts, 1976) which is assumed not to be completely suppressed in the spot, but rather to take the form of convective 'overstability' (Chandrasekhar, 1961; Savage, 1969). This instability grows in an oscillatory fashion with each successive fluid particle displacement in the opposite direction and its kinetic energy acquiring greater and greater amplitude. But Galloway (1978) finds that in nonlinear convection of an incompressible fluid in a magnetic field, nonlinear MHD waves excited in the region of strong magnetic field typically have no more than 10% of the kinetic energy density of the convective circulation which produced the concentration in field. However, Parker (1975b) points out that if the convection extends over several scale heights in a compressible fluid, more energy can be convected into fluid motion than transported as heat. This fluid motion, in turn, might excite larger amplitude waves.

These oscillations are probably not Alfvén waves (Cowling, 1976) since they are unlikely to be purely transverse to the magnetic field. They are more likely slow mode MHD waves, but no theory of convective overstability for the compressible case currently exists. In any case, Alfvén waves propagating upward in a sunspot should be almost completely reflected (Thomas, 1978), so if they are to cool the spot, they must instead escape downward into the interior of the Sun, and, as Cowling (1976) points out, their energy must be dissipated over a broad enough area far enough below the surface that no bright ring is formed.

There seems to be general agreement at present that sunspots are first formed by the gathering together of neighboring previously erupted flux tubes in a new active region (Meyer *et al.*, 1974; Zwaan, 1978). In the Meyer *et al.* (1974) scenario, supergranule motions push several flux tubes into a grouping at a point of convergence of neighboring supergranules, but the weight of observational evidence − for example, Vrabec, (1974), Zwaan (1978) − indicates that normal supergranular flow is not the primary cause of the coalescence. Whatever the mechanism, the gathering of several flux tubes results in an easily visible surface darkening or pore, which may continue to grow into a larger sunspot. The downdraft due to convergence (whether from supergranules or not) contributes to the further compression of the field to typical sunspot strength as well as the evacuation required for darkening. At this point further convection is largely suppressed, or converted into Alfvén waves in Parker's version. Meyer *et al.* (1974) argue that if the new sunspot magnetic flux is not vertical, it may quickly disintegrate. But with a central vertical field which fans out in the surface into a penumbra, the neighboring supergranulation will be partially damped, resulting in local heating below, which could reverse the local sense of circulation compared to what was present when the spot first formed. The result could be the annular outflow or moat cell seen at the surface. This flow should lead to a much more stable spot, but one which will still decay slowly as small flux tubes from the umbra are occasionally swept across the moat to the surrounding photosphere, as observed.

In the scheme of Meyer *et al.* (1974), the typical sunspot depth is assumed to be as much as 12 000 km. The upper part of the field which flares outward through the photosphere is found to be stable (Meyer *et al.*, 1977) due to the buoyancy of the light fluid inside the spot. But at whatever level below which the field lines flare out again into the convection zone, the configuration becomes unstable. Parker (1979a) argues that this means that the spot exists as a single flux tube only in a very shallow layer, perhaps only 10^3 km in depth. He postulates that below this level, the single flux tube breaks into many, which are held together in a tight network by a subsurface inflow and downflow between the tubes. Such a downflow could be driven by cooling from the interspersed cold fingers of fluid inside the flux tubes. The return flow underneath might carry away the heat built up under the spot, alleviating the need to invoke Alfvén waves to keep the spot cool. Sunspot umbras do show fine structure in brightness, which might be evidence of the warm fluid in between neighboring tubes reaching the surface of the spot (Parker, 1979f). Also, Parker (1979e) argues that the darkness of spot umbra, being almost independent of umbral area, speaks against a single central flux tube extending to any great depth. However, there is no observational evidence of the inflow and downdraft. Both the moat flow and Evershed flow are outward in the photosphere. Nevertheless, one could at least imagine that the moat cell sits on top of Parker's inflow, with some inflow returned to the surrounding fluid in the moat above without penetrating the spot (since it is a single intense tube in the photosphere) and the rest of it penetrating the spot and turning downward between the individual tubes present below. Nonlinear compressible convection calculations with a magnetic field might help to determine what circulations are preferred near a large concentration of magnetic flux such as a sunspot.

Such calculations would also hopefully lead naturally to an explanation for penumbral structure and dynamics. The penumbral filaments are still thought to be the result of convective rolls whose axis is aligned with a nearly horizontal magnetic field diverging radially outward from the umbra (Danielson, 1961). Galloway (1975) has taken this approach further with a nonlinear, but phenomenological, model to show that, as observed, the magnetic field should concentrate in the dark filaments, where the Evershed flow is also concentrated. In this model, the Evershed flow is driven by a time-dependent $j \times B$ force, rather than being a convective flow itself. An alternative model, by Meyer and Schmidt (1968), produces steady Evershed flows by assuming the gas pressure at the outer end of a penumbral flux tube to be less than at the inner end at the same level. In this model, the flow becomes sonic at the highest point in the flux tube, and shocks are required in the outer, descending part of the tube. It does not appear that observations yet allow a choice between these two models.

The running penumbral waves commonly observed are probably gravity modified magneto-acoustic waves (so-called fast-mode waves), models for which have been constructed principally by Nye and Thomas (1974, 1976). Their most likely means of excitation is overstable convection in the umbra and penumbra at lower levels (Moore, 1973; Galloway, 1978).

In summary, we are nearly ready to make significant advances in the theory of an individual sunspot based on nonlinear modeling of compressible convection in a magnetic field which can be transported and can react back on the flow. Such calculations would complement, and in some cases test, simpler conceptual models, but certainly not replace them. Such calculations would be extremely expensive if we require three space dimensions, for example, in order to see if Parker is correct that below the photosphere the spot may contain many separate flux tubes. But an axisymmetric calculation appears to be within reach now. It is not clear in detail how to connect the spot to the subsurface toroidal field, but such a connection must occur, if Hale's sunspot polarity law is to be satisfied.

5.3. SOLAR DYNAMO THEORY

A hydromagnetic dynamo is a conducting fluid in motion which maintains electric currents and therefore magnetic fields permanently against dissipation by Joule heating. The predominant theory of solar magnetism developed over the past 30 years is predicated on the hypothesis that the Sun is a hydromagnetic dynamo. For other recent reviews of the solar dynamo problem, see Stix (1981a) and Weiss (1981d).

Mathematically, dynamos usually represent solutions of some form of the induction equation (1.4). A dynamo results if the induction term $\nabla \times (\mathbf{v} \times \mathbf{B})$ is able to sustain the field $\mathbf{B}$ against diffusion and dissipation of field, represented by $-\eta \nabla \times \nabla \times \mathbf{B}$. If only the induction equation is solved, the dynamo is 'kinematic' in that the motions $\mathbf{v}$ are specified independently. The full dynamo problem involves the simultaneous solution of the induction Equation (1.4) along with the equations of motion (1.1), mass continuity (1.2), and thermodynamics (1.3). This is a much more difficult problem, on which progress has only recently been made, but it is the only approach likely to allow realistic estimates of magnetic field amplitudes, which will be limited by the feedback of the $\mathbf{j} \times \mathbf{B}$ body force on the inducing motions. In the kinematic case, since Equation (1.4) is linear in magnetic field, solutions will be exponentially growing when dynamo action is found, and exponentially decaying when ohmic diffusion wins. A large literature exists on the solutions of Equation (1.4) for various velocity fields without reference to detailed application to the Sun or other dynamos occurring in nature. For an exposition of basic concepts, mathematical developments, and results, see Moffatt (1978).

For typical laboratory fluids the diffusion term $\eta \nabla \times \nabla \times \mathbf{B}$ in Equation (1.4) is important for all scales of magnetic field and motion that can be realized. On the other hand, in a large body like the Sun, η is sufficiently small (assuming typical conductivities for an ionized gas) and the typical length scales are sufficiently large that the induction term $\nabla \times (\mathbf{v} \times \mathbf{B})$ dominates in Equation (1.4) for a wide range of motions. As a consequence, the flow drags the magnetic field lines around, unless the electromagnetic body force becomes so great as to resist. The magnetic field is said to be 'frozen' to the fluid. One result of this property is that, if the motion field is slightly complex the magnetic field will become much more complex, and highly twisted or compacted fields can result. On the other hand, a rather simple motion field can transform one relatively simple magnetic field into another, almost equally as simple. The classic example of this which is important for dynamo theory is the action of a pure differential rotation on a 'poloidal' magnetic field purely in meridional planes of a sphere. In general, the shears in the rotation will induce additional 'toroidal' field, wrapped about the axis of rotation. This is a fundamental process invoked in virtually all solar dynamos.

As we have described in more detail above, the basic features of the solar field to be explained are its rather regular magnetic reversals, with opposite magnetic polarities in the north and south hemispheres, the migration of the zones of occurrence of sunspots and active regions towards the equator with time (the butterfly diagram), the change in amplitude of the envelope of the solar cycle, and the appearance and behavior of all the many magnetic structures, from sector fields down to flux tubes. Formal dynamo theory applied to the Sun so far has focused primarily on global aspects, and not attempted to predict individual magnetic structures in detail.

Solar dynamo theory has roots that reach back to early work by Parker (1955b). Parker demonstrated that a combination of differential rotation and what he called 'cyclonic turbulence' could give rise to migratory dynamo waves that would behave much like the observed butterfly diagram. Starting from a dipole field, differential rotation generates a toroidal field, as mentioned above; then a combination of local lifting motions and twisting about a local radial axis, the so-called 'helicity' of the flow, produces new field loops in the meridional plane. The twisting is expected to arise because of the influence of Coriolis forces on convective motions. These loops coalesce into a new dipolelike field. With the right amount of twisting, this new field is of opposite sign to the old one. Stretching by the differential rotation then produces a new toroidal field of opposite sign to the old. In Parker's model, only a radial gradient in rotation is considered, and he showed that if the rotational velocity increases inward, new toroidal field of opposite sign forms on the poleward side of the old field, and the old field pattern migrates towards the equator. Reconnection of magnetic fields is implicit in the model, which allows the coalescence of poloidal loops, and also cancels old toroidal field with new. The rate at which the peak toroidal field migrates towards the equator is proportional to the square root of the product of shear in the rotation and the magnitude of the twisting. According to the theory, stars with motions more strongly influenced by rotation and with more differential rotation than the Sun should experience shorter dynamo periods, unless periodic reversals disappear altogether.

Parker's model is highly idealized compared to the real Sun, and Babcock (1961) and Leighton (1964, 1969) built models which were more heuristic but closer in some respects to the detailed phenomenology of solar magnetic fields. In these models, the magnetic field erupts to the surface in sunspot groups, and these fields are subsequently diffused across the solar surface. In Leighton's version this takes place as a random walk produced mainly by supergranules. More field of the following polarity, than of the leading polarity, reaches the pole because the spot groups are tilted in latitude. This new polar field reconnects with the polar field of opposite sign already there and cancels it out. This new field then gives rise to the new toroidal field of the next half-cycle, and the dynamo process repeats. But note that Howard and La Bonte (1981) see evidence fields migrate to the poles in a few discrete events in a solar cycle, rather than by diffusion.

The theory of dynamos driven by the combination of differential rotation and helical turbulence was placed on a more formal footing in a series of papers by Krause, Rädler, and Steenbeck (see Krause, 1976; Rädler, 1976; and Krause and Rädler, 1980, for the earlier references). They developed the theory of 'mean field electrodynamics', in which the total field is separated into a global 'mean' part and all the fluctuations about it, of smaller scale. Turbulence theory is then invoked to estimate the influence of the smaller scales on the global ones. In this system, the induction Equation (1.4) is modified to a form

$$\frac{\partial \overline{\mathbf{B}}}{\partial t} = \nabla \times (\overline{\mathbf{v}} \times \overline{\mathbf{B}} + \alpha \overline{\mathbf{B}}) - \eta_t \nabla \times \nabla \times \overline{\mathbf{B}} \tag{5.3.1}$$

in which $\overline{\mathbf{B}}$ is now the mean field that is predicted and $\overline{\mathbf{v}}$ is the mean velocity. The parameter α, usually taken to be a scalar, represents, the 'cyclonic turbulence' or helicity Parker invoked, and η_t is a turbulent diffusivity for magnetic fields.

Many solutions of Equation (5.3.1) have been studied, for different assumed motion fields, profiles of α, and values of η_t. Most of these are reviewed succinctly in Stix (1976a). The long series of dynamo calculations by Yoshimura (e.g. Yoshimura, 1972, 1975a, 1978a, 1978b) are also basically of this type, although Yoshimura identifies his regeneration action (essentially α) with the effect of global-scale convection, rather than small-scale convection. Leighton's model can also be interpreted as a mean field dynamo, if one identifies the random walk with η_t, and the tilt of sunspot groups with α (Stix, 1974).

The class of solutions to Equation (5.3.1) most often applied to the the sun are the so-called $\alpha{-}\omega$ dynamos, in which v contains only differential rotation ω (which is a function of latitude and radius). In such solutions, stretching by differential rotation of the poloidal field is the primary mechanism for producing toroidal field, while the 'α effect' produces new poloidal field from the toroidal. Best agreement with the main features of the solar cycle is obtained when the angular velocity increases inward, and the magnitude of α is small. Then migration of the mean toroidal field is towards the equator. More generally, the direction of propagation is along surfaces of isorotation, as proved by Yoshimura (1975b). By suitable choice of the magnitudes of the angular velocity gradient and α, this migration rate can be made equal to the observed one. If the α and differential rotation effect on the magnetic field are largest in low latitudes, then magnetic fields antisymmetric about the equator are favored (Stix, 1981) as observed.

Additional features have been added to $\alpha{-}\omega$ dynamo models to make them conform even more closely to the observations. For example, varying the profile of α with latitude and depth and the amount of differential rotation allows one to introduce a second branch of poloidal field which migrates towards the pole in high latitudes (Yoshimura, 1975a). Adding *ad hoc* nonlinear feedback to the induction equation allows the field amplitude to be bounded (Yoshimura, 1975a, 1978a) and if a time delay is allowed in this feedback, an envelope to the cycle can be generated, including intervals of several cycles in length that look like Maunder Minima (Yoshimura, 1978b).

The success of $\alpha{-}\omega$ dynamo theory in reproducing many features of the observed solar cycle suggests it has captured the primary processes responsible for maintaining the solar magnetic field (Weiss, 1981d). However, when one examines the assumptions behind these models more carefully, and considers the theory in the broader context of the full (as opposed to kinematic) dynamo problem, it is evident that serious difficulties remain, and it is not clear that all these difficulties can be resolved without major modifications to the theory, perhaps even to the point of scrapping it altogether.

For example, a basic assumption of $\alpha{-}\omega$ dynamo theory is that a rather low-order closure of the turbulence equations, needed to evaluate α and η_t, is valid. In its simplest form, this essentially requires that either the local magnetic Reynolds number R_m of the turbulence responsible for α and η_t be small compared to unity, or that the correlation time for the most energetic eddy be very short compared to its own turnover time. On the Sun, R_m is in fact $\gg 1$ ($\sim 10^5$ or more) for even the smallest observable scales, and, in common with most turbulent flows, the correlation time and eddy turnover times are essentially the same for all observed solar eddies. Clearly much more sophisticated closures and MHD turbulence theory are needed to deal with the much more intermittent magnetic field that results when R_m is much greater than unity. There are also serious questions as to whether η_t is even always positive (e.g. Kraichnan, 1976a, b; Knobloch, 1978).

A closely related problem is that the magnitude of α which 'works' for the solar case is much smaller than we should expect based in hydrodynamic arguments. Köhler (1973) showed that mixing-length theory predicts an α perhaps 10^3 times larger than normally used, and Gilman and Miller (1981) found that α of this large magnitude is typical for global convection which drives an equatorial acceleration of the observed magnitude.

Gilman and Miller (1981) did full hydromagnetic dynamo calculations for nonlinear Boussinesq convection in a rotating spherical shell, and found that the effective α was so large that the model did not even produce magnetic field reversals. As they point out, the reason Yoshimura (1975a) was able to mimic solar cycle behavior so well was largely because he had not consistently solved for coupled global convection and differential rotation together. The global convection he assumes is unable to sustain the differential rotation he assumes, both in terms of magnitude and sign of the latitudinal gradient. This sort of difficulty is hidden in kinematic dynamo treatments, when the equations of motion are not solved.

Subsequent nonlinear dynamo calculations by the author (unpublished) indicate that whether such basic magnetic field reversals occur or not can depend quite sensitively on the relative strengths of convection and differential rotation amplitudes. In Gilman and Miller (1981), convection contained about two-thirds of the total kinetic energy of the system, differential rotation driven by the convection about one-third. In his new calculations, the author finds that when differential rotation becomes two-thirds to three-quarters of the total kinetic energy, magnetic fields can reverse although the typical period is still much too short for the Sun. In the former case, this nonlinear dynamo is more of the 'α^2' type, in which the α effect is responsible for maintaining both the poloidal and toroidal fields, while in the latter case, it is more nearly an $\alpha-\omega$ dynamo. α^2 dynamos are known to be much less likely to produce magnetic field reversals.

The problem of α being much too large to give the right reversal period for the Sun is probably intimately related to the highly intermittent nature of the solar magnetic field. Gilman and Miller (1981) have argued that such intermittency may allow the field to escape much of the induction effects. Put another way, a solar magnetic field confined largely to the boundaries between convection cells may not feel the full effects of twisting and lifting due to the helicity in the flow. None of the current solar dynamo models has addressed this point carefully, but Childress (1979) has illustrated what may happen with an idealized model which predicts that α acting on a very intermittent field may be reduced by a factor of order $R_m^{1/2}$. For a typical magnetic flux tube in the solar photosphere, this factor could easily reach 10^2, which would bring predicted dynamo periods much closer to the observed range. But clearly, such an effect must be demonstrated for more realistic circumstances than Childress employed.

High concentrations of magnetic field in localized regions in a compressible fluid such as the solar convection zone also lead to much stronger effects of magnetic buoyancy. Parker (1977) demonstrated that its presence can require a substantially larger α or radial gradient in rotation to achieve dynamo action. Alternatively, if α is too big already, the resulting dynamo with magnetic buoyancy may run at more nearly the correct speed. Schüssler (1980b) has carried out heuristic calculations similar to those of Leighton (1969) to show that a solar dynamo model driven by differential rotation plus magnetically buoyant flux tubes can give reasonable results.

Even if all the above effects are properly incorporated into a solar dynamo model, another major difficulty may still remain. Virtually all successful kinematic solar dynamos require the angular velocity to increase inward, while as reviewed in Section 3, differential rotation models which take full account of the influence of rotation upon global convection predict angular velocity decreasing inward, and being nearly constant on cylindrical surfaces parallel to the axis of rotation. Thus in the latest (unpublished) nonlinear dynamo calculations of the author alluded to above, although magnetic field reversals are found the migration of the toroidal field patterns with time is towards the poles rather than the equator. These dynamo calculations are for an incompressible fluid, however, and the situation may change in the compressible case. In particular, Glatzmaier and Gilman (1982) find that compressible convection in a deep rotating spherical shell may drive a differential rotation whose angular velocity reaches a maximum at an intermediate depth, giving the desired increase of angular velocity with depth in the upper layers. Such profiles still possess equatorial acceleration, and angular velocity nearly constant on cylinders near the bottom as in the incompressible case. With such a profile one might imagine that the migration of the toroidal field with time is towards the poles in the deepest layers, but also towards the surface along the lines of constant angular velocity (in accordance with Yoshimura's (1975b) theorem). When the field reaches the upper layer of reversed angular velocity gradient, it then could reverse its direction of migration and proceed back towards the equator, giving rise to the observed sunspot cycle. At this time, however, such a scenario is only untested speculation (the author plans to test it). It is not clear whether this upper branch can withstand the expulsion of flux by magnetic buoyancy, as Parker (1975a) argues.

It seems clear that some form of nonlinear feedback determines the evolution of the envelope of the solar cycle, but what form this takes is unknown. Several *ad hoc* forms have been used, but Yoshimura's (1978b) time delay mechanism, requiring a lag in reaction of the $j \times B$ force of $\sim$20–30 years, seems rather implausible. One important effect this assumption takes no account of is the natural randomness likely to be present in the global convection itself. Gilman and Miller (1981) found that even when the magnetic field was extremely weak, its presence could cause the time history of the convection amplitudes to diverge away from the corresponding solution when magnetic field was absent, on a relatively short time-scale. This is a common characteristic of nonlinear turbulent fluid systems with many degrees of freedom, and itself could lead to substantial variations in the strength of successive magnetic cycles. Barnes *et al.* (1980) have shown that it is quite easy to produce a purely statistical simulation of typical sunspot cycle envelope modulations with random noise superimposed on a basic 11 yr periodicity. It may well turn out that such envelope modulations are rather unpredictable. On the other hand, the nonlinear oscillator that is the solar dynamo may behave in some respects like a 'strange attractor' (e.g. Ruzmaikin, 1981) which means there are certain points in the mathematical space, one perhaps corresponding to a 'normal' solar cycle, and another to 'Maunder Minimum' type behavior, near which the solar dynamo spends considerable time before evolving into a new state. This whole subject is relatively unexplored, and may become much more important for understanding the solar dynamo in the future.

The dynamo problem as a problem in MHD turbulence has been studied most intensively by Frisch and colleagues, e.g. Frisch *et al.* (1975), Pouquet *et al.* (1976), Pouquet and Patterson (1978), and Meneguzzi *et al.* (1981). Using conservation and scaling arguments as well as direct numerical simulation, they demonstrate a novel effect that magnetic helicity, defined as the scalar product of magnetic field and its vector potential, can, once introduced at small spatial scales in a turbulent fluid, cascade in reverse to the very largest scales allowed. Thus it can generate magnetic energy on large scales just as does the α-effect from kinetic helicity we have described above. The effect can be quite large, resulting in much larger magnetic than kinetic energy in large eddies (Meneguzzi *et al.*, 1981).

These dynamo calculations have not been applied to the Sun or any other real dynamo in detail, and are generally done for single periodic plane geometry. Nevertheless, the magnetic helicity cascade effect could be quite important in such systems, and should be explored further. In those systems, such as the Sun, where differential rotation is important, presumably this reverse cascade has to compete with direct production of magnetic helicity on the global scale which comes about simply by the differential rotation stretching axisymmetric poloidal fields into toroidal ones.

In addition to the primary thrust in solar dynamo theory, there have also sprung up in the past few years a number of alternative proposals for the origins of solar magnetic fields, some quite radical, but none yet subjected to quantitative testing with theoretical models. The most plausible of these proposals (to the author) is that of Galloway and Weiss (1981), who suggest the dynamo may actually be largely below the convection zone, in the boundary layer that forms the transition to the radiative interior. Here they expect the motions and magnetic fields to be much more regular than in the convection zone above, and therefore more conducive to producing the many regular features of the solar cycle we see. In this view, the turbulent convection zone acts largely to grind up fields produced below, and partially obscure the true regularities in the dynamo.

La Bonte and Howard (1981) go much further and propose that the magnetic field is sustained by the weak torsional oscillations they have found. This seems quite unlikely to this author, because somehow the induction effects of the much larger mean differential rotation and convection have to be shut off to allow the torsional oscillations to compete. Also a purely longitudinal motion provides no mechanism for regenerating the poloidal field. Dicke (1978) and Layzer *et al.* (1979) invoke other forms of torsional oscillations but with no supporting theoretical model. Still more extreme is the notion that tidal oscillations induced in the Sun by the planets may drive the solar cycle. For critiques of this notion see, for example, Condon and Schmidt (1970) and Smythe and Eddy (1977). Wolff (1974, 1976) noted that solar cycle length periods could be obtained as beats among neighboring global oscillations. But these oscillations are even lower in typical amplitude than the torsional oscillations of Howard and La Bonte. A common thread for all these suggestions involving oscillations is that an oscillating motion needs to be invoked to get an oscillating magnetic field in which the magnetic oscillation will be of the same period. But there is no particular reason to expect this to be true, and dynamo theory has amply demonstrated that periodic oscillations in a magnetic field will commonly result from completely steady motion fields. And what oscillations the field might induce in the motions by feedback will be at half the period of the magnetic oscillation, because the $\mathbf{j} \times \mathbf{B}$ is invariant to a change in sign of all field components. Such oscillator models must be taken far beyond the coincidence of periods to be convincing.

Finally, in a long series of papers, Piddington (e.g. Piddington, 1978) has rejected dynamo theory altogether and replaced it with the notion that the current solar magnetic field is actually primordial, not in need of maintenance against dissipation. This field is highly twisted and somehow virtually insulated from interactions with fluid around it. Overall, his arguments seem quite unconvincing to this author (and in many cases not clear enough to be testable). One crucial point Piddington seems to have missed altogether is that high electrical conductivity is no barrier to dissipation of electric currents, since in a turbulent fluid, such as solar convection, magnetic energy can be cascaded out to arbitrarily small spatial scales in one or two convective turnover times (Frisch, 1977). Such a process precludes a primordial field in the solar convection zone no matter how large the electrical conductivity.

To summarize, dynamo action remains the most likely origin of the Sun's magnetic field, but many serious difficulties have yet to be resolved. In the near future, the physics of solar dynamo models will undoubtedly be made significantly more realistic. The author himself hopes to be computing dynamo solutions from compressible convection in a rotating spherical shell by the time this article appears in print. Other such models should also be appearing. Much more attention will need to be paid in all these efforts to the role of MHD turbulence, particularly on spatial scales the model calculations cannot resolve. Eventually, such dynamo calculations may produce predictions of more subtle changes, such as in solar diameter and luminosity, but this is further in the future.

Applications of dynamo models have been made to other stars in a preliminary way, — e.g. Belvedere *et al.* (1980c), Durney and Robinson (1981), Durney *et al.* (1981), Gilman (1981), Knobloch *et al.* (1981), and Weiss (1981d), — but given the current uncertainties in solar dynamo theory, large theoretical efforts on the stellar dynamo problem seem unwarranted at this time. But an expanded effort to observe stellar dynamo action through calcium emission variability and other diagnostics would be very worthwhile (see Chapter 19 by Noyes).

6. Concluding Remarks

We have commented at the end of each section on what research we see as particularly important in the near term future in the various areas we have discussed. On the observational side, the most important measurements to be made on small spatial scales are of the interactions between individual magnetic flux tubes and the velocity fields around them. The necessary spatial resolution can only be obtained from a space platform such as the shuttle, using the Solar Optical Telescope. It is also particularly important to determine just what the relation is between flux tube occurrence on the Sun and X-ray bright points, including the variation of both with the solar cycle.

On the global scale, it seems to us it is most important to determine once and for all whether giant cells exist at the solar surface. Global oscillations need to be exploited systematically to determine the rotation rate with depth, and any remaining questions concerning the validity of the newly discovered torsional oscillations need to be resolved. Variations in solar diameter and luminosity on solar cycle time-scales are also particularly important to measure and we have high expectations such measurements will be made. Finally, the solar dynamo is the only stellar dynamo studied in detail. Observations of activity patterns with time on other stars should be vigorously pursued, as well as stellar rotation and hopefully, in at least a few cases, differential rotation measured for the same stars. This will allow a lexicon of stellar dynamo types to be built up, against which solar-stellar dynamo models of the future can be compared.

On the theoretical side, it appears we are on the verge of being able to do rather detailed two- and three-dimensional nonlinear fluid dynamical and MHD calculations with moderately realistic physics for solar application over a wide range of spatial scales, from granulation and flux tubes to global circulation and the dynamo. We judge that the most serious problem to be faced in most of these calculations is how to represent and calculate the role played by turbulent processes, both on the spatial scales explicitly calculated, and scales too small to resolve. A variety of approaches are needed, and practitioners need to draw more on fundamental developments in turbulence theory then they have to date. In many cases, detailed calculations of simpler turbulent systems would be useful to develop more reasonable representations of eddy transport coefficients (tensors varying in latitude and radius), as well as helicity or α-effects. Nonlinear calculations of dynamical interactions of isolated flux tubes with the surrounding plasma would also be particularly helpful.

Some problems may not require a large input from turbulence theory, even though they are nonlinear. Particularly ripe for attack seems to us to be nonlinear, but steady and perhaps axisymmetric, dynamical models of sunspots, as an outgrowth of the theory of convection of a compressible fluid in a reacting magnetic field. Such calculations should be done in parallel with, and as quantitative tests of, less elaborate conceptual models developed using more intuitive physical arguments. Much more can and should also be done on basic theory of dynamos driven by nonlinear convection in rotating spherical shells, particularly allowing for compressibility. But it is not clear how close such models can come to simulating the solar cycle without major improvements in the treatment of turbulence, particularly involving interactions of velocities and an intermittent magnetic field.

Acknowledgements

I wish to thank many people who helped me in this effort. Betsy Alves, Juanita Crane, and Bobbie Morse typed the manuscript, and Birdi Rogers xeroxed a mountain of reference material. Kathryn Strand, the HAO librarian, was of great help in searching for and assembling the references (originally more than 2500 titles in *Astronomy and Astrophysics Abstracts*). The editors chose three very good reviewers, Robert Howard, Hendrik Spruit, and Michael Stix, all of whom provided many useful criticisms and corrections. Any errors of fact or interpretation which remain are mine, not theirs. Finally, I thank the staff of *Astronomy and Astrophysics Abstracts* for providing the superb service they do. I had some occasion to use other abstracting services in preparation of this review, and found *Astronomy and Astrophysics Abstracts* by far the best source.

This work has been sponsored by the National Science Foundation.

References

Abarbanell, C. and Wöhl, H.: 1981, *Solar Phys.* 70, 197.
Aime, C., Martin, F., Grec, G., and Roddier, F.: 1979, *Astron. Astrophys.* 79 1.
Alfvén, H.: 1971, *Astrophys. J.* 168, 239.
Altrock, R. C. and Canfield, R. C.: 1972, *Solar Phys.* 23, 257.
Altrock, R. C. and Musman, S: 1976, *Astrophys. J.* 203, 533.
Ambroz, P., Bumba, V., Howard, R., and Sykora, J.: 1971, *Proc. IAU, Solar Magnetic Fields*, p. 696.
Babcock, H. D.: 1959, *Astrophys. J.* 130, 364.

Babcock, H. W.: 1961, *Astrophys. J.* **133**, 572.

Balthasar, H. and Wöhl, H.: 1981, *Astron. Astrophys.* **98**, 422.

Barnes, J. A., Sargent, H. H. III, and Tryon, P. V.: 1980, in R. O. Pepin, J. A. Eddy, and R. B. Merrill, (eds.). 'The Ancient Sun', *Geochim. Cosmochim. Acta Suppl.* **13**, 159.

Beckers, J. M.: 1968, *Solar Phys.* **5**, 309.

Beckers, J. M.: 1975, 'New Views of Sunspots', AFCRL Environmental Research Papers No. 499.

Beckers, J. M.: 1976, *Astrophys. J.* **203**, 739.

Beckers, J. M.: 1979, in *Plasma Instabilities in Astrophysics* Gordon and Breach, New York, p. 139.

Beckers, J. M.: 1979, *Proc. Workshop on Solar Rotation*, Osservatorio Astrofisico de Catana Pubblicazione No. 162, p. 166.

Beckers, J. M. and Brown, T. M.: 1979, *Oss. Mem. Oss. Astrofis.* Arcerti, Fasc. **10b**, 189.

Beckers, J. M. and Morrison, R. A.: 1970, *Solar Phys.* **14**, 280.

Beckers, J. M. and Nelson, G. D.: 1978, *Solar Phys.* **58**, 243.

Beckers, J. M. and Schröter, E. H.: 1969, *Solar Phys.* **10**, 384.

Beckers, J. M. and Schultz, R. B.: 1972, *Solar Phys.* **27**, 61.

Beckers, J. M. and Taylor, W. R.: 1980, *Solar Phys.* **68**, 41.

Belvedere, G., Godoli, G., Motta, S., Paterno, L., and Zappala, R. A.: 1976, *Solar Phys.* **46**, 23.

Belvedere, G., Godoli, G., Motta, S., Paterno, L., and Zappala, R. A.: 1977, *Astrophys. J.* **214**, L91.

Belvedere, G. and Paterno, L.: 1976, *Solar Phys.* **47**, 525.

Belvedere, G. and Paterno, L.: 1977, *Solar Phys.* **54**, 289.

Belvedere, G. and Paterno, L.: 1978, *Solar Phys.* **60**, 203.

Belvedere, G., Paterno, L., and Stix, M.: 1980a, *Geophys. Astrophys. Fluid Dyn.* **14**, 209.

Belvedere, G., Paterno, L., and Stix, M.: 1980b, *Astron. Astrophys.* **88**, 240.

Belvedere, G., Paterno, L., and Stix, M.: 1980c, *Astron. Astrophys.* **86**, 40.

Belvedere, G., Zappala, R. A., D'Arrigo, C., Motta, S., Pirronello, V., Godoli, G., and Paterno, L.: 1978, *Proc. Workshop on Solar Rotation*, Osservatorio Astrofisico de Catania Pubblicazione No. 162, p. 189.

Biermann, L.: 1941, *Vierteljahrsch. Astr. Gesells.* **76**, 194.

Biermann, L.: 1951, *Z. Astrophys.* **28**, 304.

Böhm-Vitense, E.: 1958, *Z. Astrophys.* **46**, 108.

Bones, J. and Maltby, P.: 1978, *Solar Phys.* **57**, 65.

Bray, R. J., Loughhead, R. E., and Tappere, E. J.: 1976, *Solar Phys.* **49**, 3.

Bray, R. J. and Loughhead, R. E.: 1964, *Sunspots*, Wiley, New York.

Bretherton, F. P. and Turner, J. S.: 1968, *J. Fluid Mech.* **32**, 449.

Brown, T. M. and Harrison, R. L.: 1980, *Astrophys. J.* **236**, L169.

Bruning, D. H.: 1981, *Astrophys. J.* **248**, 274.

Bumba, V. and Howard, R.: 1969, *Solar Phys.* **7**, 28.

Busse, F.: 1970, *Astrophys. J.* **159**, 629.

Busse, F.: 1973, *Astron. Astrophys.* **38**, 27.

Chandrasekhar, S.: 1961, *Hydrodynamic and Magnetohydrodynamic Instability*, Oxford.

Childress, S.: 1979, *Phys. Earth Pl. Int.* **20**, 172.

Christensen-Dalsgaard, J., Dziembowski, W., and Gough, D. O.: 1980, in *Nonradial and Nonlinear Stellar Pulsation*, Springer-Verlag New York; *Lecture Notes in Physics* **125**, 313.

Claverlie, A., Isaak, G. R., McLeod, C. P., Van der Raay, H. B., and Roca Cortes, T.: 1981, *Nature* **293**, 443.

Clark, A., Jr: 1979, *Solar Phys.* **62**, 305.

Cocke, W. J.: 1967, *Astrophys. J.* **150**, 1041.

Condon, J. J. and Schmidt, R. R.: 1970, *Solar Phys.* **42**, 529.

Cowling, T. G.: 1976, *Monthly Notices Roy. Astron. Soc.* **177**, 409.

Danielson, R. E.: 1961, *Astrophys. J.* **134**, 289.

Davis, J. M., Golub, L., and Kreiger, A. S.: 1977, *Astrophys. J.* **214**, L141.

Dearborn, D. S. P. and Blake, J. B.: 1980, *Astrophys. J.* **237**, 616.

Dearborn, D. S. P. and Newman, M. J.: 1978, *Science* **201**, 150.

Deubner, F. L.: 1972, *Solar Phys.* **22**, 263.

Deubner, F.-L., Ulrich, R. K., and Rhodes, E. J.: 1979, *Astron. Astrophys.* **72**, 177.

Dicke, R. H.: 1976a, *Phys. Rev. Lett.* **37**, 1240.
Dicke, R. H.: 1976b, *Solar Phys.* **47**, 475.
Dicke, R. H.: 1977, *Astrophys. J.* **218**, 547.
Dicke, R. H.: 1978, *Nature* **276**, 676.
Dicke, R. H. and Goldenberg, H. M.: 1967, *Phys. Rev. Lett.* **18**, 313.
Dravins, E., Lindgren, L., and Nordlund, A.: 1981, *Astron. Astrophys.* **96**, 345.
Dunham, D. W., Sofia, S., Fiala, A. D., Herald, D., and Muller, P. M.: 1980, *Science* **210**, 1243.
Dunn, R. B. and Zirker, J. B.: 1974, *Solar Phys.* **33**, 281.
Durney, B. R.: 1970, *Astrophys. J.* **161**, 1115.
Durney, B. R.: 1971, *Astrophys. J.* **163**, 353.
Durney, B. R.: 1981, *Astrophys. J.* **244**, 678.
Durney, B. R., Mihalas, D., and Robinson, R. D.: 1981, *Publ. Astron. Soc. Pacific.* **93**, 537.
Durney, B. R. and Robinson, R. D.: 1981, *Astrophys. J.* **253**, 290.
Durney, B. R. and Roxburgh, I. W.: 1971, *Solar Phys.* **16**, 3.
Durney, B. R. and Latour, J.: 1978, *Geophys. Astrophys. Fluid Dyn.* **9**, 241.
Durney, B. R. and Spruit, H. C.: 1979, *Astrophys. J.* **234**, 1067.
Duvall, T. L.: 1979, *Solar Phys.* **63**, 3.
Duvall, T. L.: 1980, *Solar Phys.* **66**, 213.
Eddy, J. A.: 1976, *Science* **192**, 1189.
Eddy, J. A. and Boornazian, A. A.: 1979, *Bull. Amer. Astron. Soc.* **11** (2), 437.
Eddy, J. A., Gilman, P. A., and Trotter, D. E.: 1976, *Solar Phys.* **46**, 3.
Eddy, J. A., Gilman, P. A., and Trotter, D. E.:1977, *Science* **198**, 824.
Eddy, J. A., Noyes, R. W., Wohlbach, J. G., and Boornazian, A. A.: 1978, *Bull. Amer. Astron. Soc.* **10**, 400.
Edmonds, F. N.: 1974, *Solar Phys.* **38**, 35.
Foukal, P.: 1972, *Astrophys. J.* **173**, 439.
Foukal, P.: 1976, *Astrophys. J.* **203**, L145.
Foukal, P.: 1979, *Astrophys. J.* **234**, 716.
Foukal, P.: 1980, in *Proc. Conf. on Sun and Climate, Toulouse*, p. 275.
Foukal, P., Duvall, T., Jr, and Gillespie, B.: 1981, *Astrophys. J.* **249**, 394.
Foukal, P. and Jokipii, J. R.: 1975, *Astrophys J.* **199**, L71.
Foukal, P., Fowler, L., and Livshits, M.: 1983, *Astrophys. J.* **267**, 863.
Foukal, P. V., Mack, P. E., and Vernazza, J. E.: 1977, *Astrophys. J.* **215**, 952.
Foukal, P. and Vernazza, J.: 1979, *Astrophys. J.* **234**, 707.
Frazier, E. N. and Stenflo, J. O.: 1972, *Solar Phys.* **27**, 330.
Frazier, E. N. and Stenflo, J. O.: 1978, *Astron. Astrophys.* **70**, 789.
Frisch, U., Pouguet, A., Leorat, J., and Mazure, A.: 1975, *J. Fluid Mech.* **68**, 769.
Frisch. U.: 1977, 'Problems of Stellar Convection', *IAU Coll.* **38**, 325.
Galloway, D. J.: 1975, *Solar Phys.* **44**, 409.
Galloway, D. J.: 1978, *Monthly Notices Astron. Soc.* **184**.
Galloway, D. J. and Moore, D. R.: 1979, *Geophys. Astrophys. Fluid Dyn.* **12**, 73.
Galloway, D. J., Proctor, M. R. E., and Weiss, N. O.: 1977, *Nature* **266**, 686.
Galloway, D. J., Proctor, M. R. E., and Weiss, N. O.: 1978, *J. Fluid Mech.* **87**, 243.
Galloway, D. J. and Weiss, N. O.: 1981, *Astrophys. J.* **243**, 945.
Gilliland, R. L.: 1980, *Nature* **286**, 838.
Gilliland, R.: 1981, *Astrophys. J.* **248**, 1144.
Gilliland, R. L.: 1982, *Astrophys. J.* **253**, 399.
Gilman, P. A.: 1972, *Solar Phys.* **27**, 3.
Gilman, P. A.: 1974, *Solar Phys.* **36**, 61.
Gilman, P. A.: 1975, *J. Atmos. Sci.* **32**, 1331.
Gilman, P. A.: 1976, 'Basic Mechanisms of Solar Activity', *IAU Symp.* **71**, 207.
Gilman, P. A.: 1977a, *Geophys. Astrophys. Fluid Dyn.* **8**, 93.
Gilman, P. A.: 1977b, *Coronal Holes and High Speed Wind Streams*, Colorado Assoc. Univ. Press, Boulder, Chap. 8.
Gilman, P. A.: 1978, *Geophys. Astrophys. Fluid Dyn.* **11**, 157.

Gilman, P. A.: 1979, *Astrophys. J.* **231**, 284.

Gilman, P. A.: 1980, *IAU Coll.* **51**, 19.

Gilman, P. A.: 1981, *Proc. 2nd Cambridge Workshop on Cool Stars, Stellar Systems, and the Sun* Smithsonian Astrophysical Observatory Special Report 392, p. 165.

Gilman, P. A. and Foukal, P.: 1979, *Astrophys. J.* **229**, 1179.

Gilman, P. A. and Glatzmaier, G. A.: 1980, *Astrophys. J.* **241**, 793.

Gilman, P. A. and Glatzmaier, G. A.: 1981, *Astrophys J. Suppl.* **45**, 335.

Gilman, P. A. and Miller, J.: 1981, *Astrophys. J. Suppl.* **46**, 211.

Giovanelli, R. G.: 1972, *Solar Phys.* **27**, 71.

Giovanelli, R. G., Livingston, W. C., and Harvey, J. W.: 1978, *Solar Phys.* **56**, 347.

Giovanelli, R. G. and Slaughter, C.: 1978, *Solar Phys.* **57**, 255.

Glackin, D. L.: 1975, *Solar Phys.* **43**, 317.

Glatzmaier, G. A. and Gilman, P. A.: 1981a, *Astrophys. J. Suppl.* **45**, 351.

Glatzmaier, G. A. and Gilman, P. A.: 1981b, *Astrophys. J. Suppl.* **45**, 381.

Glatzmaier, G. A. and Gilman, P. A.: 1981c, *Astrophys. J. Suppl.* **47**, 103.

Glatzmaier, G. A. and Gilman, P. A.: 1982, *Astrophys. J.* **256**, 316.

Gnevyshev, M. N.: 1977, *Solar Phys.* **51**, 175.

Golub, L.: 1980, *Phil. Trans. Roy. Soc. London. A* **297**, 595.

Golub, L., Kreiger, A. S., Harvey, J. W., and Vaiana, G. S.: 1977, *Solar Phys.* **53**, 111.

Golub, L. and Vaiana, G. S.: 1978, *Astrophys. J.* **219**, L55.

Golub, L. and Vaiana, G. S.: 1980, *Astrophys. J.* **235**, L119.

Gough, D. O.: 1969, *J. Atmos. Sci.* **26**, 448 .

Gough, D. O.: 1977a, *IAU Coll.* **36**, 3.

Gough, D. O.: 1977b, *IAU Coll.* **38**, 349

Gough, D. O.: 1977c, *IAU Coll.* **38**, 15.

Gough, D. O.: 1978, in G. Belvedere and L. Paterno (eds), *Proc. Catania Workshop on Solar Rotation*, Univ. of Catania, Italy.

Gough, D. O., Moore, D. R., Spiegel, E. A., and Weiss, N. O.: 1976, *Astrophys. J.* **206**, 536.

Gough, D. O. and Weiss, N. O.: 1976, *Monthly. Notices. Roy. Astron. Soc.* **176**, 589.

Graham, E.: 1975, *J. Fluid Mech.* **70**, 689.

Graham, E. and Moore, D. R.: 1978, *Monthly Notices Roy Astron. Soc.* **183**, 617.

Hart, A. B.: 1956, *Monthly Roy. Astron. Soc.* **116**, 38.

Harvey, J. W. and Hall, D.: 1975, *Bull. Amer. Astron. Soc.* **7**, 459.

Harvey, K. and Harvey, J.: 1973, *Solar Phys.* **28**, 61.

Harvey, K. L. and Martin, S. F.: 1973, *Solar Phys.* **32**, 389.

Harvey, J. W. and Sheeley, N. R.: 1979, *Space Sci. Rev.* **23**, 139.

Herr, R. B.: 1978, *Science* **202**. 1079.

Hill, H. A., Clayton, P. D., Patz, D. L., Healy, A. W., Stebbins, R. T., Oleson, J. R., and Zanoni, C. A.: 1974, *Phys. Rev. Lett.* **33**, 1497.

Hill, H. A., and Stebbins, R. T.: 1975, *Astrophys. J.* **200**, 471.

Hill, H. A., and Caudell, T. P.: 1979, *Monthly Notices Roy. Astron. Soc.* **186**, 327.

Howard, R.: 1972, *Solar Phys.* **24**, 123.

Howard, R.: 1974, *Solar Phys.* **38**, 283.

Howard, R.: 1976, *Astrophys. J.* **210**. L159.

Howard, R.: 1977, *Ann. Rev. Astron. Astrophys.* **15**, 153.

Howard, R.: 1978, *Rev. Geophys. Space Phys.* **16**, 721.

Howard, R.: 1979, *Astrophys. J.* **228**, L45.

Howard, R. and Harvey, J.: 1970, *Solar Phys.* **12**, 23.

Howard, R. and La Bonte, B. J.: 1980, *Astrophys. J.* **239**, 738, L33.

Howard, R. and La Bonte, B. J.: 1981, *Solar Phys.* **74**, 131.

Howard, R. and Stenflo, J. O.: 1972, *Solar Phys.* **22**, 330.

Howard, R. and Yoshimura, H.: 1976, *IAU Symp.* **71**, 19.

Hundhausen, A. J.: 1977, in *Coronal Holes and High Speed Wind Streams*, Colorado Assoc. Univ. Press, Boulder, p. 225.

Hundhausen, A. J., Hansen, R. T., and Hansen, S. F.: 1981, *J. Geophys. Res.* **86**, 2079.

Jones, C. A. and Moore, D. R.: 1979, *Geophys. Astrophys. Fluid Dyn.* **11**, 245.

Karvaguchi, I. and Kitai, R.: 1976, *Solar Phys.* **46**, 125.

Keil, S. L.: 1980, *Astrophys. J.* **237**, 1024.

Kippenhahn, R.: 1963, *Astrophys. J.* **137**, 664.

Kippenhahn, R.: 1966, *Proc. Symp. Solar Magnetic Fields and High Resolution Spectroscopy*, Atti Dei Conv. Tomo 4, 291.

Knobloch, E.: 1978, *Astrophys. J.* **225**, 1050.

Knobloch, E., Rosner, R., and Weiss, N. O.: 1981, *Monthly Notices Roy. Astron. Soc.* **197**, 45 p.

Köhler, H.: 1970, *Solar Phys.* **13**, 3.

Köhler, H.: 1973, *Astron. Astrophys.* **25**, 467.

Köhler, H.: 1974, *Solar Phys.* **34**, 11.

Kömle, N.: 1979, *Solar Phys.* **64**, 213.

Kraichnan, R. H.: 1976a, *J. Fluid Mech.* **75**, 657.

Kraichnan, R. H.: 1976b, *J. Fluid Mech.* **77**, 753.

Krause, F.: 1976, *IAU Symp.* **71**, 305.

Krause, F. and Rädler, K.-H.: 1980 *Mean Field Magnetohydrodynamics and Dymamo Theory*, Akad. Verlag, Berlin and Pergamon Press, Oxford.

Kuznetsov, V. D. and Syrovatskii, S. I.: 1979, *Sov. Astron.* **23**, 715.

La Bonte, B. J. and Howard, R.: 1981, *Solar Phys.* **73**, 3.

La Bonte, B. J., Howard, R., and Gilman, P. A.: 1981, *Astrophys. J.* **250**, 796.

Latour, J., Spiegel, E. A., Toomre, J., and Zahn, J.-P.: 1976, *Astrophys. J.* **207**, 233.

Latour, J., Toomre, J., and Zahn, J.-P.: 1981, *Astrophys. J.* **248**, 1081.

Layzer, D., Rosners, R., and Doyle, H. T.: 1979, *Astrophys. J.* **229**, 1126.

Leighton, R. B.: 1964, *Astrophys. J.* **140**, 1547.

Leighton, R. B.: 1969, *Astrophys. J.* **156**, 1.

Leighton, R. B.: Noyes, R. W., and Simon, G. W.: 1962, *Astrophys. J.* **135**, 474.

Leorat, J., Pouquet, A., and Frisch, U.: 1981, *J. Fluid Mech.* **104**, 419.

Levine, R. H.: 1977, *Solar Phys.* **54**, 327.

Levine, R. H.: 1979, *Solar Phys.* **62**, 277.

Levine, R. H., Attschuler, M. D., Harvey, J. W., and Jackson, B. V.: 1977, *Astrophys. J.* **215**, 636.

Livingston, W. and Duvall, T. L.: 1979, *Solar Phys.* **61**, 219.

Loughhead, R. E., Bray, R. J., and Tappere, E. J.: 1979, *Astron. Astrophys.* **79**, 128.

Low, B. C.: 1980, *Solar Phys.* **67**, 57.

Lubow, S. J., Rhodes, E. J. and Ulrich, R. K.: 1980, *Lecture Notes in Physics,* **125**, 300.

Macris, C. J.: 1979, *Astron. Astrophys.* **78**, 186.

Martin, S. F. and Harvey, K. L.: 1979, *Solar Phys.* **64**, 93.

Massaguer, J. M. and Zahn, J.-P.: 1980, *Astron. Astrophys.* **87**, 315.

Mattig, W.: 1980, *Astron. Astrophys.* **83**, 129.

Mattig, W. and Mehltretter, J. P.: 1968, *Proc. IAU Symp.* **35**, 187.

McIntosh, P.: 1979, 'World Data Center A for Solar-Terrestrial Physics', Rept UAG-70, NOAA, Boulder, Colorado.

Mehltretter, J. P.: 1974, *Solar Phys.* **38**, 43.

Mehltretter, J. P.: 1978, *Astron. Astrophys.* **62**, 311.

Meneguzzi, M., Frisch, U., and Pouquet, A.: 1981, *Phys. Rev. Lett.* **47**, 1060.

Meyer, F. and Schmidt, H. V.: 1968, *Z. angew. Math. Mech.* **45**, 218.

Meyer, F., Schmidt, H. V., Simon, G. W., and Weiss, N. O.: 1979, *Astron. Astrophys.* **76**. 35.

Meyer, F., Schmidt, H. V., Weiss, N. O., and Wilson, P. R.: 1974, *Monthly Notices Roy. Astron. Soc.* **169**, 35.

Meyer, F., Schmidt, H. V., and Weiss, N. O.: 1977, *Monthly Notices Roy. Astron. Soc.* **179**, 741.

Moffatt, H. K.: 1978, *Magnetic Field Generation in Electrically Conducting Fluids*, Cambridge Univ. Press, New York.

Moore, R. L.: 1973, *Solar Phys.* **30**, 403.

Musman, S.: 1972, *Solar Phys.* **26**, 290.

Nakano, T., Fukushima, T., Unno, W., and Kondo, M.: 1979, *Publ. Astron. Soc. Japan* **31**, 713.

Nelson, G. D.: 1978, *Solar Phys.* **60**, 5.

Nelson, G. D. and Musman, S.: 1977, *Astrophys. J.* **214**, 912.

Nelson, G. D. and Musman, S.: 1978, *Astrophys. J.* **222**, L69.

Newton, H. W. and Nunn, M. L.: 1951, *Monthly Notices Roy. Astron. Soc.* **111**, 413.

Nordlund, A.: 1974, *Astron. Astrophys.* **32**, 407.

Nordlund, A.: 1978, in A. Reiz and T. Andersen (eds.), *Astronomical Papers Dedicated to Bengt Ströngren*, Copenhagen Univ. Obs. **95**.

Nordlund, A.: 1980, *IAU Coll.* **51**, 17.

November, L. J., Toomre, J., Gebbie, K. B., and Simon, G. W.: 1979, *Astrophys. J.* **227**, 600.

November, L. J., Toomre, J., Gebbie, K. B., and Simon, G. W.: 1981, *Astrophys. J.* **245**, L123.

Nye, A. H. and Thomas, J. H.: 1974, *Solar Phys.* **38**, 399.

Nye, A. H. and Thomas, J. H.: 1976, *Astrophys. J.* **204**, 582.

Öpik, E. J.: 1950, *Monthly Notices Roy. Astron. Soc.* **110**, 559.

Pardon, L., Woden, S. P., and Schneeberger, T. J.: 1979, *Solar Phys.* **63**, 247.

Parker, E. N.: 1955a, *Astrophys. J.* **121**, 491.

Parker, E. N.: 1955b, *Astrophys. J.* **122**, 293.

Parker, E. N.: 1971, *Astrophys. J.* **168**, 239.

Parker, E. N.: 1974a, *Astrophys. J.* **189**, 563.

Parker, E. N.: 1974b, *Astrophys. J.* **190**, 429.

Parker, E. N.: 1974c, *Solar Phys.* **36**, 249.

Parker, E. N.: 1975a, *Astrophys. J.* **198**, 205.

Parker, E. N.: 1975b, *Solar Phys.* **40**, 275.

Parker, E. N.: 1976, *Astrophys. J.* **204**, 259.

Parker, E. N.: 1977, *Astrophys. J.* **215**, 370.

Parker, E. N.: 1978a, *Astrophys. J.* **221**, 368.

Parker, E. N.: 1978b, *Astrophys. J.* **222**, 357.

Parker, E. N.: 1979a, *Cosmical Magnetic Fields*, Oxford Univ. Press, New York.

Parker, E. N.: 1979b, *Astrophys. J.* **233**, 1005.

Parker, E. N.: 1979c, *Astrophys. J.* **231**, 270.

Parker, E. N.: 1979d, *Astrophys. J.* **231**, 250.

Parker, E. N.: 1979e, *Astrophys. J.* **230**, 905.

Parker, E. N.: 1979f, *Astrophys. J.* **234**, 333.

Parkinson, J. H., Morrison, L. V., and Stephenson, F. R.: 1980, *Nature* **288**, 548.

Peckover, R. S. and Weiss, N. O.: 1978, *Monthly Notices Roy. Astron. Soc.* **182**, 189.

Pepin, R. O., Eddy, J. A., and Merrill, R. B. (eds): 1980, 'The Ancient Sun', *Geochem. Cosmochim. Acta Suppl.* **13**.

Perez Garde, M., Vazquez, M., Schwan, H., and Wöhl, H.: 1981, *Astron. Astrophys.* **93**, 67.

Piddington, J. H.: 1978, *Astrophys. Space Sci.* **55**, 401.

Pouquet, A., Frisch, U. and Leorat, J.: 1976, *J. Fluid Mech.* **77**, 321.

Pouquet, A. and Patterson, G. S.: 1978, *J. Fluid Mech.* **85**, 305.

Proctor, M. R. E. and Galloway, D. J.: 1979, *J. Fluid Mech.* **90**, 273.

Rädler, K. H.: 1976, *IAU Symp.* **71**, 323.

Raisbeck, G. M. and Yiou, F.: 1980, 'The Ancient Sun', *Geochem. Cosmochim. Acta Suppl.* **13**, 185.

Ramsey, H. E., Schoolman, S. A., and Title, A. M.: 1977, *Astrophys. J.* **215**, L41.

Rhodes, E. J., Deubner, F.-L., and Ulrich R. K.: 1979, *Astrophys. J.* **227**, 629.

Rhodes, E. J., Ulrich, R. K., and Simon, G. W.: 1977, *Astrophys. J.* **218**, 901.

Ricort, G. and Aime, C.: 1979, *Astron. Astrophys.* **76**, 324.

Roberts, B.: 1976, *Astrophys. J.* **204**, 268.

Rüdiger, G.: 1980, *Geophys. Astrophys. Fluid Dyn.* **16**, 239.

Ruzmaikin, A. A.: 1981, *Comments on Astrophys.* **9**, 85.

Savage, B. C.: 1969, *Astrophys. J.* **156**, 707.

Schatten, K. H.: 1973, *Solar Phys.* **32**, 315.

Scherrer, P. H. and Wilcox, J. M.: 1980, *Astrophys. J.* **239**, L89.

Scherrer, P. H., Wilcox, J. M. and Svalgaard, L.: 1980, *Astrophys. J.* **241**, 811.

Schröter, E. H., Wöhl, H., Soltun, D, and Vazques, M.: 1978, *Solar Phys.* **60**, 181.

Schüssler, M.: 1977, *Astron. Astrophys.* **56**, 439.

Schüssler, M.: 1979, *Astron. Astrophys.* **71**, 79.
Schüssler, M.: 1980a, *Astron. Astrophys.* **89**, 26.
Schüssler, M.: 1980b, *Nature* **288**, 150.
Schüssler, M.: 1981, *Astron. Astrophys.* **94**, L17.
Shapiro, I. I.: 1980, *Science* **208**, 51.
Shaviv, G. and Salpeter: 1973, *Astrophys. J.* **184**, 191.
Sheeley, N. R., Jr: 1969, *Solar Phys.* **9**, 347.
Sheeley, N. R., Jr: 1972, *Solar Phys.* **25**, 98.
Sheeley, N. R., Jr: 1976, *J. Geophys. Res.* **81**, 3462.
Sheeley, N. R., Jr and Bhatnagar, A.: 1971, *Solar Phys.* **19**, 338.
Simon, G. W. and Leighton, R. B.: 1964, *Astrophys. J.* **140**, 1120.
Simon, G. W. and Weiss, N. O.: 1968, *Z. Astrophys.* **69**, 435.
Siscoe, G. l.: 1980, *Rev. Geophys. Space Phys.* **18**, 647.
Smithson, R. C.: 1973, *Solar Phys.* **29**, 357.
Smythe, C. M. and Eddy, J. A.: 1977, *Nature* **266**, 434.
Sofia, S., O'Keefe, J., Lesh, J. R., and Endahl, A. S.: 1979, *Science* **204**, 1306.
Spiegel, E. A. and Weiss, N. O.: 1980, *Nature* **287**, 616.
Spruit, H. C.: 1974, *Solar Phys.* **34**, 277.
Spruit, H. C.: 1976, *Solar Phys.* **50**, 269.
Spruit, H. C.: 1977a, Ph. D. Thesis, Utrecht.
Spruit, H. C.: 1977b, *Astron. Astrophys.* **55**, 151.
Spruit, H. C.: 1977c, *Solar Phys.* **55**, 3.
Spruit, H. C.: 1979, *Solar Phys.* **61**, 363.
Spruit, H. C.: 1981a, in S. D. Jordan, (ed.), *The Sun as a Star*, NASA-CNRS, p. 385.
Spruit, H. C.: 1981b, *Space Sci. Rev.* **28**, 435.
Spruit, H. C.: 1981c, Preprint of paper presented at Sunspot Workshop, Sacramento Peak Observatory.
Spruit, H. C. and Zweibel, E. G.: 1979, *Solar Phys.* **62**, 15.
Stenflo, J. O.: 1972, *Solar Phys.* **23**, 307.
Stenflo, J. O.: 1973, *Solar Phys.* **32**, 41.
Stenflo, J. O.: 1976 in *IAU Symp.* **71**, 69.
Stenflo, J. O.: 1977, *Astron. Astrophys.* **61**, 797.
Stenflo, J. O.: 1978, *Rep. Prog. Phys.* **41**, 865.
Stenholm, L. G. and Stenflo, J. O.: 1977, *Astron. Astrophys.* **58**, 273.
Stix, M.: 1971, *Astron Astrophys.* **13**, 203.
Stix, M.: 1974, *Astron. Astrophys.* **37**, 121.
Stix, M.: 1976a, *IAU Symp.* **71**, 367.
Stix, M.: 1976b, *Astron. Astrophys.* **47**, 243.
Stix, M.: 1977, *Astron. Astrophys.* **59**, 73.
Stix, M.: 1981a, *Solar Phys.* **74**, 79.
Stix, M.: 1981b, *Astron. Astrophys.* **93**, 339.
Straus, J. M., Blake, B., and Schramm, D. N.: 1976, *Astrophys. J.* **204**, 481.
Stuiver, M. and Grootes, P. M.: 1980, in 'The Ancient Sun', *Geochim. Cosmochimi. Acta Suppl.* **13**, 165.
Suess, S.: 1975, *Am. Inst. Aeron. Astron. J.* **13**, 443.
Svalgaard, L. and Wilcox, J. M.: 1978, *Ann. Rev. Astron. Astrophys.* **16**, 429.
Tarbell, T. D. and Title, A. M.: 1977, *Solar Phys.* **52**, 13.
Tarbell, T. D., Title, A. M., and Schoolman, S. A.: 1979, *Astrophys. J.* **229**, 387.
Thomas, J. H.: 1978, *Astrophys. J.* **225**, 275.
Thomas, J. H.: 1979, *Nature* **280**, 662.
Toomre, J. Zahn, J.-P., Latour, J., and Spiegel, E. A.: 1976, *Astrophys, J.* **207**, 545.
Tsinganos, K. C.: 1979, *Astrophys. J.* **231**, 260.
Ulrich, R. K.: 1970a, *Astrophys. Space Sci.* **7**, 71.
Ulrich, R. K.: 1970b, *Astrophys. Space Sci.* **7**, 183.
Ulrich, R. K.: 1976, *Astrophys. J.* **207**, 564.
Ulrich, R. K. and Rhodes, E. J.: 1977, *Astrosphys. J.* **218**, 521.

Ulrich, R. K. and Rhodes, E. J., Jr, and Deubner, F.-L.: 1979, *Astrophys. J.* **227**, 638.

Unno, W. and Ando, H.: 1979, *Geophys. Astrophys. Fluid Dyn.* **12**, 107.

Unno, W. and Ribes, E.: 1976, *Astrophys. J.* **208**, 222.

Van der Borght, R.: 1974, *Austral. J. Phys.* **27**, 481.

Van der Borght, R.: 1975, *Austral. J. Phys.* **28**, 437.

Van der Borght, R.: 1979, *Monthly Notices Roy. Astron. Soc.* **188**, 615.

Vauclair, S., Vauclair, G., Schatzman, E., and Michaud, G.: 1978, *Astrophys. J.* **223**, 567.

Vaughan, A. H.: 1980, *Publ. Astron. Soc. Pacific* **92**, 392.

Vaughan. A. H. and Preston, G. W.: 1980, *Publ. Astron. Soc. Pacific* **92**, 385.

Vaughan, A. H., Baliunas, S. L., Middelkoop, F., Hartmann, L. W., Mihalas, D., Noyes, R. W., and
    Preston, G. W.: 1981, *Astrophys. J.* **250**, 276.

Vrabec, D.: 1974, *IAU Symp.* **56**, 201.

Wagner, W. J.: 1975, *Astrophys. J.* **198**, L141.

Wagner, W. J. and Gilliam, L. B.: 1976, *Solar Phys.* **50**, 265.

Ward, F.: 1965, *Astrophys. J.* **141**, 534.

Ward, F.: 1966, *Astrophys. J.* **145**, 416.

Ward, F.: 1973, *Solar Phys.* **30**, 527.

Weart, S. R.: 1970, *Astrophys. J.* **162**, 987.

Webb, A. R. and Roberts, B.: 1978, *Solar Phys.* **59**, 249.

Weiss, N. O.: 1981a, *J. Fluid Mech.* **108**, 247.

Weiss, N. O.: 1981b, *J. Fluid Mech.* **108**, 273.

Weiss, N. O.: 1981c, *J. Geophys. Res.* **86**, 11689.

Weiss, N. O.: 1981d, in R. M. Bonnet (ed.), *Solar Phenomena in Stars and Stellar Systems*, Reidel,
    Boston, p. 449.

White, O. R. (ed): 1977, *The Solar Output and its Variation*, Colorado Assoc. Univ. Press, Boulder.

White, O. R. and Livingston, W. C.: 1981, *Astrophys. J.* **249**, 798.

Wiehr, E.: 1978, *Astron. Astrophys.* **69**, 279.

Wiehr, E.: 1979, *Astron. Astrophys.* **73**, L19.

Willson, R. C., Gulkis, S., Janssen, M., Hudson, H. S., and Chapman, G. A.: 1981, *Science* **211**, 700.

Wilson, O. C.: 1978, *Astrophys. J.* **226**, 379.

Wittmann, A.: 1979, in 'Small Scale Motions on the Sun', Proc. of a Colloquium held on the Occasion
    of the Change of the Name of the former Fraunhofer-Institut into 'Kiepenheuer-Institut Für
    Sonnenphysik', Frieburg Mitteilungen aus dem Kiepenheuer-Institut #179, 29.

Wittmann, A.: 1980, *Solar Phys.* **66**, 223.

Wittmann, A. and Schröter, E. H.: 1969, *Solar Phys.* **10**, 357.

Wolff, C. L.: 1974, *Astrophys. J.* **194**, 489.

Wolff, C. L.: 1976, *Astrophys. J.* **205**, 612.

Worden, S. P.: 1975, *Solar Phys.* **45**, 521.

Worden, S. P. and Simon, G. W.: 1976, *Solar Phys.* **46**, 73.

Yoshimura, H.: 1972, *Astrophys. J.* **178**, 863.

Yoshimura, H.: 1975a, *Astrophys. J. Suppl.* **29**, 467.

Yoshimura, H.: 1975b, *Astrophys. J.* **201**, 740.

Yoshimura, H.: 1976, *Solar Phys.* **47**, 581.

Yoshimura, H.: 1978a, *Astrophys. J.* **220**, 692.

Yoshimura, H.: 1978b, *Astrophys. J.* **226**, 706.

Yoshimura, H.: 1981, *Astrophys. J.* **247**, 1102.

Yoshimura, H. and Kato, S.: 1971, *Publ. Astron. Soc. Japan* **23**, 57.

Zirin, H. and Stein, A.: 1972, *Astrophys. J.* **178**, L85.

Zwaan, C.: 1978, *Solar Phys.* **60**, 213.

Zwaan, C.: 1981, Ch. II.D.1. in S. D. Jordan, (ed.), *The Sun as a Star*, NASA-CNRS publication
    (NASA SP-450) p. 163.

High Altitude Observatory,
National Center for Atmospheric Research,
Boulder, CO 80307,
U.S.A.

CHAPTER 6

SOLAR INTERNAL STRESSES: ROTATION AND MAGNETIC FIELDS

ROGER K. ULRICH

1. Introduction

The theory of rotating, self-gravitating objects is very old, going back to much names as Newton, Clairaut, and Maclaurin in the seventeenth and eighteenth centuries. Important applications to stars were made in the 1920s by von Zeipel (1924a, b), Eddington (1925, 1926, 1929), and Vogt (1925). Although the problems of rotating stars can be stated relatively easily, realistic solutions are enormously difficult to obtain. The study of the subject of rotating stars has been greatly aided by the recent publication of an excellent monograph by Tassoul (1978) which gives a very complete review of the field and makes an extensive introduction to this chapter unnecessary. Tassoul's book has been heavily drawn upon in the preparation of this chapter, those problems which bear directly on the theory of solar structure are emphasized.

Much interest in the problem of the rotation of the solar core was sparked by the experiment by Dicke and Goldenberg (1967, 1974) which yielded a measurement of 5×10^{-5} for the solar oblateness. This result has important implications for the theory of general relativity; hence, it generated a number of theoretical and observational papers. Dicke (1970a, b) has reviewed the theoretical work which closely followed the Dicke and Goldenberg paper. The oblateness problem will be discussed only briefly, and the reader is referred to the Dicke reviews for details.

The title of this chapter includes magnetic fields since both provide internal stress for the Sun if they are present with a large amplitude. Very little theoretical work has been done on magnetic fields in the deep solar interior so that the review of this topic will be brief. The book by Parker (1979) contains a chapter devoted to this topic; however, dipole-like fields of the sort studied by Cowling (1945) are not discussed. The long life of large-scale field models makes them of potential importance, and a brief account is given of the theory. The consequences of magnetic stresses for solar and stellar interior theory are similar to those due to rotation.

There are a number of reasons that solar and stellar internal stresses are important both to solar physics and to our understanding of stellar structure and evolution. The review by Kippenhahn and Thomas (1981) is of interest in this context. The rate of rotation of the solar interior can provide an important constraint of the theory of the origin of the solar system. It is difficult to account for the slowness of the solar surface rotation without some spin-down through the braking action of the solar wind. There is a possibility that the core of the Sun has not been spun down as much as the surface.

Peter A. Sturrock (ed.), Physics of the Sun, Vol. I, pp. 161–175.

The oblateness of the solar gravitational figure alters the perihelion advance of Mercury and can influence one of the critical tests of general relativity. These two points are discussed by Dicke (1970a). The meridional circulation currents possibly associated with rapid rotation could cause mixing. This point is discussed in some detail in Section 2. Dicke (1978a) has suggested that magnetic stresses could also cause circulation currents. Piddington (1971) has proposed that a core magnetic field could play a role in the solar cycle. Although this point of view is not universally accepted (Parker, 1979), it is a possibility which should not be overlooked. The transport of angular momentum out of a rapidly rotating core could also modify theories of the solar cycle. Several studies of the solar neutrino problem have suggested that either rapid rotation or a large magnetic field could reduce the predicted neutrino flux (see Chapter 17 by Newman in Volume III). A final topic is the possible role that internal rotation and magnetic fields could play in such late stages of stellar evolution as supernovae explosions and pulsar formation. Presumably, the progenitors of pulsars were once main sequence stars of rather ordinary characteristics. Fowler and Hoyle (1964) pointed out that rotation or magnetic fields could play a role in turning a core collapse into an outgoing wave. The proximity of the Sun permits searches for internal rotation and magnetic fields in ways not possible for other stars; hence, our best chance of testing the pulsar and supernova ideas may be in a solar context.

2. Theory

2.1. EDDINGTON–SWEET CIRCULATION

At the heart of the theory of rotating stars is von Zeipel's (1924a, b) theorem which states that a uniformly rotating star in hydrostatic equilibrium cannot also be in strict radiative equilibrium. The crucial property driving the circulation is the variation with position of the flux of energy crossing level surfaces. This flux is proportional to the local effective gravity. In a radiative zone of a star where the total flux across the level surface is independent of the value of the potential defining the surface, the variation of flux over the surface causes the flux divergence to be different from zero. Section 7.2 of Tassoul's book gives a good discussion of this result. The stellar matter must circulate in order to make up the local heat flux deficit or surplus. Mestel (1966) has given a discussion of the circulation pattern which is worth repeating.

In cylindrical coordinates ($\bar{\omega}$, z, ϕ) the effective potential Ψ is defined by:

$$\nabla^2 \Psi = 2\Omega^2 - 4\pi G\rho, \tag{1}$$

where Ω is the rotation angular frequency and other notation is standard. The radiative flux $\mathbf{F}$ is:

$$\mathbf{F} = -\frac{4}{3}\frac{acT}{\rho\kappa}\nabla T = -f(\Psi)\nabla\Psi, \tag{2}$$

where

$$f(\Psi) = \frac{4acT^3}{3\rho\kappa}\frac{dT}{d\Psi}. \tag{3}$$

In a model where the state variables are constant on a level surface (Ψ = const), $f(\Psi)$ is independent of $\bar{\omega}$ as indicated by Equations (2) and (3). The magnitude of the effective gravity g_{eff} is $|\nabla\Psi|$, so that $f(\Psi)$ is roughly inversely proportional to g_{eff} in a radiative zone without nuclear energy generation. Mestel shows that the energy equation can be written:

$$\rho A(\Psi)u \cdot \nabla\Psi = f'(\Psi)\left(g_{eff}^2 - \frac{\langle g_{eff}\rangle}{\langle g_{eff}^{-1}\rangle}\right),\tag{4}$$

where $\langle g_{eff}\rangle$ and $\langle g_{eff}^{-1}\rangle$ are the averages of g_{eff} and g_{eff}^{-1} over level surfaces, and

$$A(\Psi) = C_V\left[\frac{dT}{d\Psi} - (\gamma - 1)\frac{T}{\rho}\frac{d\rho}{d\Psi}\right].\tag{5}$$

$A(\Psi)$ is related to the difference between the actual and adiabatic temperature gradients. Mestel shows that the form of Equation (4) requires there to be a surface where the crossing velocity vanishes Hence, the circulation pattern consists of two distinct zones. Öpik (1951) and Gratton (1945) independently predicted this result.

If the expression for $A(\Psi)$ is not altered by rotation, as in the case of a slowly rotating body such as the Sun, we can use $dP/d\Psi = -P/(gH)$, where H is the unperturbed pressure scale height, and obtain

$$A(\Psi) = \frac{\rho T C_P}{P}(\nabla_{ad} - \nabla) \approx 2.5\,(\nabla_{ad} - \nabla),\tag{6}$$

where ∇_{ad} and ∇ are the adiabatic and true logarithmic temperature gradients with respect to pressure. This function is shown in Figure 1 for a standard solar model from Ulrich and Rhodes (1983). When $\epsilon_{nuc} = 0$ as in the solar envelope we can estimate the circulation velocity following Baker and Kippenhahn (1959) as:

$$v_r = \left(\rho A\,\frac{\partial\Psi}{\partial r}\right)^{-1}\left(a\,\frac{L\rho}{M_r}\chi + b\,\frac{L\bar\rho}{M_r}\chi\right)$$

$$\approx \frac{r^2 L\chi}{GM_r^2 A}\left(a + b\,\frac{\bar\rho}{\rho}\right),\tag{7}$$

where

$$\chi = \frac{\Omega^2}{4\pi G\bar\rho}.\tag{8}$$

$\bar\rho$ is the mean density of the Sun and a and b are functions of Ψ of order unity which depend on the rotation law.

For the surface rotation period of 25 days, $\chi = 7.2 \times 10^{-6}$ and, as is well known, the circulation velocity comes out to be 4×10^{-10} cm s^{-1} in the middle of the solar radiative interior which requires a time of over 10^{12} yr to produce any significant displacement.

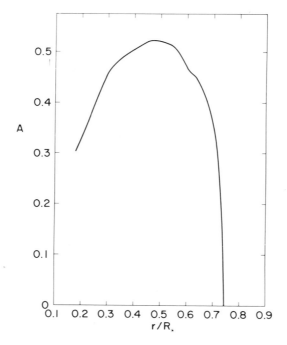

Fig. 1. The dimensionless stability parameter A which relates the circulation velocity to the local acceleration of gravity.

However, we should be cautious about dismissing meridional circulation as a potential cause of mixing because Ω could be a factor of 3 or more larger in the core than the surface and because of the $\overline{\rho}/\rho$ dependence in Equation (7) which has a value of 10 just below the convection zone. Thus, it is marginally possible for some meridional mixing to play a role in the light element depletion problem (see Strauss *et al.*, 1976, and the article by Press in this volume for a discussion). Consequently, it is worth presenting the two terms which can drive circulation by rewriting Equation (7) as:

$$v_r = (av_1 + bv_2) \left(\frac{\Omega}{\Omega_0}\right)^2,$$

(9)

where

$$v_1 = \frac{r^2 L \chi_0}{GM_r^2 A}, \quad v_2 = v_1 \frac{\overline{\rho}}{\rho},$$

χ_0 is 7.2×10^{-6} and $\Omega_0 = 2.9 \times 10^{-6}$ s^{-1}. The functions v_1 and v_2 are shown in Figure 2 along with the temperature which is included to indicate potential zones of light element burning. The velocity of 3×10^{-8} cm s^{-1} just below the convective interface permits mixing to a temperature of 2.3×10^6 K in 4.5 billion years even without any increase in Ω.

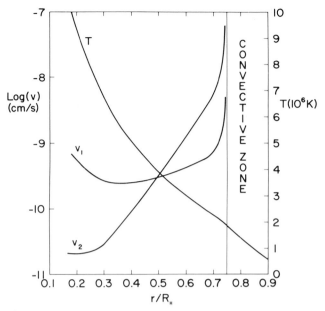

Fig. 2. Two terms in an expansion of the circulation velocity given by Baker and Kippenhahn (1959) evaluated for Ω = const. The actual circulation velocity is a linear combination of these two terms with coefficients of order unity. Present theory does not provide these coefficients.

The theory of meridional circulation I have discussed thus far is very crude. The detailed rotation law governs the dependence of the right-hand side of Equation (4) on both $\bar{\omega}$ and z or r. The flow in a radial direction implied by Equation (4), and in a more approximate way by Equation (9) and Figure 2, requires a flow in a meridional direction as well in order to conserve mass. A given parcel of matter in general must change its value of ω as a result of the flow satisfying Equation (4). Consequently, in order to conserve its angular momentum, the value of Ω following the parcel will be different at a fixed point in space as a result of the flow unless all parts of the star have the same specific angular momentum. Because of the assumed axial symmetry, there are no torques available to maintain $\Omega(z, \bar{\omega})$ at the initially prescribed value. Thus, meridional circulation implies a nonstationary state of the star. Randers (1941) initially pointed out this internal inconsistency in the Eddington–Sweet theory.

Recently, Busse (1982) has discussed some consequences of the above point. He shows that the change in $\Omega(z, \bar{\omega})$ implied by the advection of angular momentum in the circulation currents produces an imbalance in the centrifugal force which tends to restore the original position of the fluid element. Consequently, the fluid undergoes oscillations with a period comparable to the stellar rotation period about a position which circulates more slowly than would be implied by the Eddington–Sweet theory. Busse finds that the deviation of the equilibrium velocity U_e from the Eddington–Sweet velocity U_E is given approximately by:

$$U_e = U_E \left(1 + 4 \; \frac{\Omega^2 C_P}{g \, \nabla S}\right)^{-1}, \tag{10a}$$

where S is the entropy. Defining g_R and g_r to be the gravitational acceleration at the solar surface and at r, we can rewrite Equation (10a) as:

$$\frac{U_e - U_E}{U_e} = -12 \frac{R_\odot}{GM_\odot} \frac{C_P T}{A} \left(\frac{g_R}{g_r}\right)^2 \chi. \tag{10b}$$

Typical values of the coefficient of χ are 5.0 at $T = 3 \times 10^6$ K and 4.3 at $T = 5.7 \times 10^6$ K. Thus, the equilibrium velocity differs from the Eddington–Sweet velocity by an amount of order χ. However, this deviation produces a deficiency in the divergence of heat flux of order χ which is not balanced by advection. Consequently, the divergence in flux must decay on a Kelvin–Helmholtz time-scale divided by χ; i.e. on the Eddington–Sweet time-scale. Busse points out that a steady rotational state is possible only for special rotation laws as originally discussed by Schwarzschild (1947) and Roxburgh (1964). Vogt (1947) and Busse have shown that under restricted conditions such a rotation may have the form

$$\Omega^2 = g(\tilde{\omega}) \frac{RT}{\mu} , \tag{11}$$

where $g(\tilde{\omega})$ is an arbitrary function of $\tilde{\omega}$. Busse suggests that the decay of the divergence of flux is accompanied by a decay of the rotation law to a form like Equation (11). It is noteworthy that Equation (11) permits a variety of rotation laws including those with a rapidly rotating core. Also, because the time for decay to a rotation law given by Equation (11) is so long, the actual rotation law is dominated by either initial conditions or more rapid instabilities which are discussed below.

2.2. ROTATIONALLY DRIVEN INSTABILITIES

The rotation law must satisfy a sequence of stability tests in order to be an acceptable description of the interior of a middle-aged main sequence star like the Sun. The most rapid instabilities occur on a dynamical time scale and their absence requires as a necessary but possibly not sufficient condition that the rotation law satisfies the Høiland criterion (Høiland, 1941): "A baroclinic star in permanent rotation is dynamically stable with respect to axisymmetric motions if and only if the following two conditions are satisfied: (1) the entropy per unit mass S never decreases outward and (2) on each surface S = constant, the angular momentum per unit mass $\tilde{\omega}\Omega$ increases as we move from the poles to the equator."

The first part of this condition is simply the usual convective stability criterion. The restriction to axisymmetric perturbations is what prevents the Høiland criterion from being a sufficient condition. The time-scale for growth of disturbances driven by violation of the Høiland criterion can be very short. The geometry of the surfaces of constant S is nearly spherical in a subadiabatic region so that instabilities driven by violations of the Høiland criterion will not produce a modification to the rotation law averaged over a spherical shell.

Goldreich and Schubert (1967) have derived a pair of conditions for instability on a longer time-scale. Fricke (1968) independently derived the same conditions. The essential characteristic of the Goldreich—Schubert—Fricke instability is the exchange of mass between layers of different entropy. The different angular momentum content of the exchanged matter persists longer than the altered entropy content and provides a source of energy to drive a slow circulation. The speed of the motion is governed by the rate of the energy leakage by means of radiative transport. The instability is closely related to the 'salt-finger' instability found in the Earth's oceans where a salty river places a warm but high mean molecular weight layer above a cooler but lower mean molecular weight region (Stern, 1960). Although the possibility of motion is demonstrable from a linear analysis such as that used by Goldreich and Schubert (1967) and Fricke (1968), the effectiveness of the instability in redistributing the overall angular momentum of the Sun or star depends on nonlinear limiting of the motion through secondary instabilities. The problem is complicated by the competition between the tendency to favor small disturbances because they can come to thermal equilibrium most rapidly and the tendency to favor large disturbances because they will undergo nonlinear limiting after a larger displacement. Goldreich and Schubert (1967) did not impose any nonlinear limiting on very narrow disturbances and concluded that the mixing time is ten years. Fricke (1968) and Kippenhahn (1969) in contrast estimated the mixing time to be the Kelvin—Helmholtz time (t_{KH}) of 3×10^7 years. Since the theory of nonlinear limiting is imprecise, it has not been possible to develop a definitive theory of the efficiency of the Goldreich—Schubert—Fricke instability in mixing the Sun. Colgate (1968) used a Reynolds number hypothesis to estimate the nonlinear limiting and reached a conclusion similar to that of Fricke and Kippenhahn. James and Kahn (1970) argued that the mixing time is as long as the Eddington—Sweet circulation time t_{ES}. Kippenhahn et al. (1980) have studied the stream lines generated by explicit model perturbations and find that the identifiable perturbed region dissolves as soon as it moves a distance comparable to its own size. They then obtain a time-scale for a global angular velocity readjustment of $t_{ES}R/H_p$, where H_p is the pressure scale height. Although the nonlinear limiting arguments even in the case of this last paper are not rigorous, it seems most likely that the Goldreich—Schubert—Fricke instability is *not* able to spin down the solar core.

An important related mechanism for redistributing angular momentum in the solar interior is the process of Ekman pumping which results in the 'spin-down' of rotating fluids inside a more slowly rotating container. The 'spin-down' process is described in a hydrodynamic context by Prandtl (1952) and Greenspan (1968) and was introduced to solar physics by Howard et al. (1967) and Bretherton and Spiegel (1968). Spiegel (1972) gives a clear summary of the essential mechanism as applied to the Sun. Many features of the circulation which results from Ekman pumping are similar to meridional circulation, although the driving force is different. Ekman pumping occurs because two adjacent spherical shells of mass rotating at a different rate cannot both balance the centrifugal acceleration without there being a pressure gradient along a level surface. If, as we expect for the solar interior, the outer shell rotates more slowly than the inner shell, the pressure by the outer shell on the inner shell at the equator will exceed the pressure on the inner shell at the poles and matter will be forced to flow from the equator toward the poles. In hydrodynamic situations, the force imbalance is counteracted by viscous stresses in a boundary layer. For the solar case, Bretherton and Spiegel (1968) have suggested that the stresses occur in the convective envelope and that eddy viscosity dominates. Without the stable temperature gradient of the solar interior, the time-scale of the spin-down would be about one month. In fact, the stable stratification prevents the flow from penetrating the interior and sets up a thermal wind in which the Coriolis force deflects the horizontal flow into a circulation pattern where the deviation from the pre-existing pressure due to centrifugal acceleration is balanced by buoyancy forces due to the radial displacement of the matter. After this flow pattern is established, it can further evolve by radiative heat transport in much the same way as the meridional circulation currents. Spiegel (1972) estimates the time-scale for mixing due to spin-down to be

$$t_{SD} = (N^2/\Omega^2) \, t_{KH}, \tag{13}$$

where N is the Väisälä–Brunt gravity-wave frequency of the radiative region. At a distance of about 10^9 cm below the boundary of the solar convection zone where Spiegel indicates the spin-down currents should be deflected onto level surfaces, the value of N^2 is 7×10^{-7} s^{-2}. Thus, t_{SD} is comparable to t_{ES} and is too long to be relevant to the problem of mixing in the Sun unless Ω is substantially larger than 3×10^{-6} s^{-1}.

2.3. AN INTERNAL MAGNETIC FIELD

The possibility that the Sun could retain an internal magnetic field as a relic from the time of formation was first recognized by Cowling (1945). This idea has not been generally popular since, as stated by Cowling himself: "I do not feel altogether happy in making this suggestion, since the assumption that the magnetic field is explained by some unassigned cause operative a long time ago is, essentially, a confession of ignorance." Our understanding of the role magnetic fields play in the star formation process has improved since Cowling's time (see the review by Mouschovias, 1981); however, we do not have a way of constraining the possible relic field on the basis of these considerations. Parker (1974) has argued that a strong relic field would rise due to magnetic buoyancy and be lost from the solar interior in less than 10^8 years. His discussion applied to flux tubes rather than a large-scale field of the Cowling type. The possibility of a relic field in the solar core is probably still viable as long as the field has a large-scale order. There are two categories of effects which could be attributed to such a relic field: (a) it could serve as a seed field for the solar dynamo as suggested by Piddington (1971); and (b) it can modify the state of the solar interior through its associated magnetic stress. I shall confine this discussion to the second category since Gilman provides a review of the solar dynamo in Chapter 5 of this volume.

Cowling (1945) shows that the Equation for the ohmic decay of the relic electromagnetic field can be written

$$\nabla^2 \, \mathbf{E} = -\frac{4\pi\sigma}{\tau} \, \mathbf{E} \tag{14}$$

with the magnetic field given by

$$\mathbf{B} = \tau \nabla \times \mathbf{E}, \tag{15}$$

where the time dependence of both $\mathbf{E}$ and $\mathbf{B}$ is assumed to have the form $\exp(-t/\tau)$ and σ is the conductivity given by Spitzer (1962). The vector Equation (Equation (14)) can have either a poloidal or a toroidal solution. In either case, the decay time τ can have only a discrete set of eigenvalues which depend on the outer boundary condition. Cowling (1945) studied the slowest decaying mode with a poloidal **B** field which joins onto an external dipole. Bahcall and Ulrich (1971) found that for this mode $\tau = 2.5 \times 10^{10}$ yr. Dicke (1978b) discussed the solution to Equation (14) which involve both poloidal and toroidal components. He subsequently argued (Dicke, 1978a, 1982) that differential rotation will cause a poloidal core field to tip over and wind up a toroidal field. Wrubel (1952) calculated the structure of several overtone poloidal configurations which he found to have τ values of 0.4 and 0.24 times that of the longest-lived mode. Consequently, a poloidal field can have a persistent structure more complicated than a pure Cowling mode.

The currents implied by the **E** field interact with **B** to produce a magnetic force. $(\nabla \times \mathbf{B}) \times \mathbf{B}/4\pi$. The equation of hydrostatic equilibrium is modified by the addition of this force to become:

$$\nabla P + \rho \nabla \Phi - \frac{1}{4\pi} (\nabla \times \mathbf{B}) \times \mathbf{B} = 0. \tag{16}$$

where Φ is the gravitational potential and P is the pressure. Equation (16) combined with the Poisson Equation for Φ, determine the oblateness or higher order distortion of the solar surface for specified magnetic field configurations. Ulrich and Rhodes (1983) used a surface averaged value of the force term to study the effect of a Cowling field on the rate of neutrino emission. In contrast to the work by Bahcall and Ulrich (1971), Ulrich and Rhodes (1983) found that the neutrino flux decreases when the Sun possesses a Cowling-type relic field. The strength of field required to alter the neutrino flux is of order 3×10^8 gauss and is probably incompatible with the observed solar oblateness. Dicke (1982) finds that a toroidal field with a strength of order 10^7 gauss produces the degree of surface distortion required by his observations (Dicke, 1981). No calculation is yet available which gives the relationship between oblateness and the strength of a Cowling field, although Bahcall and Ulrich (1971) estimated that field must be of order 10^7 gauss to be compatible with the observed oblateness.

Ulrich and Rhodes (1983) have estimated the effect of a Cowling relic field on the frequencies of global solar oscillations. They used an approximate formulation which adds in an r.m.s. fashion the surface averaged Alfvén velocity to the sound velocity to obtain the modified acoustic wave phase velocity. They found that a central field strength must be of order 3×10^8 gauss to produce a significant frequency shift. The frequencies for modes with differing values of m, the azimuthal eigennumber, will be split by an amount comparable to the shift. The observed spectrum is not consistent with a field large enough to shift the theoretical frequencies into agreement with the observed frequencies (Grec et al., 1980). A full analysis of the magnetohydrodynamic equation of motion will be desirable to verify these conclusions.

3. Observations

Astronomy in general is a field where we are forced to rely on passive observation of remote objects and events. The solar interior is not only remote but is also completely obscured by the opaque outer layers. Only solar neutrinos, global solar oscillations, and the solar gravitational figure provide ways of obtaining information directly about the present solar interior. The neutrino fluxes are more sensitive to the deep internal thermal structure rather than the outer stress structure focused on in this review. The oscillations and gravitational figure are capable of providing precise diagnostics for the stress state of the solar interior but present observations are not definitive. The Dicke and Goldenberg (1967) result has been challenged on several observational grounds (see Dicke, 1974, for a review) and contradicted by the Hill and Stebbins (1975) observations which use a finite Fourier transform definition of the solar edge. Dicke (1976) discovered a 12-day periodicity in the solar figure as determined during the 1966 observations. Dicke (1981) has interpreted this periodicity as a rotating distortion which he attributes in his 1982 article (Dicke, 1982) to an off-axis toroidal magnetic field. Further observations by Dicke and his collaborators would be most helpful in confirming the existence of the rotation distortion. The direct measurement of the J_2 component of the solar gravitational figure with an accurately tracked space probe which closely approaches the Sun, as described by Reasenberg et al. (1982), would provide a definitive diagnostic of the global nonspherical component of the solar internal stress.

The global solar oscillations can measure frequency splitting caused by rotation, as was first noted by Ledoux (1951). A related method was applied by Deubner et al. (1979) to the measurement of the velocity gradient with depth near the solar surface using the highly nonradial oscillations in the 5 min band. The severely limited depth probed by these measurements makes it impossible to place any significant constraints on the solar internal stress. Claverie et al. (1981) have measured the splitting of modes described by the spherical harmonics $Y_l^m(\theta, \phi)$ with $l = 1$. They concluded that the solar core has an average rotation rate twice the surface rate if the rapid rotation region extends over most of the Sun. Their conclusion is not firm because they had to superimpose several different modes to increase the signal-to-noise ratio and the registration of different modes could cause spurious effects in the summed spectrum. Also, the mode with $m = 0$ moves predominantly perpendicular to the line of sight and should not have contributed a line to the observed triplet. Taking note of this last problem, Isaak (1982) has proposed that the splitting observed by his group is a consequence of a core magnetic field. Using the technique developed to measure the solar oblateness, Hill et al. (1982) have observed oscillations in the position of the solar limb at six different points around the solar circumference. Bos and Hill (1983) have found that the oscillations are coherent at the six points and possess symmetry properties expected for global solar oscillations. The observed power spectrum is extremely rich, making the identification of modes difficult. Hill et al. (1982) searched through the spectrum for equally spaced peaks which they then attribute to rotationally split components of an eigenmode described by single values of l and n but multiple values of m. They determined the value of l from the number of equally spaced peaks. There is some risk that the value of l will be underestimated using this method, since not all values of m need be excited. There is also a possibility that the identification of the equally spaced peaks is spurious since there is a large number of speaks altogether and most are not identified. Given the tentative identifications of the modes and the rotational splitting, it is possible to estimate the interior rotation law. Hill et al. (1982) find that their observations are best fit by a rotation law which rises from $\Omega = 4 \times 10^{-7}$ s^{-1} at the solar surface to 3×10^{-6}

in the inner 20% of the solar radius. Some range in this result is undoubtedly possible due to the problems of mode identification and the small number of modes (seven) used in the inversion of the frequency splitting—rotation rate integral equation.

In addition to the above direct methods for observing the solar interior stress, we can fall back on inferential methods as astronomers are frequently obliged to use. In the case of the solar interior, young stars provide an indication of the rotational state that may have existed at the time of the Sun's formation. The evolution following the arrival of the Sun on the main sequence probably involved loss of angular momentum through the solar wind, as was originally proposed by Schatzman (1962). Kraft (1967) has measured rotational velocities in the Pleiades and Hyades and for field stars and finds that the average velocity decreases from 39 $km\,s^{-1}$ for the Pleiades to 18.5 $km\,s^{-1}$ for the Hyades and 6 $km\,s^{-1}$ for field stars. He assigns ages of 3×10^7, 4×10^8, and 5×10^8 to 5×10^9 yr, respectively, for these three groups. Skumanich (1972) points out that Kraft's result as well as Ca II indications of stellar activity are consistent with decay laws where rotation and activity are proportional to $(age)^{-1/2}$. These results are generally in agreement with Schatzman's model. The degree of coupling between the convection zone and the radiative interior is critical to the interpretation of the rotation rate versus age observations. Schatzman's model assumed that stars rotate as a rigid body, whereas the summary of theory in the preceding section indicates that coupling by hydrodynamic instabilities occurs on too slow a time-scale to provide the coupling.

Kraft (1970) has given a summary figure which presents the rotation problem nicely. He shows the angular momentum per unit mass of old main sequence stars as a function of total mass, assuming the stars rotate rigidly. He used the rotation velocities from Boyarchuk and Kopylov (1964) as analyzed by Abt and Hunter (1962) to calculate the average angular momentum per unit mass J for field main sequence stars as a function of mass. The faintness of cooler main sequence stars limits the range of masses for which Kraft's relation is reliable to masses slightly greater than a solar mass. I have reproduced part of Kraft's (1970) Figure 3 as Figure 3 of this paper. The line marked Kraft main sequence is his specific angular momentum curve. Figure 3 also shows the specific angular momentum for pre-main sequence stars taken from Vogel and Kuhi (1981). These cluster around the main sequence curve and do not suggest a major loss of angular momentum. The points labeled P and H are the average results for the Pleiades and Hyades from Kraft (1967). They are in the range of the rotation rates for the pre-main sequence stars but show a decrease with age. Finally, I show three solar system points; the largest value point includes the planets, the mid-value point for the Hill *et al.* (1982) rotation law, and the lowest value point for rigid rotation. Not even the pre-main sequence stars seem to have a high enough specific angular momentum to account for the planets while the Hill *et al.* (1982) rotation curve is consistent with the young stars. The newer results are supportive of the Schatzman picture of solar wind spin-down of the envelope with a radiative core somewhat decoupled from the envelope.

4. Summary

Perhaps the most remarkable aspect of studying the state of the solar internal stress field is the number of opportunities we have to measure these fields directly. Without

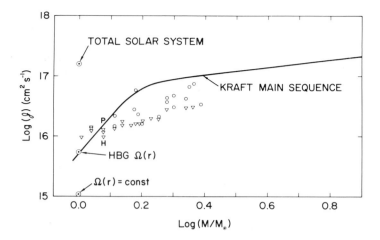

Fig. 3. The specific angular momentum per unit mass J as a function of stellar mass for the Kraft main sequence (see the text for the origin of this curve) and pre-main sequence objects observed by Vogel and Kuhi (1981). The triangles represent upper limits to the rotation velocity and the circles represent measurements of the velocity. The points designated by P and H are averages for the Pleiades and Hyades determined by Kraft (1967). The point marked $\Omega(r)$ = const is the result for a rigidly rotating Sun and the point marked HBG $\Omega(r)$ is the result for the Sun using the Hill, Bos, and Goode (1981) rotation law. The point marked total solar system includes the planets.

knowledge of the new techniques for measuring the solar figure and deducing internal structure through the global oscillations, one might think the study of solar internal stress fields would be an entirely theoretical topic. It is also clear that as an experimental science, the study of the solar internal stress is quite immature and an area for substantial future growth. The discrepancy between the rotating nonspherical figure discussed by Dicke (1981) and the small oblateness measured by Hill and Stebbins (1975) needs to be resolved. Hopefully, new observations by the Princeton group could confirm or deny the rotating figure model. The Arizona group could also look for the signal of the rotating figure in their data. It is conceivable that the amplitude of the integrated anisotropic stress field is variable due to some long-term beat phenomenon. Also, the measurement of the shape of the apparent solar disk will always require some assumptions about the structure of the solar atmosphere before the shape of the gravitational potential field can be deduced. Dicke (1970a) has discussed this problem at length and concludes that the atmospheric effects are small. Specifically, the effect of the surface magnetic field and meridional circulation currents were eliminated as possible contributors to the oblateness signal on the grounds that they would also have altered the limb brightness in a detectable way. The complexity of the argument is great enough that the inhomogeneity and dynamic state of the upper solar atmosphere could invalidate the conclusion. The ultimate test of the shape of Ψ will come through the accurate tracking of a space probe which approaches near the Sun.

   The deduction of the internal stress of the Sun through the study of global oscillations is still quite far from being a reliable method. The lowest degree modes as observed by Claverie *et al* (1981) and Grec *et al*. (1980) have been identified because the full disk observations filter out all but a few modes. Although the single-to-noise ratio for these observations is very high, it is not high enough to permit the clear observation of the frequency splitting due to internal stress fields in individual modes. The method of folding in frequency space used by these authors introduces some uncertainty due to the unknown and possibly variable frequency spacing. The observations reported by Hill *et al.* (1982) involve a technique which does not filter the modes as selectively as the full disk observations. Consequently, the spectrum is very rich and the mode identification is difficult. Also, there remain unresolved questions about the insensitivity of the method to oscillations in the 5 min band. The application of alternate outer boundary conditions as advocated by Hill and Caudel (1979) is justified only by appealing to our ignorance of the exact state of the upper solar atmosphere. A model based on inhomogeneities, non-LTE or magnetic fields — complications known to exist in the upper atmosphere — which showed the properties needed to select against the 5 min oscillations would be much more satisfying.

An extension of the spatially resolved observations to the lower frequency or to the lower degree modes could provide important confirmation of the Hill *et al.* (1982) observations. The study of internal solar stress fields by means of global oscillations places severe constraints on the observations: (1) the duration of the observations must be two to six weeks in order to provide adequate frequency resolution; (2) the duty cycle probably must be near 60% to permit the cleaning of side lobes of the solar frequencies; and (3) the detection of a systematic variation like that expected from the Dicke (1982) magnetic rotator or from solar cycle related effects will require the observations to be repeated regularly. Point (2) has not yet been studied in adequate detail to establish a definite lower limit to the duty cycle; however, reference to the spectrum published by Bos and Hill (1983) suggests 'CLEAN'ing techniques such as described by Schwarz (1978) might be hard to apply. This procedure used in radio astronomy to remove unwanted sidelobes of a dirty beam requires the assumption that the true points are separated by regions where the power is essentially zero. The richness of the solar oscillation spectrum makes this a doubtful assumption.

References

Abt, H. and Hunter, J.: 1962, *Astrophys. J.* **136**, 381.
Bahcall, J. N. and Ulrich, R. K.: 1971, *Astrophys. J.* **170**, 593.
Baker, N. and Kippenhahn, R.: 1959, *Z. Astrophys.* **48**, 140.
Bos, R. J. and Hill, H. A.: 1983, *Proc. IAU Coll.* 66 Solar Physics, 82, 89.
Boyarchuk, A. and Kopylov, I.: 1964, *Bull. Crimean Astrophys. Obs.* **31**, 44.
Bretherton, F. P. and Spiegel, E.: 1968, *Astrophys. J.* **153**, L77.
Busse, F. H.: 1982, *Astrophys. J.* **259**, 759.
Claverie, A., Isaak, G. R., McLeod, C. P., Van der Raag, H. B., and Roca Cortes, J.: 1981, *Nature* **293**, 443.
Colgate, S. A.: 1968, *Astrophys. J.* **153**, L81.
Cowling T. G.: 1945, *Monthly Notices Roy. Astrophys. Soc.* **105**, 166.
Deubner, F.-L., Ulrich, R. K., and Rhodes, E. J., Jr: 1979, *Astron. Astrophys.* **72**, 177.
Dicke, R. H.: 1970a, *Ann. Rev. Astron. Astrophys.* **8**, 197.
Dicke, R. H.: 1970b, in A. Slettebak (ed.), 'Stellar Rotation', *IAU Coll.* 4, 289. *Coll.* 4 289.
Dicke, R. H.: 1974, *Science* **184**, 419.
Dicke, R. H.: 1976, *Solar Physics* **47**, 475.

174   ROGER K. ULRICH

Dicke, R. H.: 1978a, in G. Friedlander (ed.), *Proc. Informal Conf. on the Status and Future of Solar Neutrino Research*, Vol. II, Brookhaven National Lab., Upton, New York, BNL Rept 50879, p. 109.
Dicke, R. H.: 1978b, *Astrophys. Space Sci.* **55**, 275.
Dicke, R. H.: 1981, *Proc. Nat. Acad. Sci. USA* **78**, 1309.
Dicke, R. H.: 1982, *Solar Phys.* **78**, 3.
Dicke, R. H. and Goldenberg, H. M.: 1967, *Phys. Rev. Lett.* **18**, 313.
Dicke, R. H. and Goldenberg, H. M.: 1974, *Astrophys. J. Suppl.* **27**, 131.
Eddington, A. S.: 1925, *The Observatory* **48**, 73.
Eddington, A. S.: 1926, *The Internal Constitution of Stars*, Dover, New York, p. 282.
Eddington, A. S.: 1929, *Monthly Notices Roy. Astrophys. Soc.* **90**, 54.
Fowler, W. A. and Hoyle, F.: 1964, *Astrophys. J. Suppl.* **9**, 201.
Fricke, K.: 1968, *Z. Astrophys.* **68**, 317.
Goldreich, P. and Schubert, G.: 1967, *Astrophys. J.* **150**, 571.
Gratton, L.: 1945, *Mem. Soc. Astron. Ital.* **17**, 5.
Grec, G., Fossat, E., and Pomerantz, M.: 1980, *Nature* **288**, 541.
Greenspan, H. P.: 1968, *The Theory of Rotating Fluids*, Cambridge Univ. Press.
Hill, H. A., Bos, R. J., and Goode, P. R.: 1982, Preprint submitted to *Phys. Rev. Lett.*
Hill, H. A. and Caudell, T. P.: 1979, *Monthly Notices Roy. Astrophys. Soc.* **186**, 327.
Hill, H. A. and Stebbins, R. J.: 1975, *Astrophys. J.* **200**, 471.
Høiland, E.: 1941, 'Avhandliger Norske Videnskaya – Akademi i Oslo, I', *Math.-Naturv. Klasse* **11**, p. 1.
Howard, L. N., Moore, D. W., and Spiegel, E.: 1967, *Nature* **214**, 1297.
Isaak, G. R.: 1982, *Nature* **296**, 130.
James, R. A. and Kahn, F. D.: 1970, *Astron. Astrophys.* **5**, 232.
Kippenhahn, R.: 1969, *Astron. Astrophys.* **2**, 309.
Kippenhahn, R., Ruschenplatt, G., and Thomas, H.-C.: 1980, *Astron. Astrophys.* **91**, 181.
Kippenhahn, R. and Thomas, H.-C.: 1981, in D. Sugimota, D. I. Lamb, and D. N. Schramm (eds.), *Fundamental Problems in the Theory of Stellar Evolution*, Reidel, Dordrecht, p. 237.
Kraft, R. P.: 1967, *Astrophys. J.* **150**, 551.
Kraft, R. P.: 1970, 'Stellar Rotation', in G. H. Herbig (ed.), *Spectroscopic Astrophysics*, Univ. of California Press, Berkeley, p. 385.
Ledoux, P.: 1951, *Astrophys. J.* **114**, 373.
Mestel, L.: 1966, *Z. Astrophys.*, **63**, 196.
Mouschovias, T. Ch.: 1981, in D. Sugimoto, D. I. Lamb, and D. N. Schramm (eds.), *Fundamental Problems in the Theory of Stellar Evolution*, Reidel, Dordrecht, p. 27.
Öpik, E. J.: 1951, *Monthly Notices Roy. Astrophys. Soc.* **111**, 278.
Parker, E. N.: 1974, *Astrophys. Space Sci.* **31**, 261.
Parker, E. N.: 1979, *Cosmical Magnetic Fields*, Clarendon Press, Oxford.
Piddington, J. H.: 1971, *Proc. Astron. Soc. Austral.* **2**, 7.
Prandtl, L. J.: 1952, *Fluid Mechanics*, Blackie & Sons, London.
Randers, G.: 1941, *Astrophys. J.* **94**, 109.
Reasenberg, R. D., Anderson, J. D., De Bra, D. B., Shapiro, I. I., Ulrich, R. K., and Vesoot, R. F. C.: 1982, in J. H. Underwood and J. E. Randolph (eds.), 'Starprobe Scientific Rationale', *JPL Publ. 82–49*, JPL, Pasadena, Chap. 1.
Roxburgh, I. W.: 1964, *Monthly Notices Royal Astrophys. Soc.* **128**, 157.
Schatzman, E.: 1962, *Ann. d'Astrophys.* **25**, 18.
Schwarz, U. J.: 1978, *Astron. Astrophys.* **65**, 345.
Schwarzschild, M.: 1947, *Astrophys. J.* **106**, 427.
Skumanich, A.: 1972, *Astrophys. J.* **171**, 565.
Spiegel, E.: 1972, in S. I. Rasool (ed.), *Physics of the Solar System*, NASA SP-300, Washington, D. C., p. 61.
Spitzer, L., Jr: 1962, *Physics of Fully Ionized Gases*, Interscience, New York.
Stern, M. E.: 1960, *Tellus* **12**, 172.
Straus, J. M., Blake, J. B., and Schramm, D. N.: 1976, *Astrophys. J.* **204**, 481.

Tassoul, J.-L.: 1978, *Theory of Rotating Stars*, Princeton Univ. Press, Princeton.

Ulrich, R. K. and Rhodes, E. J., Jr: 1983, *Astrophys. J.* **265**, 551.

Wrubel, M. H.: 1952, *Astrophys. J.* **116**, 291.

Vogel, S. N. and Kuhi, L. V.: 1981, *Astrophys. J.* **245**, 960.

Vogt, H. 1925, *Astron. Nach.* **223**, 229.

Vogt, H. 1947, *Astron. Nach.* **277**, 49.

von Zeipel, H.: 1924a, *Probleme der Astronomie* (Festschrift für H. von Seeligh), Springer-Verlag, Berlin, p. 144.

von Zeipel, H.: 1924b, *Monthly Notices Roy. Astrophys. Soc.* **84**, 665, 684.

Department of Astronomy,
University of California,
Los Angeles, CA 90024,
U.S.A.

CHAPTER 7

SOLAR WAVES AND OSCILLATIONS

T. M. BROWN,* B. W. MIHALAS,* and E. J. RHODES, JR.

Introduction

Study of the generation and propagation of waves in the solar convection zone and atmosphere began with the realization in the late 1940s that such waves could carry mechanical energy from the turbulent convection zone into the chromosphere and corona. Dissipation of these waves could then produce the observed increase in temperature in the outer atmosphere of the Sun, and could explain much, if not all, of the observed nonthermal broadening of solar spectrum lines. Such propagating waves are difficult to observe, largely because they are transient and of small spatial scale, and one is often reduced to inferring their presence from a combination of indirect observations together with a chain of theoretical arguments. Further difficulties arise in both theory and observation because the waves become highly nonlinear in the chromosphere and corona and experience strong nonadiabatic effects in and just below the photosphere. These effects are extremely difficult to model accurately, hence the waveforms are largely unknown.

After three decades of study, the role of waves in atmospheric heating is still uncertain, as is that of the other major contender, magnetic field-related processes. On the other hand, the knowledge we have gained of the structure of the atmosphere has alerted us to the possibility of chomospheres and coronae on other stars. Observational study of stellar atmospheres, in connection with theoretical expectations (from stellar structure theory) about which stars might generate large fluxes of acoustic waves and/or have rapidly changing small-scale magnetic fields, may help in the understanding of atmospheric heating processes.

The discovery that a broad peak, centered at about five minutes, in the frequency spectrum of solar Doppler shifts (Leighton et al., 1962), really is an interference pattern produced by the superposition of hundreds of normal acoustic modes (i.e. trapped, standing, small-amplitude acoustic waves), constitutes one of the most exciting meetings between theory and observation in recent astrophysics. More recent observations have produced highly accurate frequency measurements of 5 min acoustic modes with values of the angular degree l (number of wavelengths around a circumference) ranging from 0 (strictly radial oscillations) to about 1000. A few longer period acoustic modes and possibly some gravity modes (driven by buoyancy) have also been seen. Modes with small l-values penetrate to the center of the Sun and are truly global in character. Theoretical models yield computed frequencies for all the acoustic modes that agree with observation

Peter A. Sturrock (ed.), Physics of the Sun, Vol. I, pp. 177–247.

to better than 1%. Thus the study of solar oscillations stands as one of the rare instances in which detailed observations are well explained and understood theoretically.

Even more exciting is the possibility of probing the interior structure of the Sun and other stars by using oscillation data. The fact that the oscillations are global normal modes means that the frequencies are determined by the properties of the entire star. But because each eigenfunction tends to be large only in certain ranges of the radius, each normal mode produces information primarily about those regions. Extraordinary possibilities for inference of interior structure arise because of the linearity of the normal mode eigenfunctions: each mode carries information mainly about the mean background environment rather than about its interactions with itself or with other modes. Contrast this case with other equally important signals from stellar interiors: stellar cycles and starspots, sunspots and the solar photospheric magnetic field, all of which might yield information about stellar dynamos, but which mask the information in highly nonlinear convolutions. It is precisely because the theory provides an apparently complete explanation of the observations that we are able to deduce from the exact frequencies information about solar structure and, hopefully, from the amplitudes information about driving and damping.

So far, little observational data exists on nonradial pulsations of other stars. We now can see clearly the low-l modes in full disk-integrated sunlight and disk-averaged spectral lines, which means that such a possibility must also exist for all stars that pulsate in similar modes. Current knowledge of stellar internal structure is based on astoundingly few parameters. For most stars, we have a luminosity, atmospheric composition, and variously defined temperatures which allow us to estimate the radius. For binary stars, we also have the mass, and for rapid rotators, the line of sight component of the rotational velocity. Frequencies for the lowest four l-values ($l = 0-3$), which might cover a large range in the radial order n (the latest solar disk-averaged observations detect 80 distinct frequencies for $l = 0-3$), would vastly increase the amount of data available for a star; having such data for the range of luminosities and effective temperatures that permit nonradial pulsations would open a whole new window for looking inside the stars. While low-l nonradial acoustic modes in solar-like stars have not yet been observed, the importance of observing stellar oscillations should not be underestimated. Unfortunately, atmospheric scintillation makes it extremely difficult to detect such low amplitude pulsations from the ground, so the observations will probably have to be made from space.

The Sun is our laboratory for learning what information the oscillations can yield about stellar interiors. Here we have spatial resolution, hence can unequivocally identify a very large number of modes and further can detect the moderate- and high-l modes that are lost in the process of full-disk averaging.

The organization of this chapter reflects the fact that while the *theory* of nonmagnetic waves provides essential background and physical insight for understanding trapped oscillations, the interesting *observational work* has been mainly in the field of oscillations. We first formulate the equations applicable to waves and oscillations and then develop the basic theory of waves, both the nonmagnetic waves that are related to the trapped oscillations and can contribute to atmospheric heating, and the various magnetic waves that are seen in sunspots or may play a role in atmospheric heating and dynamics. We then treat the theory of nonradial oscillations in some detail, first discussing basic properties in the context of the relatively simple Cowling approximation, then looking at the excitation and damping mechanisms and stability calculations, and finally summarizing some of the detailed computations that have been done of normal mode frequencies, frequency perturbations, and frequency splitting. We finish the theoretical section with a glimpse at the major theoretical issues that require further work in the fields of oscillations and waves.

The second section treats observations of waves and oscillations. First we discuss the techniques used, and point out some of the difficulties inherent in making the types of observations described. Then we summarize the major developments in observations of normal mode frequencies, of high- and low-l 5 min p-modes, and of longer period modes whose identities are as yet somewhat uncertain. We much more briefly discuss the less dramatic observations of waves, and finish with a look at important advances yet to be made in observing oscillations and waves.

The last section of the chapter examines the fundamental question of what can be learned about the solar interior from analysis of the observational data, drawing on the preceding discussions of both theory and observation. We discuss what constraints have been put on solar models by direct method calculations. Then, noting the limitations of simple comparisons between model calculations and observed frequencies, we examine some of the possibilities that exist for probing the solar interior by true inversion methods.

1. Theory of Waves and Oscillations

1.1. BASIC EQUATIONS

1.1.1. *Full Hydromagnetic Equations*

The full hydrodynamic equations for a pulsating star consist of the continuity, momentum, and energy equations together with an equation of state describing the gas and Poisson's equation for the gravitational potential. Maxwell's equations, together with Ohm's law and suitable constitutive relations, are used to determine the magnetic field, which appears in the fluid equations.

The equation of continuity is

$$\frac{\partial \rho}{\partial t} + \nabla \cdot (\rho \mathbf{v}) = 0, \tag{1.1.1}$$

where ρ is mass density and $\mathbf{v}$ is fluid velocity. The momentum equation is

$$\rho \left[\frac{\partial \mathbf{v}}{\partial t} + (\mathbf{v} \cdot \nabla)\mathbf{v} \right] = -\nabla p + \rho \mathbf{g} + \nabla \cdot \tau, \tag{1.1.2}$$

where p is the pressure, $\mathbf{g} = -\nabla \psi$ is the gravitational acceleration, and ψ the gravitational potential; τ is the stress tensor which includes all stresses (viscous, diffusion-regime radiation, and magnetic — assuming no macroscopic electric field) except the gas pressure, which is assumed to be isotropic and therefore diagonal; the stress tensor can be written (following Tassoul, 1978, Equation 47; compare also Ledoux and Walraven, 1958, Equation 49.45):

$$\tau_{ik} = (\eta_m + \eta_r)\left(\frac{\partial v_i}{\partial x_k} + \frac{\partial v_k}{\partial x_i} - \frac{2}{3}\delta_{ik}\frac{\partial v_l}{\partial x_l}\right) +$$

$$+ \left(\zeta + \frac{5}{3}\eta_r\right)\delta_{ik}\frac{\partial v_l}{\partial x_l} + \frac{B_i B_k}{8\pi}(2 - \delta_{ik}), \tag{1.1.3}$$

where δ_{ik} is the Kronecker delta, η_m the coefficient of dynamic molecular viscosity, η_r the coefficient of dynamic radiative viscosity, and ζ the dynamic bulk viscosity coefficient. The diagonal contributions to τ from magnetic and radiative stresses add to the effective pressure, and the magnetic field contributes a tension equal to $B^2/4\pi$ along the magnetic lines of force. Noting that $\nabla \cdot \mathbf{B} = 0$, and writing η for $\eta_m + \eta_r$, the momentum equation becomes

$$\rho\left[\frac{\partial \mathbf{v}}{\partial t} + (\mathbf{v} \cdot \nabla)\mathbf{v}\right] = -\nabla p + \rho\mathbf{g} + \eta\nabla^2\mathbf{v} + \left(\zeta + \frac{1}{3}\eta_m + 2\eta_r\right)\nabla(\nabla \cdot \mathbf{v}) +$$

$$+ \nabla\eta \cdot \nabla\mathbf{v} + \nabla\mathbf{v} \cdot \nabla\eta + \nabla\left(\zeta - \frac{2}{3}\eta_m + \eta_r\right)(\nabla \cdot \mathbf{v}) +$$

$$+ \frac{1}{4\pi}(\nabla \times \mathbf{B}) \times \mathbf{B}. \tag{1.1.4}$$

The energy equation, in the quasistatic radiative diffusion limit, is

$$\rho T\left(\frac{\partial s}{\partial t} + \mathbf{v} \cdot \nabla s\right) = \nabla \cdot (k\,\nabla T) - (\nabla \cdot \mathbf{F}_r) + \Phi_v + \rho\epsilon_{\text{nuc}} + \frac{v_m}{4\pi}(\nabla \times \mathbf{B})^2, \tag{1.1.5}$$

where s is the specific entropy (entropy per unit mass) and should include the entropy of the radiation field, k is the thermal conduction coefficient, $\mathbf{F}_r$ is the radiative flux, and ϵ_{nuc} is the rate of energy generation by nuclear reactions per unit mass and time. The magnetic term is just the Joule heating J^2/σ_e, and Φ_v, the viscous dissipation function is

$$\Phi_v = \frac{1}{2}(\eta_m + \eta_r)\left(\frac{\partial v_i}{\partial x_k} + \frac{\partial v_k}{\partial x_i} - \frac{2}{3}\delta_{ik}\frac{\partial v_s}{\partial x_s}\right)^2 + \left(\zeta + \frac{5}{3}\eta_r\right)\left(\frac{\partial v_s}{\partial x_s}\right)^2. \tag{1.1.6}$$

If composition changes from thermonuclear reactions are negligible and if the internal energy, e, can be computed in terms of p and ρ alone, the energy equation can be written in terms of p and ρ (Cox and Giuli, 1968, Equation 17.6) as

$$\frac{1}{(\Gamma_3 - 1)}\left[\frac{\partial p}{\partial t} + \mathbf{v} \cdot \nabla p - \frac{\Gamma_1 p}{\rho}\left(\frac{\partial \rho}{\partial t} + \mathbf{v} \cdot \nabla\rho\right)\right]$$

$$= \nabla \cdot (k\,\nabla T) - (\nabla \cdot \mathbf{F}_r) + \Phi_v + \rho\epsilon_{\text{nuc}} + \frac{v_m}{4\pi}(\nabla \times \mathbf{B})^2, \tag{1.1.7}$$

where

$$\Gamma_1 = \left(\frac{d \ln p}{d \ln \rho}\right)_{ad} \quad \text{and} \quad \Gamma_3 = 1 + \left(\frac{d \ln T}{d \ln \rho}\right)_{ad}$$

are the two independent adiabatic exponents. Equations for Γ_1 and Γ_3 including the effects of changing ionization and of diffusion radiation are given by Ledoux and Walraven (1958, §53) and by Cox and Giuli (1968, §9.18).

The equation of state is not, in general, the perfect gas law, and can be written $p = p(\rho, T, X_i)$, where $\{X_i\}$ denotes composition and T is the temperature. Poisson's equation for the gravitational potential ψ is

$$\nabla^2 \psi = 4\pi G\rho, \tag{1.1.8}$$

where G is the constant of gravitation.

The magnetic field is determined from Maxwell's equations. Because we deal with time-scales long compared to the equilibration time of the conduction electrons in solar material, we shall ignore displacement currents and accumulation of electric charge. We assume the solar gas to be nonmagnetic, hence take the relative permeability to be unity. Then Maxwell's equations in Gaussian c.g.s. units are:

$$\nabla \cdot \mathbf{B} = 0, \tag{1.1.9a}$$

$$\nabla \times \mathbf{B} = \frac{4\pi}{c} \mathbf{J}, \tag{1.1.9b}$$

$$\nabla \times \mathbf{E} = -\frac{1}{c} \frac{\partial \mathbf{B}}{\partial t}, \tag{1.1.9c}$$

where $\mathbf{B}$ is the magnetic field strength, $\mathbf{E}$ is the electric field, $\mathbf{J}$ is the current density, and c is the speed of light. Using the generalized form of Ohm's law, $\mathbf{J} = \sigma_e (\mathbf{E} + c^{-1} \mathbf{v} \times \mathbf{B})$, to eliminate $\mathbf{E}$, we obtain an equation for $\mathbf{B}$,

$$\frac{\partial \mathbf{B}}{\partial t} = \nabla \times (\mathbf{v} \times \mathbf{B}) - \nabla \times (\nu_m \nabla \times \mathbf{B}), \tag{1.1.10}$$

where the coefficient of magnetic diffusivity, ν_m, is given by $\nu_m = c^2/4\pi\sigma_e$, and σ_e is the electrical conductivity; σ_e and therefore ν_m are functions of position. The Lorentz force density is given by

$$\mathbf{f} = \mathbf{J} \times \frac{\mathbf{B}}{c} = \frac{1}{4\pi} (\nabla \times \mathbf{B}) \times \mathbf{B}. \tag{1.1.11}$$

1.1.2. Linearized Equations

The full fluid equations form a system of nonlinear partial differential equations which cannot be solved in most cases. For small-amplitude oscillations the dependent variables can be written as the sum of a mean value and a small perturbation. The equations are then simplified to a linear system by neglecting all products of perturbation terms. We shall designate Eulerian perturbations by primes (e.g. x') and Lagrangian perturbations by δ (e.g. δx). We combine both types of perturbations in some of the linearized equations.

The linearized continuity equation is:

$$\frac{\partial \rho'}{\partial t} + \nabla \cdot (\rho_0 \mathbf{v}' + \rho' \mathbf{v}_0) = 0. \tag{1.1.12}$$

The linearized momentum equation, again writing η for $\eta_m + \eta_r$ is:

$$\rho_0 \frac{\partial \mathbf{v}'}{\partial t} = -\rho' \nabla \psi_0 - \rho_0 \nabla \psi' - \nabla p' + \nabla \left\{ \left(\varsigma - \frac{2}{3} \eta_m + \eta_r \right) (\nabla \cdot \mathbf{v}) \right\} +$$

$$+ \nabla \eta \cdot \nabla \mathbf{v} + \mathbf{v} \nabla \cdot \nabla \eta + \frac{1}{4\pi} [(\nabla \times \mathbf{B}_0) \times \mathbf{B}'] + \frac{1}{4\pi} [(\nabla \times \mathbf{B}') \times \mathbf{B}_0]. \tag{1.1.13}$$

Here ψ' satisfies the Poisson equation $\nabla^2 \psi' = 4\pi G \rho'$, and $\mathbf{v}_0$ is assumed to be zero. The energy equation, linearized about an equilibrium state, is given by:

$$\frac{\partial p'}{\partial t} + \mathbf{v} \cdot \nabla p_0 - \frac{\Gamma_1 p_0}{\rho_0} \left(\frac{\partial \rho'}{\partial t} + \mathbf{v} \cdot \nabla \rho_0 \right)$$

$$= (\Gamma_3 - 1) \nabla \cdot (k_{th} \nabla T') - (\nabla \cdot \mathbf{F}_r') + \rho' \epsilon_{nuc} + \rho_0 \epsilon'_{nuc} +$$

$$+ \Phi'_v + \frac{v_m}{4\pi} [2(\nabla \times \mathbf{B}_0) \cdot (\nabla \times \mathbf{B}')] + \frac{v'_m}{4\pi} (\nabla \times \mathbf{B}_0)^2. \tag{1.1.14}$$

In linearized form Maxwell's equations and Equation (1.1.10) become:

$$\nabla \times \mathbf{B}' = \frac{4\pi}{c} \mathbf{J}', \tag{1.1.15a}$$

$$\nabla \cdot \mathbf{B}' = 0, \tag{1.1.15b}$$

$$\nabla \times \mathbf{E}' = -\frac{1}{c} \frac{\partial \mathbf{B}'}{\partial t}, \tag{1.1.15c}$$

$$\frac{\partial \mathbf{B}'}{\partial t} = \nabla \times (\mathbf{v} \times \mathbf{B}_0) - \nabla \times (v_m \nabla \times \mathbf{B}'), \tag{1.1.15d}$$

In the presence of turbulence, linearized equations are obtained by averaging the perturbed equations and then eliminating higher order terms; the resulting equations are given in Ledoux and Walraven (1958, §56).

The molecular transport coefficients are too small to be important, except in the corona where the gas is so tenuous that its effect on the oscillations is negligible. Turbulent transport coefficients in the convection zone are many orders of magnitude larger and, along with the radiative conduction coefficient, are the only ones likely to be relevant for oscillation calculations.

1.2. WAVES

We turn now to a discussion of the basic theory of waves, and use it as an aid in developing an understanding of the behavior of waves and oscillations in the Sun. Some other aspects of wave theory in the solar atmosphere are discussed in Stein and Leibacher (1974).

1.2.1. *Nonmagnetic Waves*

In the absence of a magnetic field, the linearized equations admit oscillatory solutions representing waves whose restoring forces are the pressure gradient and buoyancy forces. At high frequencies are gravity-modified acoustic waves; at low frequencies are pressure-modified internal gravity waves; and in a small range of intermediate frequencies are evanescent (nonpropagating) waves. In the Sun, propagating waves are interesting mainly because: (1) they transport energy from one location to another, and therefore can modify the solar (atmospheric) structure; and (2) they affect the strengths, widths, and shapes of the spectral lines which are our main probe for learning about the Sun.

When waves are trapped between two reflecting boundaries, they form a standing wave pattern, of which the solar 5 min oscillations are a fascinating example. Frequencies and other properties of standing waves are determined by the structure of the entire region between the reflecting boundaries, hence we can in principle learn about the structure of a cavity from the properties of its trapped waves. In the solar case, eigenfunctions of the various trapped acoustic and gravity waves (*p*- and *g*-modes) cover all interior depths.

In contrast, propagating waves provide information only about local conditions, hence are useful probes only for the visible atmosphere and, to a lesser extent, for the regions where they are generated. Moreover, they are much more difficult to detect and definitively identify than are trapped modes. Before dealing with the oscillations, we discuss the properties of propagating waves, because these are the simple waves out of which the more complex standing wave patterns are formed by the addition of boundary conditions. In later sections we shall see how some of the properties are altered by trapping.

GENERAL PROPERTIES OF ACOUSTIC–GRAVITY WAVES. In a uniform atmosphere, acoustic wave propagation is described in terms of the sound speed, which is a parameter characteristic of the medium. If gravitational forces are present, two more parameters play important roles in the propagation of nonmagnetic waves. First, the force of gravity causes an exponential vertical stratification of the mean density in a compressible atmosphere, with a density scale height $H(z)$ which is given locally by $H(z) = \mu(z)g/RT(z)$, where R is the gas constant, g is the gravitational acceleration, and μ and T are the mean molecular weight and temperature.

Second, the local gradient in temperature or density can be compared to the rate of change of temperature or density that would occur in a parcel of gas rising or falling adiabatically. If the density decreases with height more rapidly than the adiabatic density gradient, the gas is stable against convective motion. A displaced parcel of gas then experiences a restoring force towards its equilibrium position and the parcel tends to oscillate about this position until damped by dissipative forces; the motion is a buoyancy oscillation or internal gravity wave. We can define a parameter N^2, the square of the Brunt–Väisälä frequency, by $N^2 = -(g/\rho)(d\rho/dz - d\rho/dz|_{ad})$, which is negative in regions where the gas is convectively unstable and positive where the gas is stable and, hence, will support internal gravity waves.

The linearized continuity, momentum, and heat conduction equations with no magnetic or dissipative terms, in an isothermal atmosphere, have adiabatic wave solutions whose velocity v can be written

$$v = v_0 \, (t = 0, x = 0, z = z_0) \exp[i(\omega t - k_x x - k_z(z - z_0))] \, \exp[(z - z_0)/2H], \qquad (1.2.1)$$

with similar expressions for the other variables. Substitution of this solution form into the simplified (i.e. nonmagnetic and nondissipative) forms of Equations (1.1.12)–(1.1.14) yields the acoustic–gravity dispersion relation (Hines, 1960),

$$k_z^2 = c^{-2} \, (\omega^2 - \omega_{ac}^2) + k_x^2 \, \omega^{-2} \, (N^2 - \omega^2), \qquad (1.2.2)$$

where c is the sound speed, ω is the wave frequency and k_x and k_z are the horizontal and vertical wavenumbers respectively. The amplitude variation of v implied by the factor $\exp[(z - z_0)/2H]$ arises because the energy density $\rho v^2/2$ of an adiabatic wave in an isothermal atmosphere must remain constant with height.

In Equation (1.2.2), ω_{ac} is the acoustic cutoff frequency, the lowest frequency at which waves can vertically propagate as gravity-modified acoustic waves; it is given by $\omega_{ac} = c/2H$, and is thus seen to result from the density stratification. Note that c/H is the inverse of the travel time τ_{s-h} for an acoustic disturbance to propagate vertically through a density scale height. Thus when the wave period $P \geq 4\pi\tau_{s-h}$, it is no longer possible to propagate information via compressions across the large change in ambient density that the disturbance encounters in a single period. The Brunt–Väisälä frequency N is the greatest frequency at which internal gravity waves can propagate. Except in the presence of extreme temperature gradients, $\omega_{ac}(z)$ is always larger than $N(z)$ at the same height, though both may vary substantially with height (or radius). Frequencies between N and ω_{ac} correspond to evanescent disturbances whose energy density decreases exponentially with increasing height.

Vertically propagating acoustic waves ($k_x = 0$) obey the simple dispersion relation $k_z^2 = c^{-2} (\omega^2 - \omega_{ac}^2)$; the pressure perturbation in terms of the vertical velocity w is

$$p' = (c^2 k_z/\omega)(\overline{\rho} w) + [(i/\omega)(2 - \Gamma_1)/(2\Gamma_1 H)] \, (\overline{\rho} w). \qquad (1.2.3)$$

Where $\omega \gg \omega_{ac}$, p' and w are almost perfectly in phase, and are also in phase with ρ' and T'. This is because acoustic waves, which are driven by pressure gradients, consist of alternating compression and rarefaction regions. In the compressions p, ρ, and T are all elevated, and the fluid velocity has its maximum value and is in the direction of propagation. In the rarefactions p, ρ, and T are diminished and the velocity again has its maximum value but is opposite to the direction of propagation. But as ω approaches ω_{ac}, $k_z \to 0$ and the imaginary term dominates, giving a phase difference between p' and w of $\pi/2$ when $k_z^2 = 0$. For $\omega < \omega_{ac}$, $k_z^2 < 0$ and k_z is imaginary, hence the phase lag between p' and w is $\pi/2$ for $\omega < \omega_{ac}$ as long as $k_x = 0$. The acoustic energy flux, $\overline{p'w}$ (averaged, e.g., over a wave cycle) has its maximum value when p' and w are exactly in phase; there is no transport of energy when $\omega < \omega_{ac}$ and $k_x = 0$, as $\overline{p'w}$ is zero in that regime.

Accoustic waves are longitudinal, so that in the high-frequency (pure acoustic) limit phase and energy propagate in exactly the same direction, along the wave vector $\mathbf{k}$. The phase velocity $\mathbf{v}_p$ and group velocity $\mathbf{v}_g$ then both equal the sound speed c, and are thus independent of both frequency and horizontal wavenumber, so the waves are totally nondispersive. If high-frequency acoustic waves are generated in the convection zone with an isotropic distribution of propagation direction, those with vertical propagation vectors will be much less strongly damped than those with highly oblique $\mathbf{k}$; the total group velocity is approximately the sound speed, hence the oblique waves spend much more time in the height range where radiative damping is strong. The acoustic wave spectrum emerging at the top of the damping region thus will have greater energy density associated with the waves that propagate approximately vertically than with highly oblique waves, and we expect the acoustic waves to have preferentially vertical velocities.

When k_x is nonzero, a second regime of propagating waves occurs at frequencies below the acoustic cutoff frequency. If ω becomes small enough that $k_x^2 \omega^{-2} (N^2 - \omega^2) > c^{-2} (\omega_{ac}^2 - \omega^2)$, then k_z^2 again becomes positive in Equation (1.2.2). These waves are internal gravity waves, for which buoyancy is the dominant restoring force. As they occur only when $k_x \neq 0$, they are inherently two-dimensional and they cannot propagate purely vertically but must be directed obliquely. Physically, gravity waves can occur only if horizontal variations are present, because the waves are driven by differences in buoyancy between adjacent fluid elements.

At very low frequencies, $\omega \ll N$, internal gravity waves approximately obey the simple relation (which is exact in an incompressible medium).

$$k^2 \equiv k_x^2 + k_z^2 = N^2 k_x^2 \omega^{-2}, \tag{1.2.4}$$

from which we can derive their characteristic properties. The phase velocity is

$$v_p = \omega/k = \omega^2/Nk_x, \tag{1.2.5}$$

directed along the wave vector $\mathbf{k}$, while the group velocity is

$$v_g = |d\omega/d\mathbf{k}| = (\omega/Nk_x)(N^2 - \omega^2)^{1/2}, \tag{1.2.6}$$

and is exactly orthogonal to $\mathbf{v}_p$. The horizontal components lie in the same direction, and the vertical components are oppositely directed. Thus when energy propagates upward, phase moves downward across the motion of the wave packet; phase fronts, or lines of constant phase, are parallel to the group velocity vector, while the wave vector $\mathbf{k}$ is orthogonal to it. Internal gravity waves are highly dispersive, as v_g depends on both k_x and ω. This implies that a wave packet with a broad spectrum of frequencies or horizontal wavenumbers will spread rapidly in space, diminishing in amplitude as it does so. The fluid velocity is approximately parallel to v_g, which is directed along an angle ϕ to the horizontal, given, for pure gravity waves by $\tan \phi \approx \omega/N$. As ϕ is always less than $\pi/4$ ($\omega < N$), the horizontal velocity is always greater than the vertical.

For gravity waves, phase relations among velocity and thermodynamic variables are quite different from those for acoustic waves. In a pure buoyancy oscillation, the gas expands and cools adiabatically as it rises, and contracts and heats as it falls. Above its equilibrium position it is denser than the surrounding gas and thus experiences braking from the gravitational force as it rises. At the top of its cycle, the velocity is zero, ρ' has its maximum value, and T' its minimum value, while at bottom of the cycle the opposite holds; in an incompressible fluid, the phase lag between the vertical velocity v_z and ρ' is exactly $\pi/2$ and that between v_z and T' is $\pi/2$ in the opposite sense, so that ρ' and T' are out of phase by $\pm\pi$. The pressure perturbation is in phase with v_z, as it is for acoustic waves. Compressibility modifies all these phase relations: v_z becomes slightly less than $\pi/2$ out of phase with ρ' and T', and the lag between v_z and p' is small but not zero. Because spectral lines are sensitive not only to velocities but also to variations in temperature and density, we shall see that phase relations are of great importance in developing tactics for detecting the different varieties of propagating waves.

To identify reflection points in a real, nonisothermal atmosphere, suppose that the fractional variation of atmospheric properties is small over a vertical wavelength. A WKB-type analysis shows that at each height the isothermal dispersion relations just discussed hold approximately with the local values of c, H, ω_{ac}, and N, but that these parameters, and therefore k_z^2, are height dependent. If the temperature decreases with height, then ω_{ac} increases with height and a vertically propagating acoustic wave with frequency ω will be reflected downward at the height at which $\omega_{ac} = \omega$. Since all the energy of the wave is then likewise reflected, there is no net vertical transport of energy by the wave. An obliquely propagating acoustic wave, with $k_x \neq 0$, can also be reflected if it propagates into a region where the temperature (and therefore the sound speed) increases so much that the term $c^{-2}(\omega^2 - \omega_{ac}^2)$ becomes smaller than $k_x^2 \omega^{-2}(N^2 - \omega^2)$, the latter being negative for acoustic frequencies. This occurs at varying depths for acoustic waves propagating downward through the convection zone. Vertically propagating acoustic waves are reflected at the center of the Sun by geometry.

A gravity wave with frequency ω always reflects at (or before) a height where the Brunt–Väisälä frequency decreases to ω. Thus in passing outward through a star, all gravity waves generated in the deep interior will reflect at or below the bottom of the lowest convection zone, where N becomes first zero and then imaginary. The quantity N^2, by its definition, represents the small difference between two much larger quantities; it is thus very sensitive to small changes in the real and adiabatic gradients of density or temperature. The relative variation of $N(z)$ with height is generally much greater and more rapid than the variation of $c(z)$, $H(z)$, or $T(z)$.

A gravity wave can also reflect when propagating into a region of decreasing scale height. To see this, neglect ω^2 compared to both ω_{ac}^2 and N^2 in Equation (1.2.2.), and consider the remaining terms: $k_z^2 \approx N^2 k_x^2 \omega^{-2} - (4H^2)^{-1}$. Then for $\omega \ll N$, we see that k_z^2 can go to zero when $Nk_x \omega^{-1} \approx (2H)^{-1}$ or $2H \approx (\omega/N)k_x^{-1}$. This type of reflection is of importance only for waves with small values of k_x.

Dissipative processes can substantially alter all the wave properties we have discussed.

For propagating waves, by far the most important of these processes is radiative damping, which besides decreasing the energy flux and wave amplitude below their adiabatic values, also alters phase relations, changes the magnitude and direction of the propagation velocities, and modifies reflection properties. These changes are discussed in detail in Schmieder (1977), Mihalas (1979), and Mihalas and Toomre (1982).

ACOUSTIC–GRAVITY WAVES IN THE SOLAR ATMOSPHERE. Both acoustic waves and internal gravity waves are expected to be produced by convective turbulence in the solar envelope. Acoustic wave generation occurs within the convection zone (see Section 1.4), while gravity waves are produced at the upper boundary, where rising gas impacts the overlying convectively stable layers. Figure 1 shows the radial dependence

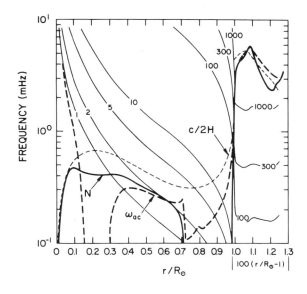

Fig. 1. Radial dependence of the acoustic cutoff frequency ω_{ac}, the Brunt–Väisälä frequency N, and the quantity S_l that determines the reflection depth for sound waves propagating into a region of increasing temperature. S_l is shown for the low-l values 1, 2, 5 and 10, and the high-l values 100, 300 and 1000. The scale of the atmosphere ($r > 1.0$) is expanded by a factor of 100 relative to that of the interior ($r < 1.0$). Acoustic waves and gravity waves reflect, respectively, at depths (or heights) where $\omega \to \omega_{ac}$ or $\omega \to N$; at small radii, however, sphericity terms become important and prevent low-l modes (e.g. l = 2, 5, 10) from being reflected upward by the rise in ω_{ac} in the deep interior.

of ω_{ac} and N, and of S_l (which reduces to ck_x in plane-parallel geometry) for several values of the angular degree l (see Section 1.3), in the solar interior and atmosphere.

　　The acoustic cutoff frequency ω_{ac} rises rapidly at the top of the convection zone, reaching a maximum value of about 3.5×10^{-2} s^{-1} ($P \approx 180$ s) near the temperature minimum of the solar atmosphere, about 500 km above the visible surface. Turbulence-generated acoustic waves with frequencies lower than this value are reflected downward and are a possible source of the 5 min oscillation modes. Those with $k_x \neq 0$ are trapped because they are also reflected upward from deeper, hotter regions of the convection zone where $(\omega^2 - \omega_{ac}^2)c^{-2} < k_x^2(N^2/\omega^2 - 1)$, as described earlier (see also Ulrich, 1970).

Acoustic waves with frequencies below 3.5×10^{-2} s^{-1} can in principle be trapped in the chromosphere, where the lower reflection results from the maximum in ω_{ac}, and the upper reflection is from the coronal temperature rise (see Figure 1). Such a cavity is not expected to produce a well-defined standing wave pattern because of pronounced horizontal inhomogeneities in the chromosphere and because there is no strong generation of acoustic waves within the cavity. However, tunneling of waves between the two cavities or coupling of resonant modes between cavities could effectively drive chromospheric modes.

Acoustic waves with higher frequencies propagate up through the atmosphere, with amplitudes that increase exponentially with increasing height. Biermann (1946) and Schwarzschild (1948) first recognized that acoustic waves could form shocks, dissipate, and heat the upper atmosphere. The compressive part of the wave, being hotter, has a greater sound speed than the rarefaction part and thus gradually overtakes it, an effect that becomes increasingly pronounced as the amplitude of T' increases. The wave forms a weak, upward propagating shock. Heating of the atmosphere by weak acoustic shocks has been studied in detail by Ulmschneider (1971, 1974). It now appears (Cram, 1977; Jordan, 1977; Athay and White, 1978) that acoustic shock heating can offset radiative losses in the lower chromosphere, but cannot do so in the upper chromosphere and corona. The heating of these regions is generally believed to result from processes involving magnetic fields.

Internal gravity waves see quite a different structure in the Sun (Figure 1). They are produced only above and below the convection zone, and are evanescent within it. In regions where radiative smoothing of temperature fluctuations occurs, Souffrin (1966) has shown that at least in the simple Newtonian cooling approximation, gravity waves cannot propagate unless the buoyancy frequency N exceeds the inverse of the characteristic radiative cooling time τ_R. Because τ_R is very short in the low photosphere, the atmospheric region in which gravity waves exist probably begins at about 100–200 km above the visible solar surface.

In the atmosphere the buoyancy frequency is roughly proportional to $T^{-1/2}$ but is more strongly affected by changes in the ionization state of the gas (Thomas *et al.*, 1971). It increases from zero near the visible surface to a maximum value of about 3.6×10^{-2} s^{-1} at around 800 km above the temperature minimum, then falls to a deep minimum in the hydrogen ionization plateau (mid-chromosphere), increases modestly again, and finally drops almost discontinuously at the transition layer.

Below the convection zone the radial dependence of the Brunt–Väisälä frequency is structured by the fact that N is proportional to the gravitational acceleration, which increases with increasing depth through most of the interior and thus partially offsets the effect on N of the increasing temperature. $N(r)$ rises from zero at the base of the convection region to a broad plateau at greater depths and near the center falls sharply as g decreases to zero.

In the atmosphere gravity waves with periods less than ≈ 550 s could, in principle, be trapped between the low photosphere ($N = 0$) and the chromospheric dip in N. Ridges for such modes have been computed by Gough (1980). In practice, heavy radiative damping in the photosphere, horizontal inhomogeneities, and the development of nonlinearities in the middle chromosphere all mitigate strongly against setting up a coherent standing-wave pattern. More likely, gravity waves generated by turbulent convection, probably with a large initial energy flux, will be heavily damped in the photosphere, then will propagate more or less adiabatically upward until they become sufficiently nonlinear to break. For gravity waves, phase relations do not permit shocks to develop; rather, highly nonlinear gravity waves tend to roll over themselves, much as do large ocean waves (surface gravity waves) and break, producing local turbulence and thus ultimately heating the gas via viscous dissipation. The height range where they are expected to dissipate energy at a rate comparable to the estimated rate of radiative loss appears to be similar to that for acoustic waves (Mihalas, 1979; Mihalas and Toomre, 1981, 1982).

Atmospheric gravity waves may make a major contribution to spectral line broadening (see Section 2.1.1). Theoretically, one expects the waves to broaden photospheric lines more at the solar limb than at disk center because of the larger horizontal than vertical velocities. In the chromosphere, broadening can result both from propagating gravity waves and from the turbulence they produce; here the velocity field is expected to be nearly isotropic.

1.2.2. *Magnetohydrodynamic waves*

In the presence of magnetic fields, there are two kinds of two-force waves. Including compressibility but not buoyancy yields the familiar magnetoacoustic or magnetohydrodynamic (MHD) waves. The analysis of these waves has been very important for our understanding of wave motions in the Sun, particularly insofar as the features of wave propagation derived in this way have become paradigms for propagation of *all* waves influenced by magnetic fields. We shall therefore consider MHD waves in some detail. Including buoyancy forces but not compressibility yields magnetogravity (MG) waves. We shall not devote much space to MG waves; they are more complicated than MHD waves, and there are few circumstances in solar physics where the conditions leading to their presence clearly occur.

MHD waves have been discussed at length by, for example, Bazer and Fleischman (1959), and Ferraro and Plumpton (1966). To derive the properties of MHD waves in their usual form, we assume that waves propagate in an unstratified planar medium with infinite electrical conductivity, permeated by a uniform magnetic field $\mathbf{B}$. We neglect viscous, radiative, and gravitational forces, and take the unperturbed fluid velocities to be zero. The linearized equations of continuity, momentum, flux conservation, and energy then become:

$$\frac{\partial \rho'}{\partial t} = \rho_0 \nabla \cdot \mathbf{v}, \tag{1.2.7a}$$

$$\frac{\partial \mathbf{v}}{\partial t} = -\frac{1}{\rho_0} \nabla p' + \frac{1}{\rho_0} \frac{1}{4\pi} (\nabla \times \mathbf{B}') \times \mathbf{B}_0, \tag{1.2.7b}$$

$$\frac{\partial \mathbf{B}'}{\partial t} - \nabla \times (\mathbf{v} \times \mathbf{B}_0) = 0, \tag{1.2.7c}$$

$$p' = c^2 \rho'. \tag{1.2.7d}$$

To obtain a dispersion relation, we assume wave solutions of the form $(\mathbf{v}, \mathbf{B}') = (\mathbf{v}(0),$ $\mathbf{B}'(0)) \exp[i(\omega t - \mathbf{k} \cdot \mathbf{x})]$, where $\mathbf{k}$ is the wave propagation vector. Inserting these forms into Equations (1.2.7a)–(1.2.7d), we get a system of linear homogeneous equations; setting the associated determinant to zero gives the desired dispersion relation:

$$\left(\frac{\omega^2}{k^2} - V_A^2 \cos^2 \theta_{kB}\right)\left[\left(\frac{\omega^2}{k^2} - c^2\right)\left(\frac{\omega^2}{k^2} - V_A^2\right) - c^2 V_A^2 \sin^2 \theta_{kB}\right] = 0, \qquad (1.2.8)$$

where $V_A = (B_0/\sqrt{4\pi\rho_0})$ is the Alfvén speed, and θ_{kB} is the angle between $\mathbf{B}(0)$ and $\mathbf{k}$.

This dispersion relation breaks into two factors. The first corresponds to waves with fluid motions perpendicular to the plane containing $\mathbf{B}_0$ and $\mathbf{k}$. These are Alfvén waves. They are purely transverse, noncompressive, and to first order do not alter the magnitude of B_0. Formally, these waves propagate in any direction, with a phase speed equal to $V_A \cos \theta_{kB}$. However, inspection shows that this amounts to translating the wave disturbance along the magnetic field at speed V_A.

The second factor in the dispersion relation relates to the combined effect of the fast- and slow-mode waves. The fluid motions in these waves lie in the $\mathbf{k}$–$\mathbf{B}(0)$ plane, with the motions for the two modes mutually orthogonal for a given $\mathbf{k}$. Figure 2 shows a

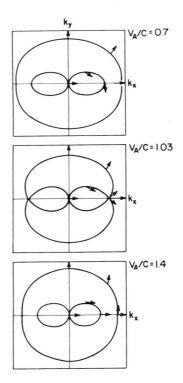

Fig. 2. Phase speed (continuous curves) and relative fluid velocity (arrows) for fast- and slow-mode MHD waves. In all cases, the magnetic field is oriented along the horizontal axis. Quantities are shown for V_A/c of 0.7, 1.03, and 1.4. Phase speeds are shown as polar plots, while the fluid velocities are shown only for propagation at angles of $0°$, $45°$, and $90°$ from the magnetic field direction.

plot of phase speed as a function of θ_{kB} for cases in which (a) $c > V_A$, (b) $c \approx V_A$, (c) $c < V_A$. As one can see, the relative motions in fast- and slow-mode waves depend considerably on whether c or V_A is the larger. In the solar atmosphere, V_A tends to dominate c at large heights, since the density falls much faster than the mean magnetic field strength. This generalization may be misleading, however, because of the very inhomogeneous nature of the solar magnetic fields. Thus, in the quiet Sun ($B \sim 1$ gauss), $V_A = c$ in the middle chromosphere, about 1500 km above the base of the photosphere; but in sunspots ($B \sim 1000$ gauss), $V_A = c$ at the bottom of the photosphere; and in small flux tubes, where the exterior gas pressure is largely balanced by interior magnetic pressure, V_A may exceed c even at great depths.

For wave propagation, consider first the limit $c \gg V_A$. Then fast-mode waves are nearly longitudinal, whereas slow-made waves are nearly transverse. Fast modes propagating parallel to $\mathbf{B}$ propagate as sound waves at speed c, while those with a component of $\mathbf{k}$ perpendicular to $\mathbf{B}$ see an additional restoring force arising from the compression of $\mathbf{B_0}$, and hence propagate at a higher speed. The motions in slow modes are nearly perpendicular to $\mathbf{k}$. If $\mathbf{k}$ is parallel to $\mathbf{B}$, the waves see the full restoring force from tension in the magnetic field lines, and the waves propagate at the Alfvén speed. But as $\mathbf{k}$ becomes perpendicular to $\mathbf{B}$, the fluid motions become parallel to $\mathbf{B}$ and the restoring force vanishes. Thus slow modes propagate at zero speed when $\mathbf{k}$ is perpendicular to $\mathbf{B}$.

In the limit $V_A \gg c$, fluid velocities in fast modes are almost perpendicular to $\mathbf{B}$, while those in slow modes are almost parallel to it. Here, fast modes are fundamentally Alfvén waves, with the restoring force augmented by compressibility when $\mathbf{k}$ is not parallel to $\mathbf{B}$. Such waves therefore propagate at the Alfvén speed or faster. The velocities in slow modes are almost parallel to $\mathbf{B}$. When $\mathbf{k}$ is parallel to $\mathbf{B}$ they are ordinary sound waves, but as $\mathbf{k}$ becomes perpendicular to $\mathbf{B}$, the waves become transverse and the restoring force from compressibility goes to zero. Since the motion is parallel to $\mathbf{B}$, there is no magnetic restoring force either, and the propagation speed perpendicular to $\mathbf{B}$ again goes to zero.

In all cases, fast modes propagate at least as fast as the larger of (c, V_A), while slow modes propagate no faster than the smaller of (c, V_A).

If V_A and c are nearly the same (as in Figure 2(b), where $V_A/c = 1.03$), one observes the peculiar effect known as conical propagation, in which the fast-mode phase velocity goes through a minimum as $\mathbf{k}$ becomes parallel or antiparallel to $\mathbf{B}$. For $V_A = c$ the fast and slow modes are degenerate for these two values of $\mathbf{k}$.

Although all three MHD wave modes can propagate phase in any direction except (sometimes) perpendicular to $\mathbf{B}$, their energy propagation, described by the group velocity $V_g = \nabla_k \omega$, is more restricted. For Alfvén waves, $V_g = V_A$, and is directed strictly along $\mathbf{B}$. For fast modes, V_g varies in a complex fashion, depending on the ratio of V_A to c. When this ratio is very large or very small, fast-mode waves can propagate energy in any direction, at speeds comparable to the phase speed. V_g for slow modes is only nonzero in a cone surrounding $\mathbf{B}$, with an opening angle that depends on c/V_A. Thus, they can only propagate energy along directions that lie near the direction of the magnetic field. For linear waves of all three kinds, the phase speed is independent of ω. Thus, the waves are not dispersive: wave pulses do not tend to spread, and monochromatic waves do not tend to steepen into shocks. These properties are not generally preserved in the finite amplitude case.

The wave properties of finite amplitude waves differ in a number of ways from those discussed above (Barnes, 1979; Barnes and Hollweg, 1974). It is still possible to construct Alfvén waves that are strictly noncompressive, but because of the contribution of **B** to the total pressure, **B** must have constant magnitude and change only in direction. Thus, the magnetic vector may move through a circle (or even oscillate through a segment of a circle), but the linear polarization given by the small-amplitude analysis is not allowed. Further, unless **k** is parallel to $\mathbf{B}_0$, the velocity vector cannot be perpendicular to both, and the wave becomes partly compressive. Thus, finite amplitude Alfvén waves must propagate strictly along $\mathbf{B}_0$. At finite amplitude, fast- and slow-mode waves retain their compressive nature and linear polarization. The compressions they apply to the fluid change the local wave propagation properties, causing different parts of a wave that is originally sinusoidal to propagate at different speeds, and leading eventually to the formation of a discontinuity (shock wave). This process occurs even in a homogeneous fluid, and is distinct from the steepening to shock waves that takes place in a gravitationally stratified fluid.

MAGNETOGRAVITY AND MAGNETOATMOSPHERIC WAVES. A gravitationally (and stably) stratified, but incompressible, atmosphere with a uniform magnetic field permits magnetogravity waves (Lighthill, 1967; Schwartz and Stein, 1975). The Boussinesq approximation, in which density variations from the mean state are not permitted except insofar as they contribute to the buoyancy force, yields the simplest form of the dispersion relation:

$$\omega^2 = V_A^2 k^2 \cos^2 \theta_{kB} + N^2 \sin^2 \theta_{kz}. \tag{1.2.9}$$

At a given ω, and for small values of k, this relation is dominated by the influence of buoyancy forces. These long-wavelength waves are thus virtually identical to pure gravity waves. At larger values of k, the magnetic restoring forces become important, and the waves take on the character of Alfvén waves. Two geometries are possible for magnetogravity waves with $\omega < N$. In the first, the orientation of **B** and the opening angle of the gravity wave cones are such that the Alfvén wave planes cut the cones in ellipses. In the second, corresponding to ω nearer to N, or to **B** more nearly horizontal, or both, the Alfvén planes cut the gravity wave cones in hyperbolas. This case is particularly interesting, because it suggests that waves propagating into a region where N vanishes need not be totally reflected; they may be converted into slightly modified Alfvén waves instead (Lighthill, 1967). Such conclusions based on two-force wave models should of course be treated with caution. However, adding compressibility to magnetogravity waves does not change the symmetry or topology of the wave normal surfaces in any essential way (Schwartz and Stein, 1975). For this reason, magnetogravity waves may provide a better qualitative guide to wave behavior in the Sun than any other two-force model.

Waves with all three restoring forces are termed magneto-acoustic-gravity (MAG) waves, or magnetoatmospheric waves. The equations governing such waves are quite complex (McLellan and Winterberg, 1968), and for the most part only special cases have been treated (Ferraro and Plumpton, 1958; Bel and Mein, 1971; Nakagawa et al., 1973; Antia and Chitre, 1978). In addition, many authors have analyzed the cases of vertical or horizontal magnetic field in attempts to explain various phenomena in sunspots and active regions.

In spite of their complexity, it is possible to say a few things about MAG waves that are at once comprehensible, useful, and true. First, the combination of all three restoring forces gives rise to only three wave modes, just as in the MHD case. One of these is identifiable as an Alfvén wave, but except in special cases, the other two modes have properties that are a mixture of those appropriate to fast mode, slow mode, and gravity waves. This admixture of properties is particularly important when $\omega < N$. Second, as with acoustic–gravity waves, the inclusion of both gravity and compressibility implies a vertically stratified atmosphere. This leads to two important effects: the existence of cutoff frequencies below or above which certain wave modes will not propagate vertically, and the tendency for propagating wave amplitudes to increase with increasing height in the atmosphere, in order to maintain a constant energy flux. Both effects have been described in the discussion of acoustic–gravity waves (Section 1.2.1) but they are somewhat modified in the magnetic case.

MAGNETIC WAVES IN INHOMOGENEOUS MAGNETIC FIELDS. We must now ask whether the approximations of small amplitude and homogeneous magnetic field have anything to do with magnetically influenced waves on the Sun. The accuracy of the homogeneous fields approximation depends on the relative scale lengths of the variations in the magnetic field and the waves themselves. Many of the waves supposed to be important in solar processes have periods in the range 10–300 s. Taking c to be a characteristic wave speed, this implies wavelengths of 70–2100 km in the photosphere, with even larger values in some regions of the Sun. Since most of the magnetic field outside of sunspots is now thought to be concentrated in narrow (radius less than 500 km, and perhaps much less) flux tubes, it is unlikely that any but the shortest period waves will see a magnetic field that is homogeneous over distances comparable to a wavelength. Similarly, observed wave energies are so large that shock waves must occur in some parts of the solar atmosphere. The linearity assumptions are then certainly violated, at least in places. Thus, an accurate treatment of magnetic waves in the Sun must deal with inhomogeneous fields and nonlinear waves, and treatments that do not do so probably cannot provide more than qualitative description of the real waves. Attempts to do this have followed one of two paths. The first is to accept the MHD or MAG analysis as being locally correct, and compute the coupling between wave modes brought about by inhomogeneities in the field or nonlinear interactions. The second is to assume an unperturbed magnetic field configuration that is dynamically consistent to lowest order and mimics what is seen on the Sun (the thin flux tube approximation), and then find the corresponding wave modes.

In the first approach, the greatest stress has been placed on the treatment of interactions between waves. These have been described in a series of papers by Melrose (e.g. Melrose and Simpson, 1977), and similar ideas have been applied to wave dissipation in the chromosphere and corona by Wentzel (1978) and by Petrukhin and Fainshtein (1976). One principal result of these studies is that it is rather easy to convert an Alfvén wave into a fast-mode wave, either through the interaction with another Alfvén wave or by propagating the wave through a region where V_A or c vary on a small scale. However, in many cases the fast mode wave produced in this way propagates almost along $\mathbf{B}$, and if $V_A \gg c$ it is almost indistinguishable from the original wave.

The second approach to waves in inhomogeneous fields is to consider the motion of fluid within and surrounding slender, isolated tubes of magnetic flux. Since the flux tube equations were first proposed by Defouw (1976), this approach has been actively pursued, and is included in a recent review by Spruit (1981, and references therein). The related problem of waves in a thin two-dimensional flux sheath has been treated by Wilson (1978a, b). The thin flux tube approximation suffers from the absence of a demonstrated exact equilibrium solution, so that perturbations about the assumed (nonequilibrium) tube states may be suspect. It presently appears that these departures from equilibrium are unimportant where very thin tubes are concerned, but may become significant in theories that try to treat the internal structure of an oscillating tube.

To illustrate the waves that arise in the thin flux tube approximation, consider a tube parallel to the z-axis containing a magnetic field, and embedded in an atmosphere containing no field. Since we are not concerned with fine structure within the tube, assume that both $\mathbf{B}$ and $\mathbf{v}$ are constant across the width of the tube, though they may vary with z. Suppose also that the tube is so thin that pressure balance across the tube is always maintained, and that the tube diameter is much less than the scale of any waves considered. The equations describing fluid motions in the tube are then (Spruit, 1981):

$$\rho \frac{\partial v}{\partial t} + \rho v \frac{\partial v}{\partial z} = -\frac{\partial p}{\partial z} + \rho g, \tag{1.2.10}$$

$$p + \frac{B^2}{8\pi} = p_e, \tag{1.2.11}$$

$$\frac{\partial (\rho/B)}{\partial t} + \frac{\partial (\rho v/B)}{\partial z} = 0. \tag{1.2.12}$$

where $p_e(z)$ is the pressure of the atmosphere outside the tube. Equations (1.2.10) and (1.2.11) describe vertical and radial momentum balance, respectively. Equation (1.2.12) may be understood by noting that magnetic flux conservation requires that B times the area of the tube be a constant; ρ/B is therefore proportional to the mass per unit length of tube, and (1.2.12) is a combination of flux conservation and the continuity equation. An energy equation is required to complete the set (1.2.10)–(1.2.12); this is usually taken to be an adiabatic relation.

The results obtained with these equations depend on whether the external atmosphere is taken to be homogeneous or stratified. If it is homogeneous, and the tube is taken to have negligible area, the dispersion relation is

$$\left(\frac{\omega}{k}\right)^4 - \left(\frac{\omega}{k}\right)^2 (c_e^2 + c_T^2) + c_e^2 c_T^2 = 0, \tag{1.2.13}$$

where c_e is the sound speed in the external medium and c_T is defined in terms of the V_A and c inside the tube by $c_T^2 = c^2 V_A^2/(c^2 + V_A^2)$. The solutions to (1.2.13) correspond to a sound wave propagating in the external medium at speed c_e, and a wave in the tube moving with speed c_T. The tube wave can be identified as a sort of slow-mode wave; it is partly longitudinal (becoming increasingly transverse as V_A/c becomes small), and moves at a speed less than the smaller of V_A and c. It develops that these tube waves do not excite traveling waves in the external medium unless $c_T > c_e$. Thus, unless the tube interior is hotter than the surrounding atmosphere, the tube waves are not damped by the emission of sound waves, nor can they gain energy from sound waves incident on the tube.

The simplest case of tube waves in a stratified medium occurs if the tube is embedded in an isothermal atmosphere. The dispersion relation that results is similar to (1.2.13), except that it includes a cutoff frequency ω_c, below which the waves are evanescent. If the adiabatic condition is relaxed and the waves are required to propagate isothermally, then the cutoff frequency may be written $\omega_c = c_T/4H$, where H is the pressure scale height. This is similar to the propagation condition of ordinary sound waves, though the exact value of the cutoff frequency is different than for sound waves, and depends on the value of V_A/c.

There are other possible oscillation modes for flux tubes, involving flows that are not axisymmetric (Spruit, 1981). Two of these are of particular interest. The first is an Alfvén wave, with fluid velocities that are perpendicular to the tube axis and chosen so that $\nabla \cdot \mathbf{v} = 0$ and so that the radial velocity vanishes at the tube boundary. Waves of this sort propagate along the tube axis at the Alfvén speed, and do not disturb the fluid external to the tube. The second kind of wave corresponds to a transverse motion of the entire flux tube. This also is similar to an Alfvén wave, except that it necessarily involves the motion of fluid outside the tube. The restoring force per unit mass is therefore smaller than for a pure Alfvén wave, and the propagation speed depends on the mass density outside the tube. This kind of wave is potentially important because it may easily be excited by turbulent motions in the convection zone, and may thus carry energy to higher levels of the atmosphere.

1.3. GENERAL PROPERTIES OF SOLAR OSCILLATIONS

1.3.1. *Equations and Spheroidal Mode Solutions*

The general linearized Equations [(1.1.12)–(1.1.15)] can be simplified considerably for a static, spherically symmetric star. Because of the symmetry one can find separable solutions where the displacement vector has the form

$$\delta\mathbf{r} = \mathrm{Re}\left\{\left[\xi_r(r)\, Y_l^m(\vartheta, \phi)\, \mathbf{a}_r + \xi_h(r)\left(\frac{\partial Y_l^m}{\partial \vartheta}\, \mathbf{a}_\vartheta + \frac{1}{\sin\vartheta}\frac{\partial Y_l^m}{\partial \phi}\, \mathbf{a}_\phi\right)\right]\exp(-i\omega t)\right\}, \quad (1.3.1)$$

where ξ_r gives the radial dependence of the radial displacement and ξ_h differs from the horizontal displacement by a gradient operator. Similarly, the perturbation in, for example, pressure, may be written

$$\delta p = \text{Re}[\delta p(r) \, Y_l^m(\vartheta, \phi) \exp(-i\omega t)]. \tag{1.3.2}$$

Here Y_l^m is a spherical harmonic, $\mathbf{a}_r$, $\mathbf{a}_\vartheta$, and $\mathbf{a}_\phi$ are unit vectors in the r, ϑ, and ϕ directions, and ω is the complex frequency. Solutions of this form are called spheroidal eigenmodes; they are characterized by l, m, and the radial order n. The degree l corresponds to the local total horizontal wavenumber k_h, such that $k_h = \sqrt{l(l+1)}/r$; m roughly determines the orientation of the mode. In a spherically symmetric star there is no preferred orientation, and thus, as indicated, the eigenfrequency ω_{nl} and the amplitude functions $\xi_{r,nl}$, etc., do not depend on m. For simple models n is the number of zeros, excluding the centre when $l > 0$, in the displacement. For more complicated models, including the Sun, a more careful definition must be made (cf. Scuflaire, 1974; Osaki, 1975).

The perturbation equations have a second class of solutions, called toroidal modes, for which $\delta \mathbf{r}$ has no radial component. These oscillations are quite different from the p-, g-, and f-modes that are the object of most observational and theoretical studies, and will be discussed separately later in this section.

The case of spheroidal modes with $l = 0$ corresponds to radial, or spherically symmetric, oscillations. These have an extensive literature (see, e.g., Cox, 1980) in connection with 'classical' variable stars which generally appear to oscillate in one or two radial modes that have large amplitudes and thus are highly nonlinear. For the Sun, observed oscillations appear for the most part to be linear, nonradial modes; $l = 0$ modes seem to play no special role and are treated together with the nonradial (i.e. $l > 0$) modes.

From the general perturbation equations it follows that the amplitude functions for spheroidal modes satisfy the following set of equations

$$\rho' = -\frac{1}{r^2} \frac{d}{dr} (r^2 \rho \xi_r) + \frac{l(l+1)}{r} \rho \xi_h, \tag{1.3.3a}$$

$$\omega^2 \xi_r = \frac{1}{\rho} \frac{dp'}{dr} - \frac{\rho'}{\rho^2} \frac{dp}{dr} + \frac{d\psi'}{dr} \tag{1.3.3b}$$

and

$$\omega^2 \xi_h = \frac{1}{r} \left(\frac{p'}{\rho} + \psi' \right), \tag{1.3.3c}$$

$$\frac{1}{r^2} \frac{d}{dr} \left(r^2 \frac{d\psi'}{dr} \right) + \frac{l(l+1)}{r^2} \psi' = +4\pi G \rho', \tag{1.3.3d}$$

where G is the gravitational constant. If the oscillations are adiabatic, $\delta p/p = \Gamma_1 \, \delta \rho/\rho$. Thus the adiabatic eigenmodes are determined by a fourth-order system of ordinary differential equations. For radial oscillations ψ' can be eliminated analytically, reducing the order of the equations to two (e.g. Cox, 1980).

At the center, regularity of the solution imposes two boundary conditions which yield $\xi_r(r) \approx r^{l-1}$ for $l > 0$ and $\xi_r(r) \approx r$ for $l = 0$ (that the displacement vanishes at $r = 0$ for radial modes is geometrically obvious). One surface condition is obtained from the continuity of ψ' and its first derivative at the surface. A second condition often used is $\delta p = 0$, which is strictly correct at the surface of a star bounded by vacuum. More complicated conditions applied in the atmospheres of realistic stellar models are discussed later in this paper and in Christensen–Dalsgaard and Gough (1975) and Unno et al. (1981).

1.3.2. Cowling Approximation

Cowling (1941) made the approximation of neglecting the perturbation in the gravitational potential ψ' in Equations (1.3.3) and of eliminating Poisson's equation. Cowling argued that for oscillations of high radial order (or high degree) regions with opposite sign of ρ' largely cancelled in Poisson's equation, so that the overall effect of ψ' would be small. This was to a large extent confirmed by Robe (1968) who compared frequencies computed for polytropic models without and with the Cowling approximation. Cowling (1941) and Kopal (1949), using a perturbation analysis, found the effects of ψ' on ω_{nl} to decrease with increasing l for the f-modes and low-n g-modes. Robe found that, for example, with a polytropic index of 3, corrections to ω_n were about 7% for the f-mode, 1.3% for p_{10} and 0.23% for g_{10}. However, a full analysis of the validity of the approximation has yet to be published.

The Cowling approximation was initially introduced as a computational convenience. In this capacity it is no longer needed. Its main advantage now is that, by reducing the order of the equations to two, it makes them much more amenable to asymptotic analysis. In fact the Cowling approximation equations correspond to the equations for acoustic–gravity waves, where ψ' is eliminated automatically by taking g to be constant, and so many of the results of Section 1.2 can be directly applied.

Corresponding to the two classes of nonmagnetic waves (Section 1.2.1) we expect two classes of spheroidal modes, one driven predominantly by the pressure gradient restoring force, the other predominantly by buoyancy. Cowling established this division on the basis of the oscillation equations, showing that for each value of $l > 0$ there were two classes of solutions. As the radial order n tends to infinity, the frequencies of the pressure-driven class, which he called p-modes, tend to infinity, while the frequencies of the buoyancy-driven class, called g-modes, tend to zero.[1] For $l = 0$ buoyancy forces vanish and only the p-modes remain. Cowling also found, from numerical solution of the equations, a mode with no zeros in the displacement, which he called the f-mode.

PROPERTIES OF P-, F-, AND G-MODES. We shall discuss properties of the modes in terms of the properties of the corresponding waves, but we note that the same results can be obtained by asymptotic analysis of the oscillation equations (e.g. Shibahashi,

---

[1] It was shown by P. Ledoux and P. Smeyers (1966 [C.R. Acad. Sci. Paris, Serie B, **262**, 841]) that stars with both radiative and convective zones have two classes of g-modes, one which they designated the g^+-modes which are concentrated in the radiative regions and oscillate with time, the second, the g^--modes, which are concentrated in the convective regions and grow or decay exponentially with time. Here we are only concerned with the g^+-modes and we therefore drop the '+'.

1979). An excellent discussion of mode properties, with somewhat different emphasis, can be found in Leibacher and Stein (1981).

Acoustic waves can only propagate when $\omega^2 \geq c^2\, k_h^2$ and when $\omega \geq \omega_{ac} = c/2H$ (see Section 1.2.1). The p-modes are confined to the region above the radius where $\omega^2 \approx S_l^2 \equiv c^2 l(l + 1)/r^2$, which decreases in depth with increasing l (or decreasing ω) as indicated on Figure 1; beneath it the eigenfunction in general decays exponentially. At the surface the acoustic waves are reflected downward by the increase in the acoustic cutoff frequency (cf. Section 1.2.1 and Figure 1). Thus standing waves are set up between the lower turning point, where $\omega = S_l$, and the surface, where $\omega = \omega_{ac}$. Ulrich (1970) realized that this could account for the observed 5 min oscillations. Leibacher and Stein (1971) independently suggested essentially the same picture, but based on nonlinear, one-dimensional calculations which required an *ad hoc* lower reflection point. Wolff (1972a, b) first identified these oscillations as Cowling's p-modes.

Similarly, gravity waves are confined to regions where $N^2 > 0$ and where $\omega \leq N$ and $\omega \leq ck_x N/\omega_{ac}$; the buoyancy frequency N is shown on Figure 1. Interior g-modes are trapped beneath the convection zone and restricted to frequencies below the maximum N_{max} of N in the interior. Those with frequencies close to N_{max} are confined deep in the model. Interior g-modes are evanescent throughout the convection zone, hence their surface amplitudes are small even when the total energy of the mode is large. The amplitude decreases over the convection zone by roughly $\exp(-\int k_r\, dr)$, with the radial wavenumber k_r given in the WKB and Cowling approximations by $k_r^2 \approx (N^2/\omega^2 - 1)(l(l + 1)/r^2 - 1/4H^2)$. In all but the upper layers of the convection zone the scale height is very large, while N^2 is small and negative, and the dominant term in k_r^2 is $(-1)l(l + 1)/r^2$; thus the ratio of surface to base amplitude decreases rapidly with increasing l, implying that the g-modes likely to be observed are those of small degree (see, e.g., Dziembowski and Pamjatnykh, 1978; Christensen-Dalsgaard et al., 1980). It is also possible (Unno, 1975) for g-modes to be trapped in the solar core by a strong gradient in mean molecular weight, which alters the Brunt–Väisälä frequency in that region.

Gough (1980) noted that as l increases for fixed n, the motions become more nearly radial, the horizontal velocity approaches zero, and ω_{nl} gradually increases until it approaches the Brunt–Väisälä frequency. As $|n|$ increases at fixed l, the motions become predominantly horizontal, implying a decrease in the amount of the gravitational potential energy available for the mode relative to the inertia of the fluid, and ω_{nl} decreases.

On the basis of this simplified discussion one can understand the overall properties of the solar spectrum of oscillations. Figure 3 shows oscillation frequencies in a typical solar model (Model 1 of Christensen-Dalsgaard, 1982) as functions of l for selected modes. The frequencies of the g-modes tend to N_{max} as l tends to infinity. For the p-modes and the f-modes ω^2 increases approximately linearly with l. The f-mode, however, is essentially a surface gravity wave, at least for $l \geq 5$, with a frequency approximately given by the dispersion relation for deep-water surface waves, $\omega \approx (gk_h)^{1/2}$ (Gough, 1980). In no circumstance does the f-mode behave as an acoustic mode; thus the practice of labeling this mode as p_0, although fairly widespread, is misleading. Also, there is a direct correspondence between the radial modes and the p-modes with $l \geq 1$, in the

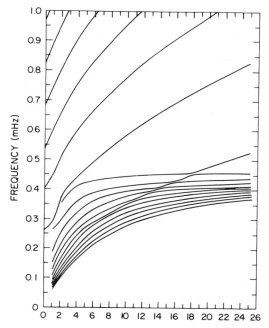

Fig. 3. Oscillation frequencies of p-, f-, and g-modes as a function of angular degree l, computed for a typical solar model. Notice that the p-modes and g-modes occupy disjoint regions of the diagram, whereas the g-modes and f-modes display avoided crossings. For further explanation, see the text.

sense that when l is varied from 0 to 1 the radial modes go over continuously into the p-modes (Vandakurov, 1967b).[2]

For fairly high-order modes, the radial component of the fluid displacement behaves, within the wave cavity, in approximately the same manner as the displacement of a simple wave (see Section 1.2). Thus for p-modes, whose group velocity is the sound speed c, energy flux conservation implies that

$$\xi_r(r) \sim \frac{1}{r\rho^{1/2}c^{1/2}} \sin \phi_p(r),$$ (1.3.4)

where the variation in the phase ϕ_p is determined by the local radial component of the wave vector. Similarly for g-modes

$$\xi_r(r) \sim \frac{1}{r^{3/2}\rho^{1/2}N^{1/2}} \sin \phi_g(r)$$ (1.3.5)

[2] Although only integer values of l have any physical meaning, the equations and boundary conditions are well defined in a mathematical sense for any value of l. When investigating the properties of the eigenmodes and eigenfrequencies it is in fact often convenient to regard l as a continuous variable.

(see also Wolff, 1979). These are essentially JWKB solutions of the oscillation equations and must be modified close to the reflection points of the modes.

Figure 4 shows $\rho^{1/2} \xi_r$ for a few p-modes with periods close to 5 min, illustrating the

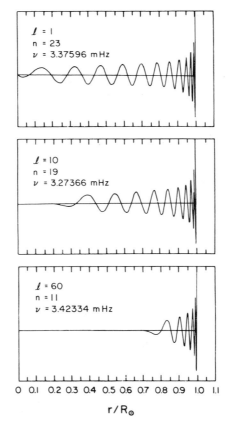

Fig. 4. Radial dependence of the eigenfunctions for three low to intermediate-l p-modes in the 5 min band. The quantity plotted is proportional to the product of the radius r with the square root of the energy density. The $l = 1$ mode has large amplitude at all radii, and the eigenfunctions are seen to become increasingly confined to the surface as l increases. At very high l-values (l of several hundred) the p-modes are confined to a very shallow layer at the top of the convection zone.

increasing confinement of the p-modes close to the surface with increasing l. The residual variation in the amplitude with r is closely approximated by $r^{-1} c^{-1/2}$. The radial mode has a significant part of its energy close to the center of the model. Similarly Figure 5 shows $\rho^{1/2} \xi_r$ for f- and g-modes with $l = 10$. The f- and g_1-modes are efficiently confined to the deep interior; the g_5-mode has a mixed character, but is predominantly the surface gravity mode, as is also suggested by its position in the frequency diagram on Figure 3.

1.3.3. *Asymptotic Behavior of p- and g-Mode Frequencies*

Although detailed numerical computations are required to determine accurate frequencies

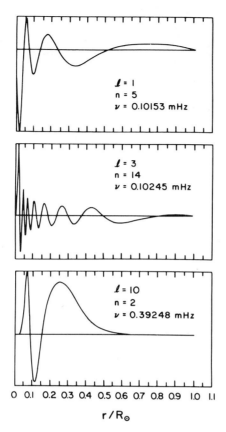

Fig. 5. Radial dependence of three low-frequency g-mode eigenfunctions with l-values of 1, 3, and 10; the quantity plotted is the same as in Figure 4. Although the modes are all evanescent throughout the convection zone, the $l = 1$ mode retains substantial amplitude out to the solar surface. The $l = 3$ mode has much smaller amplitude in the convection zone and at the surface, and the $l = 10$ mode is very strongly confined to the region below the convection zone.

for comparison with observed oscillations, a great deal of insight may be obtained from approximate, asymptotic relations for the frequencies. This has been particularly true for p- and g-modes of high n and low l, for which the asymptotic relations were first obtained by Vandakurov (1967a; see also Tassoul, 1980; Stein, 1982).

Vandakurov showed that an approximately correct analytic expression for the frequencies of high-n, low-l ($n \gg l$) modes is

$$\nu_{n,l} \sim (n + l/2 + \epsilon)\Delta\nu_0 \qquad (1.3.6)$$

where $\nu_{n,l} = \omega_{n,l}/2\pi$ is the cyclic frequency, ϵ is a parameter of order unity that is unique to the solar model, and where the frequency difference between modes with successive values of n is $\Delta\nu_0 = (2 \int_0^{R_\odot} dr/c)^{-1}$. Here $\Delta\nu_0$ depends only on the time required by an acoustic wave to travel from the surface of the Sun to the center and back again.

Christensen-Dalsgaard and Gough (1980b) used the above formula to argue that sequences of $\nu_{n,l}$ having fixed l and varying n would be approximately uniformly spaced in n, with $\Delta \nu_{n,l} \approx \Delta \nu_0$. Two such sequences with different l values would be almost coincident when both values of l were either even or odd and would be separated by roughly $\Delta \nu_0/2$ when one degree was even and the other odd:

$$\nu_{n,l} \sim \nu_{n-1,l+2}; \quad \nu_{n,l} \sim 1/2(\nu_{n-1,l+1} + \nu_{n+1,l+1}). \tag{1.3.7}$$

These approximate sequence rules begin to break down as l increases for a given order n, but have been borne out by observation for 5 min modes with $l = 0{-}3$. Tassoul (1980) obtained a second-order correction to Vandakurov's result for $\nu_{n,l}$, which can be written as $\{-[l(l+1) + \delta] A (\Delta_0)^2 / \nu_{n,l}\}$, where δ and A are parameters unique to each model. For frequencies ranging between 2.3 and 4.1 mHz the r.m.s. deviation between Tassoul's equation and frequencies calculated by Christensen-Dalsgaard and Gough (1982) was 0.8% for all modes having $l \leq 10$. Observational frequencies now have an accuracy of around 0.1%, which means that although these simple, analytic relations are surprisingly good, they cannot be used in comparisons between theoretical and observed frequencies at the level of the observational accuracy.

An approximate expression for the frequencies of p-modes of high degree which roughly reproduces the observed ridge structure for the high l solar 5 min oscillations can be obtained by noting that the solar convection zone can be approximated by an adiabatic polytropic layer (Christensen-Dalsgaard, 1980; see also Stein, 1982). The result is

$$\omega_{nl}^2 \approx (g_s/r_s) (1 + 2n/\mu)l, \tag{1.3.8}$$

where μ is the effective polytropic index at the upper reflection point.

For g-modes one obtains the approximate relation

$$\omega_{nl} \sim \frac{2\sqrt{l(l+1)}}{\pi(2n + l + \alpha)} \int_0^{r_c} \frac{N}{r} \, dr, \tag{1.3.9}$$

which is valid for small frequencies; here r_c is the radius at the bottom of the convection zone, and α is a constant that depends on the behavior at the bottom of the convection zone. This relation shows that the periods of high-order g-modes are approximately uniformly spaced, with a separation

$$\Delta \Pi_l \approx \frac{2\pi^2}{\sqrt{l(l+1)} \int_0^{r_c} (N/r) \, dr}. \tag{1.3.10}$$

There is no analogy to the near degeneracy found for the p-modes. The density of the modes in period is roughly proportional to l. Evidence has been found in differential velocity observations for the existence of sequences of long period modes uniformly spaced in period (P. Delache and P. Scherrer, private communication).

1.3.4. *Radial Oscillations*

The purely radial oscillations of Cepheid variable stars are excited to such large amplitude that significant variations in the total luminosity of the star are observed. These highly nonlinear pulsations are appropriately studied using radial, nonlinear equations. For solar oscillations, which are of small amplitude and are also nonradial, the linearized equations are probably adequate, and these are tractable in three dimensions.

Some useful insights can be gained, however, by considering the solutions for $\delta \mathbf{r}$ that are obtained from the linearized radial equations, which are much simpler than Equations (1.1.12)–(1.1.15). These solutions also become increasingly close to the correct nonradial solutions as the order n increases at fixed degree l or as l decreases at fixed n. In the nonmagnetic case the linearized radial equations can be reduced to

$$
\frac{\partial}{\partial r}\left(r^4 \Gamma_1 p_0 \frac{\partial \zeta}{\partial r}\right) + \zeta \left\{\omega^2 \rho_0 r^4 + r^3 \frac{\partial}{\partial r}\left[(3\Gamma_1 - 4)p_0\right]\right\}
$$

$$
= \frac{1}{i\omega}\left\{(\Gamma_3 - 1)\left[k_{\text{th}} \frac{\partial T'}{\partial r} - \frac{1}{r^2}\frac{\partial}{\partial r}(r^2 F_r') + \rho' \epsilon_{\text{nuc}} + \rho_0 \epsilon'_{\text{nuc}}\right]\right\}. \tag{1.3.11}
$$

For adiabatic motion the right-band side is zero and (1.3.11) becomes (see, e.g., Cox, 1980, Equation 8.6) the 'linear adiabatic wave equation'. The solutions to the adiabatic form of (1.3.11) for an isothermal atmosphere (assuming constant g) are of the form

$$
\zeta = \exp(z/2H)\{A_- \exp(-\beta_- z) + A_+ \exp(+\beta_+ z)\}\exp(-i\omega t), \qquad \omega < \omega_{\text{ac}} \tag{1.3.12a}
$$

which are purely evanescent solutions that grow (A_+) or decay (A_-) in the outward radial direction, and

$$
\zeta = \exp(z/2H)\{A_- \exp[-i\beta_- z] + A_+ \exp[+i\beta_+ z]\}\exp(-i\omega t), \qquad \omega > \omega_{\text{ac}} \tag{1.3.12b}
$$

which are upward (A_+) and donwward (A_-) propagating waves. Here

$$
\beta_\pm = \pm\left\{ABS\left[\frac{1}{4H^2} + \frac{1}{\Gamma_1 R_\odot H}\left(3\Gamma_1 - 4 - \omega^2 \frac{R_\odot}{g_\odot}\right)\right]\right\}^{1/2}. \tag{1.3.12c}
$$

In an unbounded isothermal atmosphere, the requirement that the total energy of the wave remain finite forces the A_+ solution in (1.3.12a) to be zero. If the atmospheric structure is such that an A_+ solution which is evanescent in some region can become a propagating wave at a greater value of the radius, the solution is allowed physically.

The radial approximation is excellent for interpretation of full disk observations and probably also of all the limb observations; the low l values of the modes detected by these methods imply that the modes are nearly radial in character. Hill and his colleagues (Hill *et al.*, 1978) have studied methods for estimating the relative amplitudes of the β_+ and β_- waves in their limb brightness data. These methods make use first of the fact that the relative amplitudes of T', ρ', and v are quite different for the β_+ and β_- solutions, and second of the difference in sensitivity of different observation techniques to the different perturbation quantities.

1.3.5. *Properties of Nonadiabatic Solutions*

When driving and/or dissipative terms are included in the pulsation equations, the solutions are no longer perfect standing waves. Phasing between fluid velocity and pressure, temperature, and density perturbations changes, producing a number of important effects, even when the magnitudes of the changes are small.

First, the trapped oscillations, which transport no energy in the adiabatic case, always carry at least a small energy flux if they are nonadiabatic. This allows energy to 'leak' out through the top and bottom of the cavity. In the case of the solar p-mode oscillations, energy may be carried by the oscillation up to heights in the atmosphere where the mode changes character to that of a freely propagating wave. Such modes can produce weak acoustic shocks in the upper chromosphere and transition region. Second, if the driving is not steady, but is either stochastic or impulsive, modes may appear and then decay on the dissipative time-scale appropriate to the wave cavity. Third, the structure of the eigenfunctions changes. In a weakly nonadiabatic situation, the amplitude almost goes to zero at the nodes, but never becomes exactly zero as it does in the adiabatic case. In a strongly nonadiabatic situation, the nodes tend to lose their identity altogether. Fourth, avoided crossings may, if nonadiabaticity is strong enough, become actual frequency crossings between modes.

1.3.6. *Toroidal Oscillations*

The pulsation equations allow solutions with no radial displacement, i.e. the toroidal oscillations. The displacement vector for the toroidal solutions can be written

$$\delta \mathbf{r} = \mathrm{Re}\left\{\left[(\xi_r^t = 0)\mathbf{a}_r + \xi_h^t(r)\left(\frac{1}{\sin\vartheta}\frac{\partial Y_l^m}{\partial\phi}\mathbf{a}_\vartheta - \frac{\partial Y_l^m}{\partial\vartheta}\mathbf{a}_\phi\right)\right]\exp(-i\omega t)\right\}, \qquad (1.3.13)$$

where the superscript t (for toroidal) on ξ is intended as a reminder that the functional forms are different from those for spheroidal modes. Note that there can be no toroidal modes with $l = 0$.

For purely tangential motion in a spherically symmetric star without elasticity there are no restoring forces and so these modes occur with zero frequency. Thus the restoring forces for solar toroidal modes arise from rotation (Cox, 1980) and internal magnetic fields (Ledoux and Walraven, 1958, §83; Ledoux, 1974), both of which are relatively weak effects in the case of the Sun. The geometrical appearance is described at Bolt and Derr (1969).

Applying the computation method of Ferraro and Memory (1952) to the Sun, with $B = 1000$ gauss, yields a period for the lowest degree toroidal mode of 13.4 yr, suggestive of the solar cycle. Plumpton and Ferraro (1953) found that toroidal oscillations could travel as transverse, magnetic (Alfvén) waves along the poloidal field lines, with similar periods. They noted that in a rotating star torsional oscillations produce unbalanced Coriolis forces which cause meridional motions. Plumption (1957) found that disturbances leaving the core of the Sun would eventually emerge at the surface, with the latitude of emergence drifting from 30 degrees to the equator in roughly 5 yr. Dynamo waves (Parker, 1955, 1957) were found to exist in the numerical solutions of dynamo models by Yoshimura (1975), but were not expected to be observable.

In Section 2 we report observations by Howard and La Bonte (1980) that may in fact be torsional oscillations with a period equal to the sunspot cycle. It is not possible at the present time to compute torsional modes in a realistic solar model without parametrizing some of the important dynamo effects, hence analyses of that sort are open to some question. A numerical study by Yoshimura (1981), which uses a parametrized model, indicates that the observed oscillations may be driven by the longitudinal component of the Lorentz force generated by dynamo waves. Yoshimura found that under certain input conditions, his dynamo model gave an evolutionary pattern similar to the velocity oscillation pattern found by Howard and La Bonte.

1.4. EXCITATION AND DAMPING OF SOLAR PULSATIONS

Normal modes of a star can be excited in a number of ways. The relative time-scale for excitation of a mode to finite amplitude and for damping from the effects of dissipation processes determines (1) whether the mode can exist at all, and (2) its lifetime, and therefore the width (in frequency) of the peak it produces in a power spectrum. Here we discuss the driving and damping mechanisms that have been considered in connection with solar oscillations. We then review the studies that deal with the stability of solar normal modes and finally discuss computations and the implications of mode lifetimes.

1.4.1. *Excitation and Damping Mechanisms*

KAPPA AND GAMMA MECHANISMS. The κ-mechanism (Eddington, 1941; Baker and Kippenhahn, 1962) operates in regions where the opacity increases when the gas is compressed, thereby trapping excess radiation and further heating and compressing the gas. Such a process is possible when either (1) the opacity κ is proportional to a positive power of the temperature, or (2) κ is proportional to a positive power of the density and to a negative power of the temperature, with the relative temperature fluctuations small compared to those of the density. The mechanism operates in the relatively thin ionization-recombination zones of hydrogen, helium, and perhaps certain heavy elements, where $\Gamma_3 - 1$ and hence $\delta T/T$ are small. A small periodic acoustic perturbation will gradually increase in amplitude in such a region. During the compressive part of the cycle there is a slight excess heating which causes a pressure excess, while in the rarefaction phase the gas cools excessively. The amplitude of the oscillation thus increases during both phases, and can grow until the opacity saturates so that further increases in pressure no longer increase κ.

The γ-mechanism (Cox *et. al.*, 1966) provides driving in regions where the relative Lagrangian perturbation in the luminosity $\delta L/L$ decreases with radius and where δL is approximately in phase with the density perturbtion $\delta \rho$. This happens in ionization zones, where the energy of compression acts to increase the degree of ionization rather than to heat the gas. The essence of this mechanism lies in the fact that the energy reradiated by the gas is proportional to aT^4; when the energy flowing in at a particular radius ionizes the gas instead of raising its temperature during compression, the gas reradiates less energy than it absorbs and the net trapping of energy amplifies the oscillation. Both processes are described in detail in, e.g., Cox (1980).

In regions of the Sun where κ decreases during compression or where $\delta L/L$ increases with radius and is in phase with $\delta\rho$, the inverses of the κ- and γ-mechanisms cause damping of the oscillations. These forms of damping operate outside the regions where driving by radiation trapping is effective. Because densities, temperatures, and phase lags vary greatly with radius, the driving and damping need not balance, and in some cases (e.g. classical Cepheids) the effect of the driving regions can vastly outweigh that of the damping regions.

INTERACTION WITH CONVECTION. Interactions of pulsations and waves with convection are complex and can contribute to both damping and driving. These include thermal overstability, the Lighthill mechanism, stochastic excitation of modes and damping by turbulent viscosity, and dynamical interactions which appear in time-dependent mixing-length or dynamical computations. Convective motions also generate waves outside the convection zone via overshooting and by perturbing the pressure, density and/or temperature of the external gas.

Thermal overstability. This mechanism operates in an environment where horizontal temperature variations occur. A fluid element oscillating under a destabilizing buoyancy force and a restoring force provided by rotation, magnetic fields or compressibility (∇p) may exchange energy with the surrounding gas via thermal and radiative diffusion in the horizontal direction. In a convectively unstable region, the force of buoyancy tends to drive a displaced fluid parcel farther from its equilibrium position, opposing the restoring force. If the oscillating fluid absorbs heat at the bottom of its cycle by energy transfer in the horizontal direction, and loses it at the top, the buoyancy force is decreased at both extrema, making the restoring force relatively stronger, and the oscillation amplitude will grow until limited by nonlinear effects (Moore and Spiegel, 1966). This mechanism, with a pressure gradient restoring force, has been studied as a source of driving for solar nonradial oscillations; because it requires horizontal variations, it is not relevant for radial modes.

Lighthill mechanism and stochastic excitation of modes. Turbulent flow throughout the convection zone generates a broad frequency spectrum of acoustic waves by nonlinear coupling via turbulent Reynolds stresses. Stein (1968) extended the analysis of Lighthill (1952) to the case of a medium stratified under gravity. For the dominant quadrupolar radiation, the acoustic intensity is proportional to the eighth power of the turbulent velocity (Lighthill, 1952; Stein, 1968), and is thus very sensitive to the convective model. Waves generated stochastically in this manner can be trapped within the convection zone as described in Sections 1.2.1 and 1.3; those with the appropriate relations between frequency and horizontal wavenumber can set up standing wave patterns. The stochastic interaction of the acoustic modes with turbulent eddies is largely via the Reynolds stresses (see Section 1.4.3); other second-order terms in the fluid equations in general are less important (Goldreich and Keeley, 1977b).

Turbulent viscosity. Throughout the turbulent convection zone, modes interact weakly via Reynolds stresses with eddies having turnover times that are far from the mode period. These interactions on the net are assumed to remove energy from the pulsations, and are normally parametrized by a turbulent viscosity coefficient which turns out to be orders of magnitude larger than the molecular viscosity coefficient in regions of turbulent convection.

Perturbation of the convective flux. Perturbation of the convective flux can drive or damp oscillations in much the same way as does perturbation of the radiative flux via the γ-mechanism. Driving occurs when δF_c decreases in the outward direction and is approximately in phase with $\delta \rho$. When δF_c increases with radius and is in phase with $\delta \rho$, damping occurs. As the phase lag between δF_c and $\delta \rho$ approaches $\pi/2$, the interaction between them goes to zero.

Studies which resolve convective flows (Deupree, 1975, 1977) or are based on time-dependent mixing-length theory (Unno, 1967; Gough, 1977b) produce effects which are qualitatively different from the kinematic and static stability analyses on which the rest of the convection-pulsation studies are based.

Wave generation near convective region boundaries. Convective overshooting in regions just outside convection zones can produce internal gravity waves with temporal and spatial scales comparable to those of the convective elements (Lighthill, 1967). Thus in the stable layers below the convective zone very long period gravity waves may be produced which can, in principle, become trapped in a g-mode cavity in the deep interior.

NUCLEAR BURNING. Energy released from nuclear burning can drive modes whose eigenfunctions have relatively large amplitude near the solar core. Nuclear burning is extremely sensitive to perturbations in temperature; for example, for infinitesimal fluctuations, the ^{3}He + ^{3}He energy production rate goes roughly as the 12th power of the temperature. The oscillations likely to be affected by this mechanism are the g-modes trapped beneath the convection zone and possibly the low-l, low-order p-modes. Driving by variations in nuclear burning takes place on Kelvin–Helmholtz time-scales of the order of 10^6–10^8 yr, with radiative damping in the deep interior occurring on similar time-scales.

EXCITATION BY IMPULSIVE EVENTS. It may be possible to excite normal modes in response to impulsive events like large solar flares (Wolff, 1972a), in analogy with excitation of terrestrial normal modes by volcanoes and earthquakes. This mechanism has not been studied in detail, probably because it operates only occasionally, and thus is not important for understanding the ever-present quiet Sun oscillations.

NONLINEAR MODE–MODE COUPLING. This mechanism does not actually generate waves (hence does not altogether belong in this section), but is of interest because it can cause modes which are only weakly excited by other means to develop substantial amplitudes. The interaction occurs among finite-amplitude waves whose fluctuations produce significant mean (averaged over horizontal dimension or time) second-order cross terms in the fluid equations. Mode–mode interactions readily redistribute energy among g-modes. They can also produce a g-mode from two p-modes if the p-modes have small l-values and thus have significant energy in their eigenfunctions at the inner radii where g-modes are trapped.

INTERACTION BETWEEN PULSATIONS AND THE RADIATION FIELD. The transfer of radiation in a gas acts to smooth out static spatial fluctuations in the temperature.

With a moving fluid, the radiation field interacts dynamically and alters many aspects of the motion.

In the deep solar interior the gas is dense, hence the mean-free-path of the photons is extremely small, and temperature smoothing proceeds on a diffusion time-scale, with little dynamical interaction between radiation and small-amplitude motions. Typical time-scales for radiative damping of g-modes trapped below the convection zone are $O(10^6 - 10^8)$ yr. Within the convection zone, energy is carried vertically by convective transport and by radiative diffusion; because the radiative energy density is small compared to the internal energy density, advection of radiation is a small effect in the Sun. In both of these regions the radiation field is well approximated as a local variable with a Planckian frequency distribution which depends only on the local temperature.

In the atmosphere, however, and in a small region below the visible surface where surface effects can still be felt, interaction between the matter and radiation is strong and nonlocal, with damping time-scales as short as seconds for isolated temperature perturbations in the low photosphere. Radiative interactions in the outer layers can affect oscillation amplitudes, stability, and the upper boundary condition for the pulsations.

Here the photon mean free path is generally larger than the depth of the atmosphere. Perturbations in the local radiation field which, together with the perturbation in the Planck function, control the damping rate are thus determined by emission of radiation over a very large region, and can be difficult even to estimate. Nonlocal effects can bring about significant deviations in δJ from that given by local treatments; in an inhomogeneous medium δJ may exceed δB in some circumstances, implying driving rather than damping.

An enormous simplification occurs in treating radiation in an optically thin gas for the highly idealized case of an infinite, homogeneous, isotropic gas perturbed by a sinusoidal disturbance; here the mean intensity $\bar{J}$ and the perturbation $\delta \bar{J}$ can be written in terms of local quantities, yielding a local 'Newtonian cooling time' for temperature perturbations (Spiegel, 1957). For a stratified medium, one can easily, but not very accurately, define a height-dependent cooling time in terms of local quantities. A better treatment solves the angular moments of the transfer equation self-consistently with the fluid equations and closes the hierarchy of moment equations by demanding that the ratio of the second moment to the zeroth moment, the Eddington factor K/J, must equal $1/3$. This *Eddington approximation* allows a considerable degree of generality in the solution for $\bar{J}$, though it is often used with additional constraints and approximations under the same name. One such version (Unno and Spiegel, 1966) has been employed by Ando and Osaki (1975, 1977) and by Goldreich and Keeley (1977a, 1977b) to model perturbations of the radiative flux.

Schmieder (1977), in a fully nonlocal treatment of the radiative damping of propagating acoustic waves, has shown that a height-dependent effective cooling time τ_R cannot reproduce self-consistently the effects of nonlocal radiative damping on both the amplitudes and the phases of acoustic wave perturbations, even if τ_R is calculated separately for each wave frequency. Christensen-Dalsgaard and Frandsen (1983) found the Eddington approximation adequate for radial modes. For nonradial modes the value for the radiation field obtained with the Eddington approximation diverged significantly from that obtained in a fully nonlocal treatment when $l \approx 100$, and the discrepancy increased with increasing l.

1.4.2. *Mode Lifetimes*

The lifetime of a mode is determined by the rates at which excitation and damping occur and is often discussed in terms of the 'quality' Q, defined to be the ratio of the lifetime to the period of the mode. Reliable measurements of lifetimes may eventually provide greater understanding of the excitation and damping mechanisms. The distinction between global modes and local standing wave disturbances is made on the basis of lifetimes: an envelope disturbance is local if it is stochastically driven or locally excited by an impulsive event and has a lifetime shorter than the time required for it to propagate (at its group velocity) once around the solar circumference. Modes driven by global effects, such as the κ-mechanism, and modes having long lifetimes are global.

For the high-l p-modes, Deubner *et al.* (1979) found lifetimes of at least 12 h. Five minute p-modes with $l \approx 20$ have horizontal group velocities around 100 km s^{-1}, and thus circle the Sun in about 12 h. Duvall and Harvey (1983) find lifetimes of at least 3 days for modes with $l \leq 70$, hence intermediate-l modes appear to be global in character. Subphotospheric flow variations, however, may interfere with these estimates of mode lifetimes (Hill *et al.*, 1983).

For low-l modes it is possible to resolve individual peaks in the frequency power spectrum, and a lower bound to the lifetime can be derived from the widths of the peaks for each l. For modes observed in integrated sunlight, with $l \leq 4$, Claverie *et al.*, (1979) estimated lifetimes of at least 32 h and possibly more than 96 h. Grec *et al.* (1980) found lifetimes of about 48 h, and seemed to see substantial amplitude variations on 1 day time-scales. Woodard and Hudson (1983) found 168 h while newer observations by Claverie *et al.* (1981) imply lifetimes of about 28 days. Conflicting pieces of evidence thus leave the lifetimes of these modes an unsettled issue.

The mysterious 160 min mode, whose degree and order are not known, has exhibited phase constancy over the nine years it has been observed, but has varied in amplitude. No other clearly identifiable observations of low-l g-modes exist, but theoretical lifetimes are of the order of $10^6 - 10^8$ yr.

1.4.3. *Stability of Solar Pulsation Modes*

Numerous computational studies have examined stability of the various solar normal modes using different combinations of the driving and damping mechanisms just described. Overall, the analyses suggest that many modes are of uncertain stability because of large uncertainties in the physics of the solar models; it is specifically the nonadiabatic processes that we know least about.

The primary driving of a mode generally does not occur in the same physical locale as the main damping, and thus only a global stability calculation has any meaning at all. Uncertainties in our knowledge of physical conditions within the Sun, and difficulties in treating complex physics in a computationally satisfactory way, make the results of global stability analyses more uncertain than the small margins by which modes are found to be stable or unstable. The conclusions of the studies should therefore be viewed with some skepticism.

STABILITY OF SOLAR P-MODES. Driving mechanisms of the p-modes are the κ- and γ-mechanisms, Lighthill mechanism, stochastic interactions with convective cells, thermal overstability, and (rarely) radiative transfer (see Section 1.5.1). Damping comes primarily from radiative effects and turbulent viscosity. Solar p-mode stability studies which considered only the κ- and γ-mechanisms and thermal overstability have generally found the p-modes to be unstable (Ulrich, 1970; Wolff, 1972b; Ando and Osaki, 1975), with possible effects of the chromosphere quite uncertain because the modeling of the atmo-sphere was crude (Ulrich and Rhodes, 1977; Ando and Osaki, 1977).

When interaction with convection is taken into account, the stability results for many modes change considerably. Goldreich and Keeley (1977a), treating perturbations in radiative flux by the Eddington approximation and in convective flux by a time-depen-dent prescription for the mixing-length ratio (Cox et al., 1966), found all radial modes with $P > 6$ min to be unstable if turbulent viscous damping was omitted and stable with it included. Modes with $P < 6$ min were damped but not stabilized by turbulent viscosity. In contrast, Berthomieu et al. (1980), using a time-dependent mixing-length theory (Gough, 1977b) found all p-modes they studied ($l = 200$ and 600) to be stabilized by convection. Methods for estimating the convection—pulsation interaction are highly uncertain, and both groups (see also Gough, 1980) concluded that the narrow margins of stability or instability indicate the inability of theoretical calculations to unequivocally determine the stability of solar radial and p-mode oscillations.

Gough (1980) argued that if overstability (κ, γ, and convective overstability) is the driving mechanism, amplitudes should be much larger than are observed, because this type of mechanism is only limited by finite amplitude nonlinearities. Thus he concluded that the 5 min oscillation modes of both high and low degree probably are excited stochastically by the turbulent convection. This argument fails to take into account, however, that the driving by such mechanisms could be much weaker and damping much more rapid in the Sun than in Cepheid variables where the κ-mechanism is dominant.

Goldreich and Keeley (1977b) explicitly included the interaction between turbulent convection and the oscillations to study stochastic excitation of radial modes. For modes with ω^{-1} less than the turnover time of the largest eddies (i.e. those whose characteristic size equals the pressure scale height), they concluded that the energy of each radial mode would be approximately equipartitioned with the kinetic energy of the (resonant) turbu-lent eddies whose turnover time approximately equals the inverse frequency of the mode.

Keeley (1977), from a stochastic analysis for nonradial p-modes similar to that of Goldreich and Keeley for radial modes, predicted a variation of velocity amplitude with frequency similar to that observed by Rhodes et al. (1977a). Gough (1980), using the Berthomieu et al. (1980) model and assuming exact equipartition of energy between the modes and resonant eddies, obtained results nearly identical to those Keeley found from a full stochastic excitation analysis. However (Gough, 1980), the relative surface velocities predicted by these studies for p-modes with different l-values and for f-modes, do not agree with the observations of Deubner et al. (1979).

Excitation of low-l modes was studied by Baker and Gough (Gough, 1980) using an extended solar envelope model, and employing the diffusion approximation for the radiative transfer and Gough's (1977a, b) time-dependent mixing-length theory. They found stability for all radial modes with $\omega < 4$ mHz (p_{28}). For $2.5 < \omega < 4.0$ mHz their stability coefficients were within a factor of 2 of those obtained by Goldreich and Keeley (1977a), who treated the convection in a totally different manner. The theoretical power spectrum they obtained by assuming equipartition is qualitatively similar to the observed spectra of Claverie *et al.* (1979) and Grec *et al.* (1980).

In the analyses discussed so far, radiative effects were treated in the Eddington or Newtonian cooling approximations. Christensen-Dalsgaard and Frandsen (1983) reported initial results in a study comparing mode analysis with the Eddington approximation to calculations made with a fully nonlocal transfer equation coupled to the fluid equations. The solar model was otherwise the same for the two sets of computations. Their results for the value of the radiation field (Section 1.4.1) indicate that stability of radial modes will be the same in both cases, but that stability of nonradial modes with $l \geq 100$ should be calculated using the nonlocal treatment. Using the Eddington approximation, they found all acoustic modes to be stable, in disagreement with Ando and Osaki's (1975, 1977) results obtained with the same method; they attributed the difference to Ando and Osaki's assumption that $J = B$ in the equilibrium model.

STABILITY OF SOLAR G-MODES. Whereas *p*-mode observations are now almost commonplace, the evidence for interior *g*-modes is sketchy, but tantalizing. Observations of a 160 min oscillation will be discussed in a later section, as will observations of several periods less than an hour which could be either *g*-modes or *p*-modes. Much of the impetus for theoretical studies of *g*-mode stability has arisen from the disagreement between theory and observation of neutrino production in the present Sun (see Fowler, 1972 and Newman, Chapter 17 of this work). Unstable low degree, low-order *g*-modes might be able to provide occasional mixing of the nuclear burning core, which would temporarily alter the composition, producing transients that change the nuclear energy generation rates so as to produce a much smaller neutrino flux and may have been responsible for the ice ages (Dilke and Gough, 1972). Ulrich (1974) noted that overstable *g*-modes do not necessarily cause significant mixing of the core, they will produce such mixing only if the large amplitude waves result in turbulent or convective motions. Another significant reason for studying *g*-mode stability will be discussed further in Section 3.3. It appears that the use of *g*-mode eigenfunctions may be of great importance in resolving the core structure via inversion methods, and thus it is useful to know what part of the frequency spectrum may contain unstable *g*-modes.

The dominant mechanisms for generating interior *g*-modes of low *l* appear to be perturbations in the nuclear energy generation rate ϵ and interaction between the *g*-modes and convection via the gamma mechanism. In some circumstances, interaction with radiation may produce some driving in the nuclear core. Damping results from interaction with radiation throughout the Sun and from perturbations in the convective flux in much of the convection zone. The excitation and damping of the interior *g*-modes take place on a Kelvin–Helmholtz time-scale.

Dilke and Gough's suggestion that in the presence of a strong composition gradient in the nuclear core, some low-*l* *g*-modes could become overstable and ultimately cause core mixing, prompted a series of stability studies, which have more or less converged on certain conclusions. In general, a 1 $M_\odot$ zero age main sequence (ZAMS) star is stable to interior *g*-modes; as the ^{3}He gradient builds up with time in the core, the star becomes unstable to low-*l* *g*-modes at around 2×10^8 yr, and then by an age of about 3×10^9 yr the *g*-modes are again largely stabilized. The age of the present Sun is believed to be about 4.5×10^9 yr. The stability coefficients for several low-l g-modes are extremely small, giving driving or damping times of the order of 10^6 to 10^8 yr.

Several studies considered stability without accounting for effects of the envelope convection zone (Dziembowski and Sienkiewicz, 1973; Christensen-Dalsgaard et al., 1974; Shibahashi et al., 1975). These gave conflicting results regarding the stability of low-l, low-n g-modes at intermediate stages of the Sun's evolution. However, all but Christensen-Dalsgaard et al. found all g-modes to become stable again well before the age of the current Sun. Christensen-Dalsgaard and Gough (1975) found all previously unstable g-modes to be stabilized by radiative damping in the the outer 300 km of the convection zone by the present epoch of the Sun; their analysis was fully nonadiabatic, but neglected the Eulerian perturbation in the convective flux.

Unno (1975) considered the possibility of mixing due to overstability of g-modes trapped in the nuclear burning core. He found that overstability cannot occur for core modes in a current standard solar model, but that in previous epochs the gradient in mean molecular weight could alter and radial dependence of N^2 so as to trap the g_1 ($l = 3, 4$) modes, while both higher and lower l modes and higher n modes were not trapped. Unno concluded that these modes could have produced the mixing postulated by Dilke and Gough (1972).

A group at Liege extended Unno's (1967) time-dependent mixing-length theory to estimate the perturbation of the convective flux, δL_c, in quasiadiabatic pulsation calculations. Gabriel et al. (1975) and Boury et al. (1975) found the present Sun to be stable to all g-modes only if they included the perturbation of the convective flux. Gabriel et al. (1976) considered core mixing during solar evolution, and found that unstable g-modes could occur in a fully evolved Sun with $\delta L_c \neq 0$, provided the mixed core had a sufficiently large mass fraction; moreover, core mixing could produce neutrino fluxes in the range of 1 to 2 SNU.

Saio (1980) used a fully nonadiabatic formulation which included the flux perturbation δL_c, in an evolving solar model with an unmixed core. He found the present Sun to be unstable to the g_2 ($l = 1$) mode with a period of 80 min; driving was produced by fluctuations in nuclear energy generation from the $^3\text{He} + {}^3\text{He}$ reaction in the core and from the perturbation in δL_c in the hydrogen ionization zone. In contrast to the nonadiabatic analysis by Christensen-Dalsgaard and Gough (1975), Saio did not find radiative damping in the outer envelope to be sufficient to stabilize this mode.

Driving of the 160 min oscillation remains a mystery, as none of the low-l g-modes which appear to be unstable or even marginally stable has a period near this value. Keeley (1980) found that modes with periods around 160 min could be driven by stochastic coupling to convection, but that the inferred amplitudes of these modes were orders of magnitude too small to be observed. To date, no detailed studies have been done on nonlinear mode—mode coupling that could produce an isolated resonance at 160 min. It may be that the 160 min oscillation could be produced as a subharmonic of the unstable 80 min g_2 ($l = 1$) mode found by Saio (1980). On the other hand (Gough, 1980), the oscillation need not be that of any normal mode, but could be generated by a fully nonlinear process.

1.5. Detailed solutions for frequencies and frequency splitting

Observations of high-l p-modes have been refined to the extent that errors in the ridge locations are very small. Computed ridge locations, slope and curvature, and separation between ridges are more uncertain than the observed values, and depend on several features in the solar model: (1) depth of the convection zone; (2) composition and possible composition gradients; (3) equation of state and opacities; (4) presence of chromosphere and corona; (5) radiative transfer in the outer layers of the convection zone and in the atmosphere.

The f-modes and low-l p-modes in the 5 min band and the longer period modes with low l-values are sensitive also to effects deep in the Sun, namely (6) core composition and gradients, and (7) temperature structure below the convection zone. While the high-l modes are far too densely packed in frequency to resolve individual modes, those with low l-values are relatively isolated, and provide a tool for studying the physical effects that produce frequency splitting, namely (8) solar rotation and (9) possible large-scale internal magnetic fields.

In this section we shall first discuss how variations in each of features (1)–(7) alters oscillation frequencies, and then present slightly expanded discussions of (8) and (9), with formulae for the resulting frequency perturbations. In Section 3 we shall consider what we can learn about the interior of the Sun from observations of oscillations (Section 2.2) in the light of the computational information reviewed here. The way in which the various effects (1)–(9) are constrained relative to each other by imposing the requirement that the Sun evolved from a zero age main sequence star to the present Sun (with the correct solar parameters) will also be discussed in Section 3.

1.5.1. *Effects of Structure on Unperturbed Frequencies*

Depth of the convection zone. All solar models used to calculate oscillation frequencies employ some form of mixing-length theory to describe convection. Varying the ratio of the mixing-length l_c to the density scale height H is equivalent to changing the depth of the convection zone, with the depth increasing as the ratio increases. For a given degree l the frequencies of the high-l p-modes systematically decrease as l_c/H increases. This response of the frequencies is actually to the total entropy in the convection zone (Ulrich and Rhodes, 1977), and the high-l p-mode frequencies change little if the convection zone depth increases beyond about 30% $R_\odot$. The low-l modes exhibit much less sensitivity in frequency to the convection zone depth because the eigenfunctions of these modes are much more spread out in radius.

Composition, composition gradients, and core structure. Increasing the heavy element abundance in a ZAMS Sun decreases the H abundance and in general produces a deeper convection zone in the fully evolved Sun, hence decreases the high-l p-mode frequencies (Gough, 1982c). For the low-l p-modes the frequencies and the frequency spacings between (n_1, l) and (n_2, l) modes and between (n, l) and $(n, l + 2)$ modes are altered in a more complicated manner (Christensen-Dalsgaard, 1982; Ulrich and Rhodes, 1983). Varying the initial composition also changes the present composition of the core, hence alters the nuclear reaction and neutrino production rates, the core opacities and the Brunt–Väisälä frequency. These changes shift the frequencies of the low-l p-modes and the interior *g*-modes. Composition gradients in the core significantly affect the Brunt–Väisälä frequency there, hence affect the behavior of interior *g*-modes.

Equation of state and opacities. There are significant uncertainties in the standard opacity tables, particularly in the values given for more complex atomic species, and in the atomic physics that enters the equation of state. Changes in opacity, such as the difference between use of two different opacity tables, are found by Christensen-Dalsgaard (1982) to produce measurable changes in mode frequencies. Shibahashi *et al.* (1983) found that including the Debye–Hückel correction for electrostatic forces in an ionized gas produces corrections to the gas pressure that range up to 7% in the convection zone.

Upper boundary condition, chromosphere, and corona. The simple switch between a zero vertical velocity and zero pressure perturbation at the upper reflection point causes a major change in the form of the eigenfunction, the first producing a node and the second an antinode in the vertical displacement at the upper boundary. Important changes in the eigenfunctions and eigenfrequencies are produced by several processes in the outer layers of the Sun. Many trapped modes become freely propagating waves in the upper atmosphere, thereby allowing some wave energy to leak out of the mode cavity and changing the phase relations among the velocity components and the fluctuations in T, ρ, and p at the upper boundary. A mode that becomes freely propagating at some height must have a mixture of growing and decaying solutions in the evanescent region in order to match all the necessary continuity conditions, and Hill and coworkers have pointed out that it can be important to take this mixture into account when interpreting spectrum line data. Radiative transfer introduces large nonadiabatic effects which change the phase relations and the relative magnitudes of the perturbations at the upper boundary.

Radiative transfer. Radiative transfer usually has its greatest effect on the imaginary (damping) part of the frequency. However, Christensen-Dalsgaard and Frandsen (1983) find that the Eddington approximation gives inaccurate values for the mean radiation field for $l > 100$. Resulting changes in the real part of the frequency have not yet been calculated.

Convective temperature and velocity structure. Gough and Toomre (1983) found that small perturbations in horizontal flow velocity theoretically can produce shifts in ridges which are asymmetric between the $+m$ and $-m$ modes of a given l and n, while perturbations in the temperature produce symmetric shifts. The two effects, therefore, should be separable.

1.5.2. *Effects of Rotation*

For a compressible sphere whose rotation rate is a function of radius but not of angle, i.e. $\Omega = \Omega(r)$, a very general variational principle for small amplitude displacements can be used (Lynden-Bell and Ostriker, 1967) to derive the eigenfrequencies and stability of normal modes. The formulation includes the perturbation Ψ', the pressure, and any other unspecified forces (e.g. magnetic fields). Gough (1978b) used this method to show that the frequency splitting for a slowly rotating sphere with $\Omega = \Omega(r)$ is given in the frame of a nonrotating observer in terms of the surface rotation rate Ω_0 by

$$\omega' = m(1 - C_{ln}) \, \Omega_0 \tag{1.5.1}$$

where C_{ln} is given in terms of $\Omega_1 = \Omega(r) - \Omega_0$ and the vertical and horizontal spatial eigenfunctions, $\xi \equiv \xi_r$ and $\eta \equiv \xi_h$ (see Equation (1.3.1)), as

$$C = \frac{\int \rho [\eta^2 + 2\eta\xi - (\Omega_1/\Omega_0) \{(\xi - \eta)^2 + [l(l+1) - 2]\eta^2\}] \, dz}{\int \rho \{\xi^2 + l(l+1)\eta^2\} \, dz} . \qquad (1.5.2)$$

From these expressions we can see the important qualitative features of rotational splitting: (1) the frequency perturbation is purely real, hence is nondissipative; (2) to first order the perturbation in ω is strictly linear in m when Ω is only a function of radius, hence produces $2l + 1$ evenly spaced frequencies for modes of degree l, separated by $(1 - C_{nl})\Omega_0$ in the observer's frame [if $\Omega = \Omega(r, \vartheta)$ the spacing is not uniform] ; (3) for negligible Ω_1 the dominant terms in C_{nl} do not contain the effects of differential rotation, but do depend on the eigenfunctions and on l and hence change from mode to mode; (4) to lowest order, C_{nl} is linear in Ω_1 and the contribution of Ω_1 is weighted by different combinations of the eigenfunctions and l than are the solid-body terms [when $\Omega_1 = 0$ this expression reduces to that derived for solid-body rotation (Cowling and Newing, 1949; see also Ledoux, 1951)] ; (5) when $\eta \ll \xi$ and $l \gg 1$ the differential rotation term dominates C_{nl}, even when Ω_1 is fairly small.

For comparison with observation, we need estimates of the magnitudes of C_{nl} and of the differential rotation term it contains. First we consider the case of solid-body rotation. If we set Ω_1 equal to zero and make the very crude approximation that the ratio α of η to ξ is depth independent, we obtain the simple expression $C_{nl} = (\alpha^2 + 2\alpha)/[1 + l(l+1)\alpha^2]$. For purely radial modes, $\alpha \equiv 0$ by definition, and it is small for low-l p-modes, yielding small values of C_{nl}. As l increases, the modes exhibit more nonradial motion, but the denominator of C_{nl} becomes dominated by the large $l(l+1)$ term, hence C_{nl} is again small (see, e.g., Gough, 1981). For the solar case, with $\Omega_0 = 3 \times 10^{-6}$ s^{-1}, $|C_{nl}\Omega/\omega_{nonrot}|$ has a maximum estimated value of about 10^{-4} at $l = 10$, and is about one order of magnitude smaller at $l = 1$ and two orders smaller at $l = 1000$. These values are currently below the limits of detectability, even for a continuous observing run of several days, and the observable shifts are given strictly by the $m\Omega_0$ term that arises in the coordinate transformation from the rotating to be observer's frame.

We now consider the case of $\Omega_1 \neq 0$ and $l \gg 1$, noting that for the high-l p-modes $|\xi| \approx l |\eta|$ (Gough, 1978b). The solid-body terms of C_{nl} can be neglected, and for high-l p-modes the differential rotation term becomes approximately

$$mC\Omega_0 \approx \frac{-m \int \rho\Omega_1 [\xi^2 + l(l+1)\eta^2] \, dr}{\int \rho [\xi^2 + l(l+1)\eta^2] \, dr} \qquad (1.5.3a)$$

while for the low-l modes it tends to

$$mC\Omega_0 \approx \frac{-m \int \rho\Omega_1 \xi^2 \, dr}{\int \rho\xi^2 \, dr} . \qquad (1.5.3b)$$

The measurable quantity that results from the differential rotation is thus for each mode a weighted (by the eigenfunctions and the density) average of Ω_1. The perturbation in each eigenfrequency is determined by the range of values of Ω_1 in the radial region where the corresponding eigenfunction is largest. From Equations (1.5.3) we see that the change in the frequency has opposite sign for $+m$ and $-m$ components, hence the perturbation should produce an asymmetry between eastward and westward propagating $(\pm m)$ components of the same (n, l) mode. We shall see in the section on observations that data analyzed in a manner that separates the $\pm m$ components and plots the two sets of ridges on the right and left sides of the ordinate does indeed exhibit a slight, and measurable, asymmetry.

1.5.3. *Effects of Internal Magnetic Fields*

The splitting caused by magnetic fields is much more elusive to calculate than that by rotation, because it depends on the configuration of the field and on the field orientation with respect to the rotation axis. At present we have only the vaguest hints of what the magnetic field may look like in the deep interior and there is nothing resembling consensus on the matter. There are a few theoretical results that may facilitate observational distinction between rotational and magnetic splitting. Ledoux and Walraven (1958, p. 537) note that for several simple field configurations the magnetic field splitting depends on $|m|$ rather than on m, hence produces $l + 1$ components instead of $2l + 1$; further, the frequency splitting is generally proportional to H^2/ρ, hence is very small except for enormous values of the field H. Magnetic splitting is not strictly linear in $|m|$, so the components for a given l and n are not expected to have exactly even spacing.

There are few calculations available; the results of one fairly complete study may elucidate the foregoing remarks. Using a variational method, Chanmugam (1979) studied combined differential rotation and magnetic field, with the field aligned along the rotational axis. For general axisymmetric perturbations, he obtained an equation for the eigenfrequencies of the form $-a\omega^2 + 2b\omega + c = 0$. The stable solutions ($\omega$ real) are given by $\omega = (b/a) \pm (1/a) [b^2 + ac]^{1/2}$, with $b^2 + ac > 0$. For the case studied, b contains a term linear in m and linear in Ω, a is the positive definite integral $\int \rho \xi \xi^* \, dV$, and c contains terms linear and quadratic in m and quadratic and higher order in Ω. In this configuration, m does not enter any of the magnetic terms, but the presence of the magnetic field complicates the m dependence of the rotational terms. An analysis by Goossens (1972) for the frequency splitting of high-order p-modes in a magnetic star shows frequency perturbations whose m dependence only involves $|m|$ and m^2, and is distinctly nonlinear in m (see Gough, 1982b, Equation 9).

1.6. FUTURE THEORETICAL NEEDS

In order to derive meaningful information about the solar interior from observations of normal mode frequencies, several aspects of model calculations require improvement.

First, aspects of basic microphysics need to be redone. While recent work by Shibahashi *et al.* (see Section 1.5.1) has improved the equation of state, they note that significant problems remain in the calculation of the ionization (Saha) equations. Opacities likewise need improved calculations. The 'neutrino problem' points out the presence of uncertainties in either reaction rates or composition and composition gradients of the core.

Second, a fully consistent treatment of radiative transfer is needed for the nonradial modes. Moreover, improved analysis of the interaction between radiation and convection may revise current pictures of the outer layers of the convection zone.

Third, of course, is the convection zone itself. Even time-dependent mixing-length theory fails to take into account many of the dynamical and multidimensional effects of convection. Although mixing-length theory is essential for stellar *evolution* calculations, full dynamical calculations should be possible in the future for studying the *present* behavior of the convection zone, to learn more about the structure of the region and about the interaction between convection and pulsations.

Fourth, the atmosphere and the convection zone are inhomogeneous in the horizontal dimension, thus negating the assumption of spherical symmetry on which model calculations are based. It is currently unknown how important the horizontal inhomogeneities are in perturbing the frequencies. Because the structure of inhomogeneities varies with time, resulting perturbations in the frequencies will vary also. Possible excitations or transfer of energy among existing modes via mode–mode coupling are, in principle, calculable, but have largely not been calculated for acoustic-gravity waves.

The remaining effects are those whose structure, and therefore physics, is to some degree intrinsically not deducible from first principles. These include large internal magnetic fields, the rotation curve $\Omega(r)$ or $\Omega(r, \theta)$, and the composition and composition gradients of the nuclear burning core. Exploration of these phenomena will require a combination of numerical experiments in direct model calculations, together with refinements in both the techniques and the observational input into inversion calculations.

For propagating waves, which can contribute to mechanical heating and microturbulence in the atmosphere, the theoretical needs are somewhat different. In this case one of the most pressing concerns is to know the amount of energy transported by different varieties of waves through different heights in the atmosphere. Another is the question of where and how dissipation of the nonlinear waves occurs. Thus wave studies need more reliable calculations of wave generation, some of which will rely on improved convection calculations. For waves the calculation of nonlocal (and probably non-LTE) radiative damping is critical, as damping in the outer envelope and low photosphere has a major effect on wave energy fluxes. Finally, nonlinear calculations are needed to determine the heights of energy deposition and the manner in which it occurs. The other important theoretical issue for waves concerns the complicated manner in which waves affect spectrum lines (see Section 2.1). Our ability to detect propagating waves reliably will require improvements in our ability to analyze features of spectrum lines specifically for evidence of waves.

2. Observations

2.1. OBSERVATIONAL TECHNIQUES

2.1.1. *Diagnostics of Spectrum Lines*

Velocity fields associated with temperature and density fluctuations produce effects on spectral lines that are complex and difficult to decipher. A spectral line represents an integral over several hundred kilometers of depth and an average over an area that is seldom smaller than 500 km on a side (mainly because of variations in refraction caused by motions in the Earth's atmosphere). At each point in this region, the line profile function is shifted by an amount determined by the local line-of-sight velocity, and has a depth determined by the local opacity, which is a function of the temperature and density. For lines that are not in local thermodynamic equilibrium (e.g. most chromospheric lines), the line depth is affected by temperature and density variations over a large area via their effects on the local radiation field $\bar{J}$.

We can distinguish four effects that a velocity field such as a wave, pulsation or convective mode can have on an observed spectrum line: (1) a simple Doppler shift; (2) broadening of the line, i.e. an increase in the frequency range of the line; (3) change in line strength, i.e. change in the integrated area under the emergent line profile; and (4) asymmetry in the shape of the line. A pure Doppler shift occurs when the velocity is coherent over the entire line-formation region, while line broadening becomes important when the line represents an average over a wide range of velocities, so that through local shifting of the profile function the line covers a broadened range of frequencies. Changes in line strength can be produced by fluctuations in temperature or density and also, in an unsaturated line, by vertical variations in the velocity. Asymmetry can result from the interplay between the velocity and the thermodynamic perturbations. For example, the observed line will be skewed if the temperature is elevated wherever the velocity is upward and diminished where the velocity is downward; the resulting line shape depends critically on the phase shift between velocity and temperature. Asymmetry also can occur when there are strong height gradients in the line-of-sight velocity.

A single dynamical process may produce all four effects in varying degrees. Convective overshoot velocities tend to produce velocity-gradient asymmetries and line strengthening. In the granulation overshoot, for example, the velocities are large enough, and the overall effect on the line profile is simple enough, that the basic form of the motion can be identified. Even so, there is a great range in the magnitudes of the velocities deduced from photospheric lines by different observers.

The 5 min oscillation modes are seen almost exclusively in the Doppler shift and in variations of line intensity. The vast amount of existing information about them comes, of course, from Fourier analysis of long time-strings of data taken on one and two-dimensional spatial rasters. Such analysis yields useful results only because the pulsations exhibit coherence on long spatial and temporal scales. Information about the energy density in oscillatory modes, however, is likely to require more detailed analysis of line profiles, as knowledge of phase relations may become critical.

Waves with wavelengths of, say, 1000 km or less, produce very small Doppler shifts, but may contribute to all three of the other effects. The signatures of waves are in general much more difficult to detect and decode than are those of the relatively coherent oscillation and convection velocities. The line broadening they produce cannot in any way be distinguished from that caused by random turbulence. Changes in line strength can be measured on spatial rasters over time spans long compared to relevant wave periods and Fourier analyzed, but such a procedure tends to be thwarted by, first, the small size of the fluctuations, which may be swamped by competing velocity fields, and second, observational problems that will be discussed in this section.

The study of asymmetry produced by propagating waves presents special problems. The most desirable procedure would be a type of inversion, in which intensity information obtained at many frequencies across the line profile is used to infer the run of velocity, temperature, and density over the depth range of the line formation region. Such a procedure is not only difficult and limited by errors in the intensity measurements, but also requires some *a priori* assumptions about the velocity fields that have perturbed the line. The actual velocity field probably contains a superposition of several waves at any given time, in addition to nonwave velocities.

Another, more tractable, alternative is to identify forms of asymmetries by obtaining synthetic waveforms (for different sorts of waves at varying frequencies and wavenumbers) from dynamical or kinematical calculations and then use these to obtain synthetic perturbed spectrum line profiles. This approach is currently hampered by our lack of precise knowledge of the physical environment through which the waves propagate and our inability to properly treat the interaction of the radiation field with fluid motions in the photosphere. For propagating waves, these factors have major effects on vertical wavelengths, phase relations, and relative amplitudes of the velocity and thermodynamic fluctuations, all of which are important in determining the shape of a spectrum line perturbed by the wave. Satifactory techniques for identifying and studying propagating waves in the solar atmophere have yet to be developed. However, a number of techniques which obtain partial information about propagating waves, often rather indirectly, will be discussed later in this section.

2.1.2. *Techniques for Observing Oscillations and Trapped Waves*

The methods that have been employed to observe solar oscillations fall into three categories: (1) Doppler shift measurements; (2) high-precision intensity measurements; and (3) measurements of the apparent solar diameter.

Most Doppler measurements of high-l p-modes have been made with traditional spectrographs. These instruments provide the spatial resolution required to distinguish the structure of high-degree tesseral harmonics, but generally have relatively poor noise and stability characteristics. Furthermore, spectrographs often have a field of view substantially smaller than the solar disk, limiting the wavenumber resolution available with such instruments. Studies of low-l oscillations require both low noise and full-disk coverage. Thus, these studies usually rely on differential observations made with wide-field spectrographs, or on observations of integrated flux made with atomic resonance devices. Because the displacements of p-modes are predominantly radial, useful velocity observations are limited to the central part of the solar disk, though integrated-flux measurements necessarily involve as well the Doppler shift of segments of the solar disk near the limb.

No fully three-dimensional (X, Y, T) analysis has yet been performed on spatially resolved velocities. Deubner (1975) employed a one-dimensional scan $880''$ long to obtain a two-dimensional (X, T) data set. Assuming horizontal isotropy for the oscillations, he converted his estimates of k_x along the slit into estimates of $k_h = (k_x^2 + k_y^2)^{1/2}$. This and other techniques that average along one of two orthogonal axes assume that tesseral surface harmonics can be represented as horizontally propagating plane waves of the form $\exp(ik_x x + ik_y y)$, and thus introduce small errors. These have thus far been unimportant, but may matter in future comparisons between theory and observations (Hill, 1978).

Frazier (1968b) and Rhodes *et al.* (1977a, b) obtained three-dimensional velocity data, but averaged the velocity in one spatial direction to filter out obliquely propagating waves before computing two-dimensional $k_h - \omega$ power spectra. The k_x computed in this way is thus nearly the same as the total horizontal wavenumber k_h. Deubner et al. (1979) and Hill et al. (1983) have also used this long-slit averaging technique.

Duvall et al. have used a related technique which does not require storing a full three-dimensional data set (see Rhodes et al., 1981b). A rotating glass prism averages light over one direction on the Sun onto the entrance slit of a spectrograph. The resulting line profiles then provide a Doppler shift averaged over an elongated slice of the solar disk. These Doppler shifts are then transformed to give a $k_x - \omega$ power spectrum, as in the previous cases. A disadvantage of this method is that the information lost by the optical averaging process can never be recovered.

Sectoral harmonics are sharply peaked in amplitude at the solar equator, and peak more sharply as l increases. In terms of the $k_h - \omega$ diagrams, this means that a given p-mode ridge corresponds more and more to the equatorial region as k increases. The high-l p-modes are also more confined to the surface than are the lower l modes.

While spatially resolved imaging experiments allow one to observe high-l harmonics, their small-amplitude sensitivity is low because of the drift and noise inherent in unstabilized spectrographs. This small-signal sensitivity can be improved by measuring the line profile position either differentially or with a suitably stable wavelength reference. Two variations of this idea have been applied to solar oscillations. One of these (Fossat and Ricort, 1975; Brookes et al., 1976; Grec and Fossat, 1977; Claverie et al., 1979, 1980, 1981; Grec et al., 1980) uses a sodium or potassium resonance cell to integrate the Doppler shift over the entire visible surface (i.e. views the Sun as a star). This integration averages out the small-scale variations of high-l harmonics. However (Fossat, 1981), the full-disk technique is sensitive to transparency variations in the Earth's atmosphere, because such fluctuations occur at different places in front of the rotating solar disk and so continuously alter the position and shape of the line profile.

The other large-scale Doppler method was developed at the Crimean Astrophysical Observatory (Severny et al., 1976; Kotov et al., 1978). It employs a differential, rather than an absolute, technique. An occulting mask is used to measure alternately the mean Doppler shift of the central portion of the solar disk (a circular region with $R = 0.60–0.70 \, R_\odot$) and of the surrounding annulus. Taking the difference between the two mean velocities minimizes instrumental drifts and eliminates the large trend introduced in the full-disk measurements by the Earth's rotation. Unfortunately, the differential technique is as sensitive to telluric transparency effects as the full-disk method (Fossat, 1981). This method was slightly modified for use at the Stanford Solar Observatory (Dittmer, 1978; Scherrer, 1978; Scherrer et al., 1979, 1980).

The predominantly horizontal velocities of the low-frequency g-modes suggest that they will be best observed near the solar limb. No systematic velocity study for this purpose has yet been carried out, largely because geometrical foreshortening makes the determination of spatial scales difficult for oscillations seen near the limb.

A few spatially resolved oscillation studies have used time series of two-dimensional images of intensity in either a line core (Rhodes et al., 1977b) or the continuum (Brown and Harrison, 1980a, b). Both of these studies concentrated on small portions of the disk. Recently, Harvey, Duvall, and Pomerantz (private communication) obtained time series of full-disk intensity images takes in the core of Ca II K line. These data are still being reduced, and no results are yet available. Intensity oscillations can be observed over a larger portion of the disk than velocity oscillations, because the intensity fluctuations of a mode are independent of μ, the cosine of the center-to-limb angle, while the radial velocity falls off as μ. Thus the horizontal wavenumber resolution, Δk_h, of the full-disk intensity $k_h - \omega$ diagrams can be better than that of the (necessarily) partial-disk velocity $k_h - \omega$ diagrams. The principal signal contamination for intensity measurements comes from detector noise, variations in atmospheric transmission, and intensity fluctuations in the solar granulation. Transmission effects may be minimized by comparing the intensity in the core of a solar line with that in the wings, but the other noise sources are more troublesome, and cause intensity oscillation measurements to have lower signal-to-noise ratios than do velocity measurements.

The planetary reflection measurements by Deubner (1981) were the first successful detection of solar oscillations using measurements of the integrated solar intensity. A much superior set of measurements was recently obtained with the ACRIM instrument on the SMM satellite (Willson, 1979) and used (Woodard and Hudson, 1983) to study low-l p-modes.

Astrometric techniques consist of repeated measurements of the apparent solar diameter. By measuring the intensity distribution near the extreme solar limb, independent determinations of the limb position at opposite sides of the disk are made at regular intervals; these are combined using careful astrometric methods to produce a times series of the total angular diameter. Because the intensity measurements refer to only a small area of the solar surface, this method is sensitive to oscillations with a rather large range of l-values. We shall discuss this technique further when we describe the observations.

Each of the observing techniques just described has a unique variation of sensitivity to oscillations with different degrees. Generally, the sensitivity of any observing technique is low whenever more than one horizontal wavelength of the oscillation fits within the smallest spatial scale resolvable by that technique. Thus, all techniques lose sensitivity when l is large enough, but those with high spatial resolution maintain their sensitivity out to higher l-values than those with low resolution. Using the known geometry of each method, one can calculate the sensitivity variation for each of them (Dziembowski, 1977; Hill and Rosenwald, 1978; Hill, 1978; Christensen-Dalsgaard, 1981; Christensen-Dalsgaard and Gough, 1980a, 1982). Although the dominant considerations in such calculations are geometrical, the results depend somewhat on what other assumptions one uses. Common approximations include neglecting line shifts caused by solar rotation or non-oscillatory velocity fields, neglecting variation in the radial velocity eigenfunction with height in the line-formation region, and assuming that all modes with a given n and l (but different m) are excited equally. The most complete calculation to date is that by Christensen-Dalsgaard and Gough (1982). They included the effects of solar limb darkening, and described the spatial filters of several observational techniques in terms of one filter function due to radial velocities and another due to horizontal velocities. This description shows that the full-disk filter function is most sensitive to $l = 1$, with $l = 0$ and $l = 2$ yielding slightly smaller values; $l \approx 3$ is about the largest l-value visible in full-disk measurements. The differential velocity measurements conducted at Stanford and the Crimea should be most sensitive to l-values between 2 and 6, with maximum sensitivity at $l = 4$.

2.2. OSCILLATIONS OBSERVATIONS

2.2.1. *Observations of 5 Minute Period p-Mode Oscillations*

The most obvious manifestations of resonant waves in the Sun are the *p*-mode oscillations with periods of approximately 5 min, discovered by Leighton *et al.* (1962). Later observations (e.g. Evans *et al.*, 1963; Frazier, 1968a; Tanenbaum *et al.*, 1969) showed that these oscillations are ubiquitous on the Sun outside of sunspots. The oscillations are best seen in the Doppler shift of photospheric and chromospheric spectrum lines, with inferred velocities that are almost entirely vertical. The photospheric r.m.s. velocities seen are typically 400 m s^{-1} when averaged over areas a few arcsec square, but decrease as the square root of the observed area if larger areas of the Sun are considered. The oscillations are not confined to a single period, but rather span a range between about 4 and 8 min. The result of this finite bandwidth is that any given point on the Sun *appears* to respond to distinct wave packets, with each packet containing typically three or four full cycles of oscillation. One may sample the vertical structure of the oscillations by measuring the Doppler shift in several lines formed at different heights in the atmosphere. In chromospheric lines, the frequency band containing the oscillations tends to spread out and move to higher frequencies. The amplitudes of the oscillations increase very slowly with height, and the vertical phase speed is much larger than the sound speed.

Little progress was made in understanding the physical nature of the oscillations until various theoretical studies (e.g. Ulrich, 1970; Ando and Osaki, 1975) suggested that the observed power of the 5 min oscillations should be confined to distinct 'ridges' in the kind of $k_h-\omega$ power spectrum first obtained by Frazier (1968b). Deubner (1975) obtained the first data set extensive enough to demonstrate clearly the existence of the predicted ridges. Working independently, Rhodes *et al.* (1977a) employed a longer data set which produced a more highly resolved $k_h-\omega$ diagram with better defined ridges. Stimulated by the striking qualitative resemblance between observed and theoretical ridges, Rhodes *et al.* were able to make an estimate of the depth of the convection zone based on theoretical modeling of the ridges. Rhodes *et al.* (1981b) and Duvall and Harvey (1983) have improved the techniques used to make these observations, obtaining yet more precise positions for the ridges. The techniques used for these observations are described in detail in Section 2.1; the key point is that all of these studies used observations covering many hours of time and typically half the solar diameter, and included provisions to filter out all wave motions except those propagating in certain well-defined directions. The resulting data sets allowed high resolution in both frequency and wavenumber, and showed that virtually all of the power in the oscillations is confined to narrow ridges in the $k_h-\omega$ plane (see Figure 6).

The ridges shown in the figure correspond to sets of resonant *p*-modes trapped below the photosphere. Each ridge results from oscillations with a distinct radial order n; the ridge with lowest ω at a given k_h is the *f*-mode, while at higher frequencies one finds pressure modes with increasingly many nodes in their vertical eigenfunctions. The presence of distinct ridges demonstrates that pressure waves survive longer than the vertical propagation time across the cavity in which they are trapped. For the modes to be truly global, they must survive long enough to travel at least once around the Sun. Regardless of whether this condition is met, it is often convenient to discuss the modes

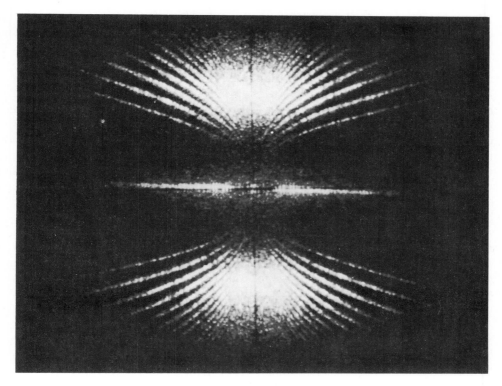

Fig. 6. An observed power spectrum of high-*l* *p*-mode oscillations, obtained by Rhodes, *et al*. (1983). Increasing power is shown as increasingly lighter shades of gray. In this spectrum ω runs from zero to 0.5 cycles per minute at the top and bottom, while k_h ranges from zero at the center to 993 cycles per solar circumference at the left and right. Evidently almost all the power is confined to a few distinct ridges, each corresponding to oscillations with a particular radial order *n*. At a given k_h, the mode with the smallest ω is the *f*-mode, while at larger ω one finds *p*-modes with radial order 1, 2, 3, etc. The left half-plane is the inverted mirror image of the right half-plane in order to better display the frequency asymmetry introduced by solar rotation in the two half-planes. Note that virtually all the power is found between periods of 4 and 7 min, and that at small k_h, the ridges are so close together that they cannot be separately resolved.

in terms of their angular degree *l*, which for large values of k_h is approximately $R_\odot$ k_h. The observing geometries used have almost invariably selected regions near the equator for study, and have filtered out waves with any N–S component of propagation. The modes selected are therefore those with $m = \pm l$, and the *l*-values represented on the figure range from $l \approx 75$ to $l \approx 1000$.

   P-modes with large *l*-values are trapped in a relatively shallow region near the top of the Sun's convection zone. For this reason, observations of high-*l* *p*-modes have been used principally to test models of the structure and dynamics of the solar envelope. Berthomieu *et al*. (1980) and Lubow *et al*. (1980) used more accurate measurements of the relation between k_h and ω to refine the estimate of convection zone depth made by Rhodes *et al*. (1977a), while Lubow *et al*. (1980) and Scuflaire *et al* (1981) demonstrated the importance and accuracy of various refinements to solar interior models. Rhodes

(1977) and Rhodes et al. (1979) showed that the difference in observed frequencies between modes with $m = +l$, $m = -l$ (eastward or westward propagating waves) leads to an asymmetry in the $k_h - \omega$ diagram that is proportional to the solar rotation rate, appropriately averaged over the part of the Sun where the waves are trapped. This asymmetry has been observed, and may give information on the variation of solar rotation as a function of depth (e.g. Duebner et al., 1979). Measurements of the intrinsic width of the ridges in ω should give an estimate of the mode lifetimes, and perhaps cast some light on the driving and damping processes acting on the oscillations. These measurements are very difficult, however, since observations to date have yielded ridge widths that are only marginally larger than the theoretical resolution of the observations. As yet no convincing results are available for high-l p-modes. Finally, minute changes in the mode positions with time may provide clues concerning the subsurface velocity and temperature structure (Hill et al., 1983). A detailed description of the way in which these inferences may be drawn from the available data will be given in Section 3.

Recently, observations using completely different techniques have succeeded in measuring the properties of p-mode oscillations with very low l-values. These methods are described in Section 2.1; they are important because the low-l modes that are accessible to them penetrate very deeply into the solar interior. Measurements of the frequencies and amplitudes of these modes thus allow one to make inferences about conditions over almost all the Sun's interior.

The first such observations to show low-l p-modes clearly were those by Claverie et al. (1979). These observations measured the Doppler shift of a solar line in light integrated over the entire disk; they were thus primarily sensitive to oscillations with l-values between 0 and 3. In the frequency band occupied by the 5 min oscillations, the power spectrum of these observations showed a large number of narrow peaks with a uniform separation of 67.8 μHz. The amplitudes of these narrow-band oscillations range from 10 to 30 cm s^{-1}, attesting to the extreme sensitivity of the technique employed. The uniform frequency spacing of the power spectrum peaks is explained by the asymptotic theory of global oscillations (cf. Section 1.3): modes with the same l-value and adjacent values of n have frequencies that differ by a constant, which for the Sun is about 135 μHz. Modes with the same n and adjacent values of l differ by half of this amount, or about 67.5 μHz. Successive peaks in the power spectrum thus correspond to modes with odd or even l, and successive values of n. Later observations from the South Pole by Grec et al. (1983) refined this picture (Figure 7). These results came from a continuous five-day run obtained during the austral summer, and are the best observations of this type currently available. They show not only the separation between odd and even l-values predicted by asymptotic theory, but also deviations from that behavior, allowing one to assign distinct frequencies to each l from 0 to 3. Thus, the mean frequency separation between modes with adjacent radial order and l-values differing by 2 are $\nu_{n0} - \nu_{n-1,2} = 9.4$ μHz and $\nu_{n1} - \nu_{n-1,3} = 15.3$ μHz. The linewidth in ω observed by Grec et al. is indistinguishable from the observational resolution (about 2 μHz) for $l = 0$, but is nearly twice this value for $l = 2$; the linewidths also seem to increase with increasing frequency. More recent observations at Stanford (Scherrer et al., 1982), using a differential technique, have extended these results out to $l = 5$, while Duvall and Harvey (1983) have used a spectrographic method to obtain frequencies for all the oscillation modes with $m = 0$ and $l < 160$. Woodard and Hudson (1983) have observed the very small variations in solar irradiance caused by oscillation modes with $l \leq 2$.

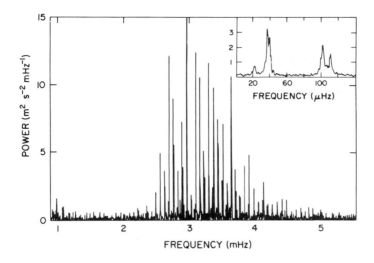

Fig. 7. An observed power spectrum of low-*l* *p*-modes, resulting from a 5 day continuous timestring of radial velocity measurements, obtained at the geographic South Pole by Grec *et al.* (1983). Most of the power is concentrated in a few dozen very narrow peaks, where each peak corresponds to an oscillation mode with particular values of radial order *n* and angular degree *l*. The largest amplitudes seen are about 20 cm s^{-1}. The inset figure demonstrates the regular spacing of features within the spectrum, as predicted by asymptotic theory. It is a superposed frequency diagram, in which the spectrum has been folded on itself with a repetition period of 0.136 mHz; the four resulting peaks (from left to right) correspond to oscillations with *l* of 3, 1, 2, and 0.

The lifetimes of the low-*l* *p*-modes are interesting for the same reasons as the high-*l* lifetimes. There is general agreement that the lifetimes are longer than about two days, but how much longer is an open question. Grec *et al.* (1983) report substantial changes in the amplitudes of several modes over times of a few days. Woodard and Hudson, however, find typical lifetimes of a week or longer.

The most intriguing and controversial results derived so far from low-*l* *p*-mode observations concern the solar rotation. Modes with the same *n* and *l* but with different *m* should have frequencies that differ from one another by roughly the solar rotation frequency, averaged over the part of the Sun where the eigenfunction is large. Attempts to observe this effect require detection of the rotational splitting of individual peaks in the power spectrum. The observation time needed to do this is about a month (less if the solar interior rotates faster than the surface). Because such long observing runs are not yet available, and because the observing methods used have been sensitive to all even *m*-values, the peaks observed are generally broadened, not cleanly split. Nevertheless, early results (Claverie *et al.*, 1981) show that peaks in the spectrum arising from *l*-values of 1 and 2 are apparently split into three and five components, respectively. The splitting between components averages about 0.75 μHz, or about twice what one would expect if the Sun were rotating at the surface rate throughout its volume. These results suggest that the solar interior may rotate substantially faster than the surface, but there are difficulties with this interpretation. First, symmetry considerations require that, at most, *l* + 1 of the 2*l* + 1 rotationally split frequency components should be visible in observations using integrated sunlight, unless the Sun's rotation axis is tipped toward the Earth at a substantial angle. Second, the width of the $l = 2$ component observed by Grec *et al*. is about 20% smaller than one would expect if the rotational splitting is as large as Claverie *et al*. suggest. Bos and Hill (1983) have reported observations of rotational splitting in power spectra of the solar diameter. They identify seven distinct sets of split components in a complicated spectrum, with splittings between adjacent *m*-values ranging from the expected surface value of 0.4 μHz for *p*-modes to 2.4 μHz for low-frequency *g*-modes. These and other applications of the low-*l* *p*-modes to helioseismology will be discussed further in Section 3.3.

2.2.2. *The 160 Minute Oscillation*

A solar velocity oscillation with a period near 160 min was reported independently (and nearly simultaneously) by two groups (Brookes *et al*., 1976; Severny *et al*., 1976). The former (Birmingham) group used an atomic resonance cell to measure wavelength shifts (averaged over the whole solar disk) of a sodium or potassium line in the solar spectrum relative to a laboratory source. The latter (Crimean) group used a modified magnetograph to observe the wavelength difference between light from the central part of the solar disk and that from the surrounding annulus. Both groups found that the largest signal in their observations was a sinusoidal oscillation with an amplitude of about 1 m s^{-1} and a period near 160 min, though the amplitude at this period was not a great deal larger than that at many nearby periods. Since one only sees four or five cycles during a long observing day, the only effective analysis technique was to combine the results of many different days, often with long gaps between days. To effect this combination, both groups used a superposed-epoch procedure, which isolates that part of the incoming signal that lies within a narrow frequency band; one always expects to see a signal of some sort from such an analysis. Thus, the most convincing aspect of these early observations was not the amplitude of the oscillation, but rather the good agreement between the two groups concerning its phase (Kotov *et al*., 1978).

The observations yielded one other perplexing fact. Both groups observed the amplitude to change dramatically over time spans of days, though long-term trends seemed evident as well (e.g. low amplitude throughout 1976) (Severny *et al*., 1978; Brookes *et al*., 1978a, b). From this, the Birmingham group inferred that the Sun is a damped oscillator with stochastic excitation, leading to rapid changes in the oscillation amplitude and, presumably, phase. The Crimean group took the measured phases to be meaningful even though the amplitude was changing, and looked for phase coherence in the oscillations over their entire body of observations. Since 160 min is exactly one-ninth of a day, any slight difference between the true period and 160 min should appear as a secular change in the local time of maximum velocity signal. Such a change of about 20 min yr^{-1} was indeed observed, implying a true period of about 160.01 min (Severny *et al*., 1978).

These results were confirmed by observations similar to those of the Crimean group, performed by a group at Stanford University (Dittmer, 1977; Scherrer, 1978). These showed oscillations with a period, amplitude, and phase in good agreement with those already reported by Severny and colleagues. Scherrer *et al*. (1979, 1980) and Severny *et al*. (1979) reported continued phase coherence of the oscillations, with a year-to-year phase shift of the best-fit sinusoid relative to 0 h UT of about 30 min. Scherrer and

Wilcox (1983) computed a combined power spectrum for the two sets of observations, finding once again that an oscillation with period near 160 min is the most significant feature.

There are a variety of less convincing observational results from other sources. These include measurements of the global velocity by Snider *et al.* (1978), as well as reported detections of the 160 min oscillation in the solar infrared center-to-limb brightness ratio (Kotov and Koutchmy, 1979), the visual (Kotov *et al.*, 1978) and radio (Eryushev *et al.*, 1980) brightness of the Sun, the global solar magnetic field (Kotov *et al.*, 1977), and the geomagnetic field (Toth, 1977). It is probably fair to say that these results have raised more doubts about the existence of the oscillation than they have quelled, chiefly because they are often contradicted by other observations.

Adding to the confusion is the absence of any satisfactory mechanism for producing or maintaining the oscillation. The period is too long for a pressure mode of the Sun; gravity modes provide a more plausible alternative, in the sense that normal modes with modest radial and angular numbers can produce periods in the correct range. However, for periods of 2–3 h, the periods vary slowly with mode number, and several modes have periods near 160 min. Attempts to explain why these other modes are not excited have invoked a sharp resonant coupling with *p*-modes (or with the beat frequency between two or more *p*-modes), e.g. Dziembowski (1983) and Kosovichev and Severny (1983). However, such attempts are so far only qualitative. A few explanations invoke more exotic phenomena. These include a solitary wave propagating around the solar core (Childress and Spiegel, 1981), a small black hole (Gough, 1978a; Severny *et al.*, 1979) or a planet of *y*-matter (Blinnikov and Khlopov, 1983) in orbit some 2×10^4 km below the solar surface. None of these models addresses the way in which the oscillation varies in amplitude while apparently maintaining phase coherence. This behavior implies at least two processes at work: a low loss (presumably interior) oscillation with good phase stability, and a variable gain process that transfers information about this oscillation to the solar surface. The nature of this second process is as yet completely unexplained.

In summary, the accumulation of evidence for the existence of a 160 min oscillation has become too compelling to ignore. But the observations contain so many contradictions, and their theoretical interpretation involves so many unanswered questions, that one can say little either about the physical process responsible for the oscillation or about the oscillation's importance for understanding the Sun's structure. It is clear that more and better observations will be required before the physical properties of the oscillation can be understood, or, indeed, before the very existence of this mode can be unambiguously demonstarted.

2.2.3. *Torsional Oscillations*

Howard and La Bonte (1980), using a technique which isolated horizontal motions in the east–west direction to analyze twelve years of Mount Wilson full-disk velocity data, found what appears to be a wave pattern consisting of alternating latitude zones of faster and slower than average rotation superimposed on the mean differential rotation. The amplitude of the difference between the mean and the perturbed differential rotation rates averages about 3 $\mathrm{m\,s^{-1}}$. Alternating zones of fast and slow rotation appear to originate at polar latitudes and drift to the equator in about 22 yr. The zones are roughly symmetric about the equator, while the equator itself is generally in a slow zone. Regions of solar activity tended to be centered on the poleward boundary of a fast zone, where the horizontal velocity shear was greatest.

If it is real, the wave pattern found by Howard and La Bonte seems to fit all the criteria for torsional oscillations. Because the pattern appears exactly on the time-scale of the solar cycle, it must likewise be driven by a combination of rotation and large-scale internal magnetic fields. The period and latitudinal drift behavior are qualitatively similar to those found in computations based on simple, incompressible spheres (studies by Ferraro and Plumpton). Moreover, because of the data reduction technique used, only horizontal motions were detected. These observations may provide a link between periodic velocity fluctuations (i.e. waves) and the solar cycle, which previously has been observed only in sunspots and other magnetic phenomena.

2.2.4. *Localized Brightness Oscillations*

Limb brightness oscillations have been studied most extensively by Hill and colleagues, who first saw them as variations in the apparent diameter of the Sun (e.g. Hill and Stebbins, 1975; Hill *et al.*, 1976; Brown *et al.*, 1978). The diameter variations seen appeared to be chiefly narrow-band oscillations with amplitudes of $5-8 \times 10^{-3}$ arcsec and periods between about 3 min and 1 h. To date, diameter observations refer to variations measured at only a few position angles, averaged over regions less than 100 arcsec in extent. They are therefore sensitive to oscillations with even l-values over a considerable range in l; oscillations with odd l simply move the solar image without affecting its diameter. If one interprets the oscillation amplitudes as motion of material in the solar atmosphere, the indicated radial velocities are about $2-10$ m s^{-1}. However, radial velocity observations (Snider *et al.*, 1974; Grec and Fossat, 1977; Severny *et al.*, 1976; Brookes *et al.*, 1976) have failed to reveal such oscillations, at least at periods significantly longer than 5 min. For this reason, the cause of oscillations in the diameter is believed to be mainly changes in the atmospheric temperature and opacity structure.

These observations immediately raised two questions: whether observed variations might be caused by refraction effects in the Earth's atmosphere, and whether the observed power spectrum might be consistent with that of broad-band solar noise (Fossat *et al.*, 1977; KenKnight *et al.*, 1977; Dittmer, 1977). Variations in the apparent diameter due to refractive inhomogeneities in the Earth's atmosphere have been estimated in two ways. One (Fossat *et al.*, 1977; KenKnight *et al.*, 1977; Grec *et al.*, 1979) uses observations of atmospheric column density or relative positions of stars to estimate (usually by invoking some model of the spatial structure of the refractive inhomogeneities) the expected variation in the observed solar diameter. The resulting estimates are typically about the same size as the observed diameter fluctuations, though the estimates by KenKnight *et al.*, which are model independent, are somewhat smaller. Another approach is to examine the diameter data itself for evidence of contamination from atmospheric sources. Atmospheric refraction should translate the solar limb darkening function without changing its shape, while changes in the solar temperature or opacity distribution should change both the shape and position of the profile. Evidence for such limb profile shape changes has been found (Hill and Caudell, 1979; Knapp *et al.*, 1980; Caudell *et al.*, 1980), suggesting that at least a significant part of the variation is of solar origin.

One can also look directly for brightness variations in the solar atmosphere. Early efforts to do this (Musman and Nye, 1977; Beckers and Ayres, 1977) failed, apparently because of inadequate sensitivity. Brown (1979b), Stebbins (1980), and Brown and Harrison (1980a, b) found power in the continuum intensity at large spatial and temporal scales, both near the limb and at disk center, with observed fluctuations large enough to account for the diameter oscillations. Moreover, the diameter and limb brightness observations produce qualitatively similar power spectra for the diameter for frequencies less than about 1.5 mHz. At higher frequencies the limb brightness observations show the p-mode oscillations quite clearly, while in the diameter power spectra the p-modes are much less evident, perhaps because of greater noise in the diameter measurements. These observations provide evidence that the diameter oscillations arise on the Sun and not in the Earth's atmosphere, since they are almost completely insensitive to atmospheric refraction errors.

If the oscillations are global, and not broad-band solar noise, they must exhibit temporal and spatial coherence. To date, methods for monitoring the diameter have produced little meaningful spatial information, hence only the temporal coherence has been tested. Brown *et al.* (1978), proposed estimating the coherence by plotting the phase of a suspected oscillation (referenced to a particular time of day) as a function of the date on which it was observed. For a coherent oscillation, the phase should advance by the same amount every day, giving a linear relation between the observed phases and the date. Using this technique with larger data sets Hill and Caudell (1979), Caudell *et al.* (1980), and Caudell and Hill (1980) found that, in a power spectrum of the solar diameter, essentially every identifiable peak in the frequency range 0.2–1.0 mHz displays phase coherence over an interval of 22 days duration. The probability of reproducing these coherence results from random data appears to be very small; Monte Carlo simulations of the period-fitting process show that, with random data, the likelihood of obtaining fits to the phases that are as good as those actually found is typically 4×10^{-3} for each frequency examined.

If correct, this analysis provides strong proof of the presence of global oscillation modes in the diameter time series data. However, most of the spectral band under consideration should be filled with an extremely dense and complicated superposition of p-modes (and possibly g-modes). Thus each bin in the observed power spectrum should be expected to contain many excited modes, with varying spatial dependences and amplitudes. However, one can show that if a frequency bin contains many discrete modes, the observed phase for that bin will advance linearly with time only if the power is distributed *symmetrically* about some point within the bin. This very strict condition is unlikely to be satisfied by every peak in a broad region of the spectrum. Suggested explanations for the observed behavior are not yet satisfactory, and thus, while the phase coherence results are persuasive from a formal point of view, their implications are so perplexing that they should be treated with caution. Further discussion of these arguments may be found in reviews by Hill (1980, 1978) and by Gough (1980).

2.3. WAVE OBSERVATIONS

Evidence for the presence of propagating short period waves comes from a variety of sources. However, short period waves ($P < \sim 100$ s) have wavelengths that are shorter than both the ground-based resolution limit and the photon mean free path in the atmosphere, hence the observations are difficult to interpret for the reasons given in Section 2.1.1. Intensity observations in the wings of strong calcium lines (Liu, 1974) show clear instances of upward-propagating impulsive disturbances, which probably are weak acoustic shocks. Deubner (1976) showed that, in power spectra of the Doppler shift seen in the lines Na I 5896 and Fe I 5383, the high-frequency tail of the spectrum shows a periodic structure in frequency out to periods as short as 20 s. Deubner interpreted this result as a spatial filtering effect, with equally spaced minima in the power spectrum corresponding to frequencies at which an integral number of wavelengths fit into the line formation region. Lites and Chipman (1979) used ground-based data to obtain power and phase spectra in three lines formed between the upper photosphere and mid-chromosphere. They found that the phase lag in velocity between different lines is almost zero for the 5 min oscillations, but gradually increases at higher frequencies, indicating propagating acoustic waves. At frequencies below the 5 min band the phase lags are random. This does not exclude the presence of gravity waves, however, since such waves propagate obliquely, whereas the observational phase comparison was made between line profiles observed at the same horizontal position.

Athay and White (1978, 1979), White and Athay (1979), and others have attempted to place limits on the role of acoustic waves in atmospheric heating. Using observations from the ground and from the OSO-8 satellite, they estimated both the radiative losses and the wave energy flux at several heights in the atmosphere. An upper limit on the wave flux may be obtained by assuming that all the wave energy represents upward-traveling waves (in effect setting $v_g = + c$), and that all unresolved atmospheric motions (i.e. line broadening) also contribute to the upward wave flux. In this way, one can estimate the wave flux to be 1.4×10^7 erg cm^{-2} s^{-1} in the mid-chromosphere, $< 10^5$ near the top of the chromosphere, and $1-2 \times 10^5$ in the transition region and low corona. The radiative losses are found to be about a factor of 10 greater than these estimates in the upper chromosphere, and a factor of 2–3 greater in the transition region. Thus the acoustic energy flux in the upper chromosphere appears inadequate to account for the radiative losses in those regions.

One of the strongest pieces of evidence for gravity waves comes from the center-to-limb and height dependence of nonthermal line broadening in the photosphere. The roughly factor of 2 ratio between limb and disk-center unresolved velocities indicates that either granulation or gravity waves are responsible. However, the convective velocities seem, on both observational and theoretical grounds, to decrease much too rapidly with height to explain the observed line broadening.

2.4. WAVES AND OSCILLATIONS IN SUNSPOTS

In active regions, unlike the quiet Sun, one expects magnetic fields to have a large influence on wave propagation throughout most of the atmosphere. However, in spite of extensive theoretical work, and in spite of a considerable body of observations, it is still not clear that we understand all the processes at work in waves in active regions. The current understanding of dynamical processes in sunspots has been summarized in a recent review by Thomas (1982).

Oscillations in sunspots were first seen as 'umbral flashes' by Beckers and Tallant

(1969). Flashes are localized chromospheric phenomena, seen as transient but repetitive intensity variations in the cores of the Ca II H and K lines with repetition periods of typically 150 s. In addition to the flashes, less localized chromospheric oscillations are present in virtually all spots, and may be seen as variations in both velocity and intensity in a number of strong lines (Giovanelli, 1972; Rice and Gaizauskas, 1973; Moore and Tang, 1975). Umbral oscillations have also been observed in lines formed in the transition region (Gurman *et al.*, 1982). Most of these observations indicate that the oscillations in any one spot tend to have well-defined periods and good time coherence. The periods vary from spot to spot, but are always in the range 145–180 s. The umbral oscillations and umbral flashes are probably related, but at this time the connection is obscure.

Sunspot penumbrae have oscillations with different properties (Zirin and Stein, 1972; Giovanelli, 1972; Musman *et al.*, 1976). Penumbral waves are traveling disturbances first seen in Hα filtergrams, but like the umbral oscillations, they can be detected in either velocity or intensity in most chromospheric lines. They differ from umbral oscillations in that they are most apparent as propagating waves, and their periods are longer (typically 250 s). Further, at least some spots do not display prominent penumbral waves (Musman *et al.*, 1976). When present, penumbral waves appear as well-defined arcs moving outward from the umbra–penumbra boundary, often encompassing as much as one-third of the spot circumference. The observed horizontal phase velocities for these waves range from 10 to 35 km s^{-1}.

It is curious that no photospheric counterparts of either umbral or penumbral oscillations have been seen, despite several attempts to detect them. Short-period photospheric oscillations in umbrae have been seen (Beckers and Schultz, 1972; Schultz, 1974), but these oscillations are apparently not found in all (or even most) spots, and are not correlated either with umbral flashes or with the chromospheric oscillations. Similarly, the photosphere in sunspot penumbrae seldom shows wavelike phenomena, and then with different wave parameters than are seen in the chromosphere (Musman *et al.*, 1976).

Perhaps the best data on umbral and penumbral oscillations available to data are provided by a particularly complete set of photoelectric observations obtained by Lites *et al.* (1982a, b). These observations show clearly that for the spot observed, flashes occur continuously, low in the umbral chromosphere, with each flash becoming the source of a rapidly propagating wave. The periods of these waves are sharply peaked near 170 s, and their phase velocities are very large (60–70 km s^{-1}). As these waves arrive at the umbra–penumbra boundary, a fraction of them proceed outward as penumbral waves, with phase speeds that decrease rapidly as the waves approach the boundary of the spot. No waves are seen propagating out into the quiet chromosphere, nor are any oscillations visible in the spot photosphere, either in intensity or line position. The phase relation between oscillations in line position and intensity are quite different from those seen in the photospheric *p*-mode oscillations. The significance of this difference is not yet clear.

A few points concerning oscillations in sunspots are plain from even a cursory examination of the observations. First, since these oscillations occur only in regions where the magnetic field strengths are large, presumably the properties of the waves are intimately connected with these same fields. Second, the well-defined oscillation frequencies observed suggest a resonant phenomenon; the evident weak dependence of the oscillation periods on spot size indicates that the resonance derives from the vertical statification of the atmosphere, rather than from the spot's horizontal structure. Third, energy densities deduced from the observed velocities in the oscillating regions indicate that the energy carried by these waves is irrelevant to the total energy balance of the spot. Finally, the large horizontal phase speeds of these oscillations do not necessarily signify large propagation speeds; if the waves propagate nearly vertically, very large horizontal phase speeds can result from modest propagation speeds.

Starting from these general notions, many workers have attempted detailed modeling of oscillations in active regions and flux tubes. Parker (1974 a, b) applied pure Alfvén waves in an attempt to understand what happens to the missing flux in sunspots. This model suggested that sunspots are cool not because they suppress convective heat transfer, but rather because they are efficient radiators of energy in the form of Alfvén waves. However, observations showed no evidence for a significant flux of Alfvén waves in the corona above spots (Beckers and Schneeberger, 1977). Further, realistic models of wave trapping (Thomas, 1978; Nye and Hollweg, 1980) show that upward propagating waves in spots are strongly reflected because of the rapid increase of Alfvén speed with height. Evidently the waves must escape downward if they are carrying any significant amount of energy. Only pure Alfvén waves are likely to escape into the interior; the other MHD modes are probably reflected by the increase in sound speed beneath a spot. Thus, the significance of wave cooling in sunspots hinges on the efficiency of Alfvén wave production, and on the ability of such waves to maintain their character in the complicated field geometries below the spot.

The physics of running penumbral waves has been considered by Moore (1973), and by Nye and Thomas (1976, and references therein), who derived the observed wave periods and horizontal phase speeds from resonances produced by trapped MAG waves. These later models treat the physics of wave propagation quite realistically: buoyancy, the static density stratification, the effects of compressibility and the magentic restoring forces are all taken into account. The only obvious idealization is the assumption that the magnetic field is uniform and horizontal. These calculations find the relevant waves to be gravity modified fast-mode waves, trapped between the rising Alfvén speed above and the rising sound speed below. The periods and wavelengths for the oscillations computed with these models are in good agreement with observations, but there may be some disagreement concerning the height dependence of the velocity amplitude.

Models for umbral oscillations are similar to those for penumbral waves, except that the magnetic fields in umbrae are taken to be vertical, rather than horizontal. As with penumbral waves, most efforts to understand umbral oscillations have invoked MAG wave resonances within the boundaries of the spot. Uchida and Sakurai (1975) modeled the oscillations as a standing 'quasi-Alfvén' wave, driven by overstable convection. In this model, most of the reflection at the upper boundary occurs because the Alfvén speed increases rapidly with height in the spot chromosphere. In spite of its somewhat restrictive assumptions, this model yields oscillation periods that are in fair agreement with observations. Scheuer and Thomas (1981) and Thomas and Scheuer (1982) have proposed a model that is similar to that of Uchida and Sakurai (1975), except that it takes full account of the effects of gravity, vertical stratification, and compressibility. Their results are similar to those of Uchida and Sakurai, although the physical interpretation is significantly different on some points. In particular, a model with compressibility shows larger vertical motions than horizontal ones at low altitudes, while the quasi-Alfvén treatment yields the opposite result. Also, it appears that the vertical velocities reach a maximum at a somewhat greater altitude in the model of Scheuer and Thomas than in that of Uchida and Sakurai. Detailed computations (Scheuer and Thomas, 1981) show that, while most of the wave energy is concentrated in the photosphere, the density stratification causes the photospheric vertical velocity to be smaller than that in the chromosphere by a factor of perhaps 40. This may explain the inability of observers to detect photospheric oscillations in sunspots.

2.5. FUTURE OBSERVATIONAL NEEDS

Although current observing techniques have provided a fascinating and informative look at many kinds of solar waves and oscillations, we have seen that these techniques are inadequate to answer many of the questions one now wants to ask. In this section, we shall try to outline goals for observational advances in a number of crucial areas.

For studies of p-mode oscillations, the principal obstacle at this time appears to be the difficulty of obtaining long, continuous timestrings of solar velocity data. Long timestrings are essential for determining such parameters of the p-mode spectrum as mode spacing and lifetime, and these parameters are, in turn, essential for an understanding of excitation mechanisms and solar structure. Unfortunately, the diurnal cycle ordinarily makes it impossible to obtain continuous data records much longer than 8 h from a single site, even if weather conditions are favorable. Recent studies have attempted to extract the required information from data runs on several independent days, or to bypass the problem by obtaining observations from the South Pole, where continuous runs of many days duration are sometimes obtainable.

The first of these techniques has had a number of significant successes (e.g., Claverie et al., 1979; Deubner, 1977), but may be intrinsically limited in its prospects (at least for linear analyses), by the unavoidable gaps in the data string (Brown, 1979a; Rhodes, 1979). Nevertheless, the crucial question of what percentage and distribution of missing data one can tolerate remains largely unanswered. Sturrock and Shoub (1981) considered this issue for randomly gapped data and linear analysis techniques, while Fahlman and Ulrych (1982) presented an analysis method, based on maximum entropy principles, that may avoid many of the problems encountered in Fourier analysis of gapped data. However, neither of these techniques has been applied to solar data or to a convincing simulation of it. Much work remains to be done in this area before one may be certain about the ultimate requirements on data continuity.

Observations from the South Pole have also produced significant results (Grec et al., 1980, 1983), but are unable to provide year-round monitoring of solar oscillations. Moreover, even during the austral summer, clear intervals of the required duration are rare. Stebbins and Wilson (1983) reported that in the last four summer seasons combined, there was only a single occasion on which clear weather persisted for more than 6.5 days, and only about ten occasions during which the sky was clear for longer than 4 days. This may prove inadequate to answer questions about driving mechanisms and damping times.

Two other methods suggested for obtaining long time series are use of a worldwide distribution of observing stations, and of a satellite instrument, either in polar orbit or placed at one of the Earth–Sun libration points (e.g. Newkirk, 1980). Problems of logistics, coordination, and compatibility have so far deterred serious consideration of a worldwide network. Unfortunately, the remaining option of a spacecraft instrument is much more expensive than any ground-based effort.

A second goal for future observations is to obtain spatially resolved velocity images of the whole Sun. If images with sufficient velocity precision were available, one could detect and identify the global oscillation modes with l-values lying between 3 and approximately 30. These intermediate l-values are essentially unobservable with present techniques, but they are the modes that tell the most about the structure of the deep interior, and about the solar rotation. It is therefore extremely important to determine their properties. Unfortunately, the amplitudes of these modes are typically less than 20 cm s^{-1}, so considerable precision and stability is required to detect them. A number of new instruments intended to achieve this goal are now under construction (Cacciani and Fofi, 1978; Title and Ramsey, 1980; Title and Rosenberg, 1980; Brown, 1980; Evans, 1980; Rhodes *et al.*, 1983), but none is yet fully operational.

Propagating waves in the solar atmosphere must have comparatively large frequencies, and correspondingly small wavelengths; this leads to many observational problems. Because the vertical wavelengths are typically comparable to the width of a line formation region, observation of these waves requires spectral resolution high enough to resolve minute changes in line shape and symmetry. In order to distinguish these shape changes from bodily shifts of the line profile, one requires an extremely stable optical wavelength reference. Finally, because of the small horizontal wavelengths, one must have very high spatial resolution to avoid seeing a superposition of signals from many waves with different phases. These requirements probably cannot all be satisfied except by observations with a large orbiting telescope; in particular, the spatial resolution required is not available from the ground. To address these problems, a number of spectroscopic experiments are now being prepared for use with SOT, the shuttle-borne Solar Optical Telescope. The data provided by these experiments should greatly improve our understanding of the nature and importance of propagating waves in the solar atmosphere.

The last requirement for a significant improvement in measurements of waves and oscillations involves the facilities needed to process the raw data. All of the observational techniques just described involve long timestrings of data, often with two-dimensional spatial resolution and/or spectral resolution. Also, to give adequate noise immunity it is usually necessary to record the incoming data at very high time resolution, with subsequent averaging. These requirements lead to very large data sets, ranging up to 10^{14} bytes for a week of SOT data. Simply displaying such a large quantity of data requires a computer system with substantial capacity, while doing serious analysis of such data sets is even more taxing. As a result, great emphasis must now be put on the development of efficient and capacious image processing facilities, that allow one to examine, edit, and reduce the results of the new observational methods. Fortunately, these problems are not unique to solar physics, and workers in other areas of astronomy are rapidly gaining experience in the necessary hardware and software techniques. However, it is clearly necessary for solar workers to take advantage of this experience.

3. Oscillations as Probes of the Solar Interior

3.1. INTRODUCTION

We have now reviewed the essential aspects of nonradial pulsation theory and the available observational information which, when combined and compared, can yield increased knowledge of conditions inside the Sun. In this concluding section we discuss the two available methods for inferring information from the oscillation data, and summarize what we have been able to establish so far with the data described in Section 2, at our present level of theoretical accuracy. We also describe some questions about interior structure and dynamics that currently lie on the horizon of our capabilities. These are problems that in principle can be solved using inversion techniques but require either data on presently unobserved modes or significantly more precise frequencies for known modes.

3.2. DIRECT METHOD

3.2.1. *Technique*

The direct method of using oscillation frequencies to place constraints on models of the interior is simply to calculate the normal modes of a particular solar model and compare with observed frequencies. The more reliable calculations of this sort also constrain the solar model by requiring that it be an evolved main sequence $1 \, M_\odot$ star with the age $(4.5 \times 10^9$ yr), luminosity, radius and atmospheric composition of the present Sun.

This technique has the disadvantage that it involves trial and error, because to some extent one can trade off, for example, depth of the convection zone (i.e. mixing-length ratio) against nonstandard composition ratios and composition gradients, boundary condition effects, and various uncertainties in the physics. The latter points up the other major problem with the method, namely that one must specify all the underlying physics, hence the accuracy of the computed frequencies is limited by uncertainties in the opacity, equation of state, modeling of convection, boundary conditions, composition gradients in the core and nonadiabatic effects, particularly radiative transfer effects.

The method has the clear advantage that it can be used with any number, however small, of measured frequencies, and can be used in conjunction with any other measurable properties of the Sun which can also place constraints on the model. As the quantity of data grows and as the input physics becomes more accurate, the limits that we can place on various aspects of internal structure become more stringent.

Some of the uncertainty inherent in the direct method can be reduced by calculating a sequence of models which differ in a single parameter and then deriving a best estimate for that parameter from a least-squares fitting procedure. This approach was used by Christensen-Dalsgaard and Gough (1981) to estimate the metal abundance.

3.2.2. *Direct Method: Results*

At present the data we have to work with, described in Section 2, consists of (1) the high-l p-mode ridges, whose absolute location in the (k_h, ω) plane, and whose separation, slope and curvature can all be determined to fairly high accuracy from observations extending over many hours; (2) frequencies for the first few low-l modes, $l = 0-5$, and possible rotational frequency splitting for $l = 0-3$ and for long period modes; (3) new data showing p-mode ridges for $l = 0-140$, which provide the first link between (1) and (2); and (4) frequencies but not l-values for a number of longer period modes seen in limb brightness fluctuations, and the period plus a lower bound to the lifetime of the 160 min oscillation. Other data in hand consist of estimates of lifetimes and amplitudes for many of the modes.

The depth of the convection zone has been a major focus of investigation by the direct method because high-l p-mode frequencies are very sensitive to the ratio of mixing length to pressure scale height (Ulrich and Rhodes, 1977; Berthomieu *et al.*, 1980; Lubow *et al.*, 1980; Scuflaire *et al.*, 1981). The consensus is that the convection zone must be about one-third of the solar radius in depth in order to reproduce the p-mode ridges with high accuracy. The low-l modes also seem to be best fit by a deep convection zone. Exception has been taken by Rosenwald and Hill (1980), who claim that nonlinearities in the atmosphere may invalidate the need for a deep convection zone. Belvedere *et al.* (1983), however, find that the observed frequencies cannot be matched by the combination of a shallow convection zone and nonlinearities in the atmosphere.

A shallower zone would also permit greater tunneling of energy by the deep g-modes. Whether a real conflict between the apparent observations of g-modes and the presence of a deep convection zone exists awaits the definitive observation and identification of the g-modes. Thus spatially resolved g-mode observations are needed.

Composition of the solar interior has also been studied extensively by this method (e.g. Christensen-Dalsgaard, 1982; Ulrich and Rhodes, 1982; Gough, 1982c; Ulrich and Rhodes, 1983). Evolved solar models have been produced with metal abundances that are smaller or larger than that observed in the atmosphere. It is possible to fit the p-mode ridges fairly accurately with both standard and nonstandard abundances by varying other parameters of the model. However, because the eigenfrequencies 'feel' a change in metal abundance primarily through the opacity of the metals (the total metal fraction is too small to affect significantly the mean molecular weight, though in some regions the metals may contribute most of the electrons and thus affect specific heats, adiabatic exponents, and ionization equilibria), the uncertainties which are known to exist in metal opacities strongly limit our ability to deduce the interior metal fraction from normal mode frequencies. In the absence of compelling reasons to believe otherwise, the abundances which correctly produce evolved solar models are generally assumed.

Attempts to fit the low-l frequency data have thus far not produced any entirely satisfactory model. To some extent the poorer fit may just be an artifact of the much greater precision obtained in the low-l frequency measurements. However, Duvall and Harvey (1983) compare observed frequencies for $l = 0-80$ with frequencies calculated from the most recent standard model of Ulrich and Rhodes (1983); they find the best agreement for the highest l-values ($l > 40$) and the poorest agreement at intermediate l and low n ($l \approx 10-30$). Discrepancies reported in the paper ranged up to about 17–18 μHz. For the low-l modes, a number of models have been obtained (by varying parameters) which can fit either one frequency exactly, or fit the frequency spacing between modes with the same l and successive values of n, but not both (e.g. Ulrich and Rhodes, 1983; Christensen-Dalsgaard and Gough, 1981). Ulrich and Rhodes find that uncertainties in treatment of the solar atmosphere are sufficient to account for the error in the computed frequencies, but that treatment of the atmosphere acts in effect as one free parameter, where two are needed.

Failure to fit the low-l frequencies with standard models thus seems to imply that the true structure of the deep interior differs from that of the standard models. There are several properties whose variation below the convection zone can significantly affect the low-l frequencies, such as composition and composition gradients, core mixing by diffusion during solar evolution, large core magnetic fields, and changes in opacity and equation of state calculations. From a series of experiments with these properties, Ulrich et al. (1982) find one model with nonstandard H, He, and heavy metal abundance interior to $M = 0.7 M_\odot$, which gives good agreement with most of the low-l p-mode frequencies. They remark that the model is almost surely not unique, and note that an acceptable model of the interior also will have to produce intermediate-l p-mode frequencies, low-l g-mode frequencies, and neutrino fluxes that agree with observed values.

The conclusion that work remains to be done on modeling of the deep interior is further supported by the ongoing failure to understand the low level of neutrino flux, and by the appearance of $2l + 1$ instead of $l + 1$ rotationally split components in the low-l p-mode observations (Claverie et al., 1981; see discussions by Isaak, 1982 and Gough, 1982a). Understanding of this region involves all the properties mentioned in the previous paragraph, as well as better modeling of the convection zone and of the interaction between convection and pulsations in the superadiabatic region, and taking into account nonadiabatic effects and the possibility of a rapidly rotating core.

Information on amplitudes (e.g. Grec et al., 1983; Duvall and Harvey, 1983), from which one hopes to derive greater knowledge of driving and damping mechanisms, is still too uncertain to be useful. Recent observations made by Harvey et al. at the South Pole may yield more certain amplitude estimates for low and intermediate-l p-modes.

Study of depth-dependent rotation rates for the outer envelope, using the high-l p-modes probably can be carried out most reliably via inversion methods which will be discussed shortly. Estimates by a direct method (Hill et al., 1982) have so far yielded somewhat uncertain results.

The recent observations by Claverie et al. (1979, 1980, 1981), in which rotational splitting of p-modes with $l = 0-3$ is reported, have been used to try to estimate the rotational velocity of the solar core (Claverie et al., 1981). They find the splitting, which is nearly twice that predicted for solid-body rotation at the photospheric rotation rate Ω_0, implies a core rotation rate of about $2\Omega_0$ if the rapidly rotating part extends though most of the solar radius, about $3\Omega_0$ if it is $0.5 R_\odot$ and $9\Omega_0$ if it is $0.15 R_\odot$. The rotational splitting of the low-l modes is still regarded as uncertain, as discussed in the section on observations.

From rotational splitting of seven long period modes, Hill et al. (1982) derived a tentative rotation curve which runs from about 3.2 μHz at the central $0.2 R_\odot$ to the surface value of 0.456 μHz at $R = R_\odot$, thus varying by a factor of more than 6 from surface to core. From the rotation curve they estimated the quadrupole moment of the Sun, J_2, to be 5.5×10^{-6}. Using the same data, Gough (1982a) estimated the quadrupole moment to be $J_2 = 3.6 \times 10^{-6}$. If J_2 can be accurately determined, it can then be used to calculate a theoretical value for the precession rate of Mercury, and hence to test the general theory of relativity. It may also be possible to measure the quadrupole moment of the Sun from a space probe that flies to within a small fraction of 1 AU, and thereby check the estimates made from the derived rotation curves.

Isaak (1982) has suggested that the appearance of all $2l + 1$ components of rotationally split $l = 1-3$ modes implies the existence of a large, obliquely rotating magnetic field in the solar interior. While Gough (1982a) questions the validity of the logic involved, he also notes that Dicke (1982) hypothesizes a similar magnetic field to explain the Princeton solar oblateness data.

3.3. INVERSION METHODS

3.3.1. *Technique*

Direct methods, even those which incorporate parameter fitting, produce nonunique solutions. Indeed, within the context of direct methods, it is not possible to demonstrate anything about the uniqueness of the solution.

Methods for deducing information about the interior in a mathematically more rigorous manner are the inversion techniques. The overall philosophy is as follows. A number of measureable quantities $\{q_i\}$ from the Sun can be shown to be functionals of some set of properties $\{S_k\}$ of the solar model, and can thus be written:

$$q_i = F_i[S_1, \ldots, S_k], i = 1, \ldots n, \tag{3.3.1}$$

where [ ] indicates the functional relationship; this expression hides multitudes of difficulties, such as the fact that F_i may represent a coupled set of partial differential equations. If one has a large number of measurements $\{q_i\}$ which are variously determined by the properties $\{S_k\}$ at all radii of the model, one may be able to use inversion techniques, which are mathematical procedures for turning (3.3.1) inside out and expressing the $\{S_k\}$ at a large number of depths as functions or functionals of the $\{q_i\}$.

In the solar case, the $\{q_i\}$ can be identified with the eigenmode frequencies $\{\omega_i\}$ or with frequency differences, the latter being either the difference between observed and calculated frequencies or the small difference between the frequencies of the components of a rotationally (or perhaps magnetically) split peak in the power spectrum.

An inversion procedure can be attempted with some hope of success if (1) a large number of frequencies (or frequency perturbations) are measured to high accuracy; (2) the eigenfunctions and the eigenfrequencies calculated from the model are approximately correct; (3) the normal mode perturbations are strictly linear (i.e. small amplitude); (4) the frequencies have measureable sensitivity to small fluctuations or to the existing gradients in the chosen variable; (5) the eigenfunctions corresponding to the measured frequencies span, collectively, the region whose properties we wish to infer; (6) the solutions are continuous (or 'stable') in the sense that varying S_k continuously to $S_k + \delta S_k$ produces a correspondingly continuous change in the frequencies. Furthermore, in practice inversion is likely to succeed only when the measured quantities are *linear* functionals of the model properties.

The important issues of uniqueness and 'stability' will not be discussed here (but see, e.g., Parker, 1977). We only remark that they are difficult to settle, and are frequently ignored for that reason, though ultimately the validity of the whole solution rests on them.

If we neglect all effects which are not spherically symmetric (e.g. rotation, magnetic fields), the eigenfrequencies of the Sun can be written in terms of the radial components of their eigenfunctions and a small set of solar properties, which we abbreviate S,

$$\omega_i^2 = \int_0^{R_\odot} dr\, f(S,\, \xi_i,\, \eta_i). \tag{3.3.2}$$

S might represent, for example, ρ, Γ_1; ξ_i and η_i contain the r-dependence of the eigenfunctions along and perpendicular to the radial direction. But ω_i is a highly nonlinear functional of the properties S, hence inversion of (3.3.2) is not really tractable.

The approach taken is to assume that a model which produces nearly correct frequencies in direct calculations is in fact a nearly correct model. Equation (3.3.2) is then perturbed about the state of the known model, giving

$$\frac{\delta\omega_i}{\omega_i} = \int dr \left[\frac{\partial \ln f}{\partial \rho}\, \delta\rho + \frac{\partial \ln f}{\partial \Gamma_1}\, \delta\Gamma_1\right] \equiv \int dr\, F_i. \tag{3.3.3}$$

The leading terms in the perturbations of ξ_i and η_i are second order in $\delta\xi_i$ and $\delta\eta_i$ (or the cross product) and are neglected. If the $\{\delta\omega_i\}$ are regarded as the difference between the observed frequencies and those calculated from the model, Equation (3.3.3) expresses these differences as linear functionals of the differences ($\delta\Gamma$ and $\delta\rho$) between the presumed correct model properties (to be determined) and those of the known model. The set of equations ($i = 1, \ldots, n$) can then be subjected to any of the various linear inversion procedures. Frequency differences among rotationally split eigenfrequencies are linear functionals of the differential rotation in radius, $\Omega_1(r) = \Omega(r) - \Omega(R_\odot)$, as given in Equations (1.5.3).

Different approaches to linear inversion problems may weight the various eigenfrequencies differently by combining them in different ways, and may treat errors in the data measurements differently. For this reason, each method has its own set of advantages and weaknesses, and it may often prove useful to apply more than one inversion procedure to the same problem.

The Backus–Gilbert (1968, 1970; see also Gough, 1982c) method constructs from the kernels F_i [of (3.3.3) or (1.5.3)] corresponding to measured eigenfrequencies ω_i a set of normalized kernels $K_j(r,\, r_j)$ which are as strongly peaked as possible at a chosen set of radii $\{r_j\}$. The integral over such a kernel then produces information about the properties of the model only over the interval where the kernel is large; i.e. if $\Omega_1(r)$ in Equations (1.5.3) is slowly varying and $K_j(r,\, r_j)$ is very narrow, then

$$\Omega_1(r_j) \approx \int K_j(r,\, r_j)\Omega_1(r)\, dr. \tag{3.3.4}$$

where $K_j(r,\, r_j) = \Sigma a_i(r_j)F_i(\xi_i,\, \eta_i)$ and F_i represents the kernel in (1.5.3). Computation of the solution consists mainly of determining the values of the $\{a_i(r_j)\}$ which maximize the narrowness of the K_j's around the r_j's. The method has the nice property that it displays explicity how well information can be resolved at each depth and totally fails in regions where the eigenfunction coverage does not permit resolution of the structure. A major disadvantage is that the linear superposition of eigenmodes (kernels) can sometimes

severely magnify the effects of errors in the frequencies; sometimes it is advantageous
to choose K_j's that are somewhat broader in exchange for a reduction in the error
magnification.

The optimal averaging method (e.g. Cooper, 1981) minimizes the difference between
the properties of the perturbed model and the original model (Equation (3.3.3)), subject
to the frequency constraints and to other constraints such as constant radius and total
mass. It always produces a solution, but in regions where the Backus–Gilbert method
fails, the solution generally will be highly inaccurate.

A spectral decomposition technique is useful for minimizing the input from data
with large errors. Here each measured quantity (e.g. ω_i or $\delta\omega_i$) is divided by its error
σ_i to produce a new set of equations in $\delta\omega_i \equiv \delta\omega_i/\sigma_i$ and $f_i \equiv f_i/\sigma_i$:

$$\delta\omega_i = \int f_i \, dr. \tag{3.3.6}$$

The f, which give highest weight to the most accurate measurements, are subjected to
a type of spectral decomposition. Specifically, the eigenvalues of $\Gamma_{ij} \equiv \int f_i(r) f_j(r) \, dr$
are obtained by diagonalizing Γ. The largest eigenvalues correspond to the spatial wavenumbers that contribute most heavily to the most accurate measurements.

Other methods which may be adaptable to inversion problems are maximum entropy
techniques and the 'clean' procedure used in radio astronomy.

3.3.2. *Applications*

Experiments with constructing kernels K_i for the Backus–Gilbert method have produced
some interesting results. For the high-l p-modes it is possible to form well-peaked kernels
over the outer layers of the convection zone, without significant magnification of the
errors (F. Hill and J. Toomre, private communication). Resolution of the deep interior,
particularly the nuclear burning core, appears to be difficult without including some
g-modes; experiments done so far with p-modes alone give very poor results in the inner
one-third to one-half of the solar radius and produce uncomfortably large error magnification (Gough, private communication). Thus, observations of g-modes may be of critical
importance for studying the structure of the solar core.

Inferring the depth-dependence of the rotation curve from observations of rotational
splitting is perhaps the least model-dependent use of the inversion methods, because
the frequency difference between rotationally split components is almost independent
of small errors in the unsplit frequency. The splitting is clearly seen in the differences in
frequency between the $+\omega$ and $-\omega$ ridges for the high-l p-modes. Deubner *et al.* (1979),
using an approach that combines aspects of the direct method and inversion, found
indications of an increase in rotation rate with depth below the solar surface in the
outer envelope. This analysis assumed a linear rotation curve, and exhibited very large
error bars; subsequent analysis by Deubner (1983) and by Rhodes *et al.* (1983) fails
to confirm the earlier result. Solution for the rotation curve is likely to require a true
inversion analysis.

So far, few observations provide information that is relevant to depths below the
outer 10% of the solar radius. Claverie *et al.* (1979, 1980) believe that they see splitting
of $l = 1-3$ p-modes; and Bos and Hill (1983) see evenly spaced peaks for seven long-period
modes that they believe to be rotationally split. Neither set of observations has been

confirmed by any other group, and many more modes, with eigenfunctions that are large at large depths, are needed to perform a reliable inversion for the rotation curve. New observations from the South Pole by Harvey, Duvall and Pomerantz, if averaged in the north–south direction, may provide rotational splitting data for the intermediate-l values, and thus may enable the use of eigenfunctions covering a much greater range of depths if the frequency splitting for those eigenfunctions can be measured accurately enough. These observations and the Duvall and Harvey (1983) observations, which are averaged in the east–west direction, are also of crucial importance in a study of the model structure, i.e. in inverting for $\delta\rho$ and $\delta\Gamma_1$.

Inversion techniques have been discussed by Gough and Toomre (1983) for diagnosing the horizontal velocities and temperature fluctuations associated with giant cells. The frequency perturbations associated with the horizontal speed U, and with perturbations in the sound speed and γ, can be written

$$\frac{\Delta\omega}{\omega} = \frac{k_x}{\omega} \int K_1 U \, d \ln z + \int K_2 \frac{\delta c^2}{c^2} \, d \ln z + \int K_3 \frac{\delta\gamma}{\gamma} \, d \ln z \qquad (3.3.7)$$

with the kernels K_1, K_2, and K_3 all independent of the sign of k_x. Thus U causes perturbations in ω that have opposite sign for waves propagating in opposite (east–west) directions, while variations in c^2 and γ produce symmetric perturbations. Thus one hopes to be able to separate the effects by taking sums and differences of the frequency perturbations. Perturbations calculated from convection theory are just at or slightly beyond the present limits of detectability. With improvements in observing techniques, the necessary measurements may be possible (Hill et al., 1983).

Another type of inversion problem is being investigated by Duvall (1982). Duvall notes that (1) an asymptotic WKB analysis gives $(n + \epsilon)\pi/\omega = \int dz/c = $ the sound travel time for a wave trapped in a cavity (with the value of ϵ set by the choice of boundary conditions; see, e.g., Leibacher and Stein, 1981), and (2) the cavity depth is very nearly the same for all waves with the same horizontal phase speed $v_p = \omega/k_x$, because the lower reflection point for acoustic waves is approximately at the depth where $\omega/k_x = c$. Taken together, these points imply that all the p-mode ridges (but not the fundamental which has $n = 0$) should collapse into a single curve on a plot of sound travel time versus horizontal phase speed. Duvall finds such a curve for $(n + 3/2)\pi/\omega$ versus ω/k_x. To the extent that the asymptotic analysis is accurate, the resulting curve can, in principle, be inverted to give the sound speed as a function of depth; there is a more than adequate number of data points available for performing the inversion.

Inversion techniques have so far hardly been used for diagnosing interior structure of the Sun, though they have been extensively employed for the Earth. This is because inversion requires accurate measurements of the frequencies and/or frequency perturbations for the modes corresponding to a large number of eigenfunctions, and morevoer to eigenfunctions that cover all the different radial regions of the Sun. Many of the necessary observations are just being made, or are just on the horizon. The information that these observations and the subsequent inversion analysis might uncover could have enormous implications for the whole field of stellar structure, on the premise that whatever our ordinary main sequence Sun does is likely to occur in many other ordinary main sequence stars.

We wish to acknowledge numerous fruitful discussions with J. Christensen-Dalsgaard, who also pointed out several important references of which we were unaware and offered helpful comments on many portions of this article.

References

Ando, H. and Osaki, Y.: 1975, *Publ. Astron. Soc. Japan* **27**, 581.
Ando, H. and Osaki, Y.: 1977, *Publ. Astron. Soc. Japan* **29**, 221.
Antia, H. M. and Chitre, S. M.: 1978, *Solar Phys.* **63**, 67.
Athay, R. G. and White, O. R.: 1978, *Astrophys. J.* **226**, 1135.
Athay, R. G. and White, O. R.: 1979, *Astrophys. J. Suppl.* **39**, 333.
Backus, G. and Gilbert, F.: 1968, *Geophys. J. Roy. Astron. Soc.* **16**, 169.
Backus, G. and Gilbert, F.: 1970, *Phil. Trans. Roy. Soc. London* **A266**, 123.
Baker, N. and Kippenhahn, R.: 1962, *Z. Astrophys.* **54**, 114.
Barnes, A.: 1979, in Parker, E. N., Kennel, C. F., and Lanzerotti, L. J. (eds.), *Solar System Plasma Physics*, Vol. 1, North-Holland, Amsterdam, p. 249.
Barnes, A. and Hollweg, J. V.: 1974, *J. Geophys. Res.* **799**, 2302.
Bazer, J. and Fleischman, O.: 1959, *Phys. Fluids* **2**, 366.
Beckers, J. M. and Ayres, T. R.: 1977, *Astrophys. J.* **217**, L69.
Beckers, J. M., and Schneeberger, T. J.: 1977, *Astrophys. J.* **215**, 356.
Beckers, J. M. and Schultz, R. B.: 1972, *Solar Phys.* **27**, 61.
Beckers, J. M. and Tallant, P. E.: 1969, *Solar Phys.* **7**, 351.
Bel, N. and Mein, P.: 1971, *Astron. Astrophys.* **11**, 234.
Belvedere, G., Gough, D. O., and Paterno, L.: 1983, *Solar Phys.* **82**, 343.
Berthomieu, G., Cooper, A. J., Gough, D. O., Osaki, Y., Provost, J., and Rocca, A.: 1980, *Lecture Notes in Physics* **125**, 307.
Biermann, L., 1946, *Naturwiss.* **33**, 118.
Blinnikov, S. I. and Khlopov, M. Yu: 1983, *Solar Phys.* **82**, 383.
Bohm-Vitense, E.: 1953, *Z. Astrophys.* **32**, 135.
Bolt, B. A. and Derr, J. S.: 1969, *Vistas in Astronomy* **11**, 69.
Bos, R. J. and Hill, H. A.: 1983, *Solar Phys.* **82**, 89.
Boury, A., Gabriel, M., Noels, A., Scuflaire, R., and Ledoux, P.: 1975 *Astron. Astrophys.* **41**, 279.
Brookes, J. R., Isaak, G. R., and Van der Raay, H. B.: 1976, *Nature* **259**, 92.
Brookes, J. R., Isaak, G. R., McLeod, C. P., Van der Raay, H. B., and Roca Cortes, T.: 1978a, in S. Dumont and J. Rösch (eds.), *Pleins Feux sur la Physique Solaire*, CNRS, Paris, p. 115.
Brookes, J. R., Isaak, G. R., McLeod, C. P., Van der Raay, H. B., and Roca Cortes, T.: 1978b, *Monthly Notices Roy. Astron. Soc.* **184**, 759.
Brown, T. M.: 1979a, in *Study of the Solar Cycle from Space*, NASA CP-2098, Washington, D.C., p. 101.
Brown, T. M.: 1979b, *Astrophys. J.* **230**, 255.
Brown, T. M.: 1980, in R. B. Dunn (ed.), *Solar Instrumentation: What's Next?*, Sacramento Peak Observatory, Sunspot, p. 150.
Brown, T. M. and Harrison, R. L.: 1980a, *Lecture Notes in Physics* **125**, 200.
Brown, T. M. and Harrison, R. L.: 1980b, *Astrophys. J.* **236**, L169.
Brown, T. M., Stebbins, R. T., and Hill.: 1978, *Astrophys. J.* **223**, 324.
Cacciani, A. and Fofi, M.: 1978, *Solar Phys.* **59**, 179.
Caudell, T. P. and Hill, H. A.: 1980, *Monthly Notices Roy. Astron. Soc.* **193**, 381.
Caudell, T. P., Knapp, J., Hill, H. A., and Logan, J. D.: 1980, *Lecture Notes in Physics* **125**, 206.
Chanmugam, G.: 1979, *Monthly Notices Roy. Astron. Soc.* **187**, 769.
Childress, S. and Spiegel, E. A.: 1981, in S. Sofia (ed.); *Proc Workshop on Solar Constant Variations*, NASA, Washington, D.C., p. 273.
Christensen-Dalsgaard, J.: 1980, *Monthly Notices Roy. Astron. Soc.* **190**, 765.
Christensen-Dalsgaard, J.: 1981, *Monthly Notices Roy. Astron. Soc.* **194**, 229.

Christensen-Dalsgaard, J.: 1982, *Monthly Notices Roy. Astron. Soc.* **199**, 735.

Christensen-Dalsgaard, J., Dilke, F. W., and Gough, D. O.: 1974, *Monthly Notices Roy. Astron. Soc.* **169**, 429.

Christensen-Dalsgaard, J., Dziembowski, W., and Gough, D. O.: 1980, *Lecture Notes in Physics* **125**, 313.

Christensen-Dalsgaard, J. and Frandsen, S.: 1983, *Solar Phys.* **82**, 469.

Christensen-Dalsgaard, J. and Gough, D. O.: 1975, *Mem. Soc. Roy. Sci. Liege* **8**, 309.

Christensen-Dalsgaard, J. and Gough, D. O.: 1980a, *Lecture Notes in Physics* **125**, 184.

Christensen-Dalsgaard, J. and Gough, D. O.: 1980b, *Nature* **288**, 545.

Christensen-Dalsgaard, J. and Gough, D. O.: 1981, *Astron. Astrophys.* **104**, 173.

Christensen-Dalsgaard, J. and Gough, D. O.: 1982, *Monthly Notices Roy. Astron. Soc.* **198**, 141.

Claverie, A., Isaak, G. R., McLeod, C. P., Van der Raay, H. B., and Roca Cortes, T.: 1979, *Nature* **282**, 591.

Claverie, A., Isaak, G. R., McLeod, C. P., Van der Raay, H. B., and Roca-Cortes, T.: 1980, *Astron. Astrophys.* **91**, L9.

Claverie, A., Isaak, G. R., McLeod, C. P., Van der Raay, H. B., and Roca Cortes, T.: 1981, *Nature* **293**, 443.

Cooper, A. J.: 1981, Thesis, University of Cambridge.

Cowling, T. G.: 1941, *Monthly Notices Roy. Astron. Soc.* **101**, 367.

Cowling, T. G., and Newing, R. A.: 1949, *Astrophys. J.* **109**, 149.

Cox, J. P.: 1980, *Theory of Stellar Pulsation*, Princeton Univ. Press., Princeton.

Cox, J. P., Cox, A. N., Olsen, K. H., King, D. S., and Eilers, D. D.: 1966, *Astrophys. J.* **144**, 1038.

Cox, J. P. and Giuli, R. T.: 1968, *Principles of Stellar Structure*, Gordon and Breach, New York.

Cram, L. E.: 1977, *Astron. Astrophys.* **59**, 151.

Defouw, R. J.: 1976 *Astrophys. J.* **209**, 266.

Deubner, F.-L.: 1975, *Astron. Astrophys.* **44**, 371.

Deubner, F.-L.: 1976, *Astron. Astrophys.* **51**, 189.

Deubner, F.-L.: 1977, *Astron. Astrophys.* **57**, 317.

Deubner, F.-L.: 1981, *Nature* **290**, 682.

Deubner, F.-L.: 1983, *Solar Phys.* **82**, 103.

Deubner, F.-L., Ulrich, R. K., and Rhodes, E. J., Jr: 1979, *Astron. Astrophys.* **72**, 177.

Deupree, R. G.: 1975, *Astrophys. J.* **198**, 419.

Deupree, R. G.: 1977, *Astrophys. J.* **211**, 509.

Dicke, R. H.: 1982, *Solar Phys.* **78**, 3.

Dilke, F. W. W. and Gough, D. O.: 1972, *Nature* **240**, 262.

Dittmer, P. H.: 1977, Ph. D. Thesis, Stanford University.

Dittmer, P. H.: 1978, *Astrophys. J.* **224**, 265.

Duvall, T. L.: 1982, *Nature* **300**, 242.

Duvall, T. L. and Harvey, J. W.: 1983, *Nature* **302**, 24.

Dziembowski, W.: 1977, *Acta Astron.* **27**, 203.

Dziembowski, W.: 1983, *Solar Phys.* **82**, 259.

Dziembowski, W. and Pamjatnykh, A. A.: 1978, in S. Dumont and J. Rösch (eds.), *Pleins Feux sur la Physique Solaire*, CNRS, Paris, p. 135.

Dziembowski, W. and Sienkiewicz, R.: 1973, *Acta Astron.* **23**, 273.

Eddington, A. S.: 1941, *Monthly Notices Roy. Astron. Soc.* **101**, 182.

Eryushev, N. N., Kotov, V. A., Severny, A. B., and Tsvetkov, L. I.: 1980, *Sov. Astron. Lett.* **5**, No. 5.

Evans, J. W.: 1980, in R. B. Dunn (ed.), *Solar Instrumentation: What's Next?*, Sacramento Peak Observatory, Sunspot, p. 155.

Evans, J. W., Michard, R., and Servajean, R.: 1963, *Ann. Astrophys.* **26**, 368.

Fahlman, G. G. and Ulrych, T. J.: 1982, *Monthly Notices Roy. Astron. Soc.* **199**, 53.

Ferraro, V. C. A. and Memory, D. J.: 1952, *Monthly Notices Roy. Astron. Soc.* **112**, 361.

Ferraro, V. C. A. and Plumpton, C.: 1958, *Astrophys. J.* **127**, 459.

Ferraro, V. C. A. and Plumpton, C.: 1966, *An Introduction to Magneto-Fluid Mechanics* (2nd ed.), Clarendon Press, Oxford.

Fowler, W. A.: 1972, *Nature* **238**, 24.

Fossat, E.: 1981, in R. M. Bonnet and A. K. Dupree (eds.), *Solar Phenomena in Stars and Stellar Systems*, Reidel, Dordrecht, p. 75.
Fossat, E., Harvey, J., Hausman, M., and Slaughter, C.: 1977, *Astron. Astrophys.* **59**, 279.
Fossat, E. and Ricort, G.: 1975, *Astron. Astrophys.* **43**, 243.
Frazier, E. N.: 1968a, *Astrophys. J.* **152**, 557.
Frazier, E. N.: 1968b, *Z. Astrophys.* **68**, 345.
Gabriel, M., Noels, A., Scuflaire, R., and Boury, A.: 1976, *Astron. Astrophys,* **47**, 137.
Gabriel, M., Scuflaire, R., Noels, A., and Boury, A.: 1975, *Astron. Astrophys.* **40**, 33.
Giovanelli, R. G.: 1972, *Solar Phys.* **27**, 71.
Goldreich, P. and Keeley, D. A.: 1977a, *Astrophys. J.* **211**, 934.
Goldreich, P. and Keeley, D. A.: 1977b, *Astrophys. J.* **212**, 243.
Goossens, M.: 1972, *Astrophys. Space Sci.* **16**, 386.
Gough, D. O.: 1977a, in R. M. Bonnet and Ph. Delache (eds.), *The Energy Balance and Hydrodynamics of the Solar Chromosphere and Corona*, G. De Bussac, Clermont Ferrand, p. 3.
Gough, D. O.: 1977b, *Astrophys. J.* **214**, 196.
Gough, D. O.: 1978a, in S. Dumont and J. Rösch (eds.), *Pleins Feux sur la Physique Solaire*, CNRS, Paris, p. 115.
Gough, D. O.: 1978b, in G. Belvedere and L. Paterno (eds), *Proc. Workshop on Solar Rotation*, Univ. of Catania, Italy, p. 255.
Gough, D. O.: 1980, *Lecture Notes in Physics* **125**, p. 273.
Gough, D. O.: 1981, *Monthly Notices Roy. Astron. Soc.* **196**, 731.
Gough, D. O.: 1982a, *Nature* **298**, 334.
Gough, D. O.: 1982b, *Nature* **298**, 350.
Gough, D. O.: 1982c, in J. P. Cox and C. J. Hansen (eds), *Proc. Pulsations in Classical and Cataclysmic Variable Stars, . . .*, J.I.L.A., Boulder, p. 117.
Gough, D. O.: 1983 (unpublished).
Gough, D. O. and Toomre, J.: 1983, *Solar Phys.* **82**, 401.
Grec, G. and Fossat, E.: 1977. *Astron. Astrophys.* **55**, 411.
Grec, G., Fossat, E., Brandt, P., and Deubner, F.-L.: 1979, *Astron. Astrophys.* **77**, 347.
Grec, G., Fossat, E., and Pomerantz, M.: 1980, *Nature* **288**, 541.
Grec, G., Fossat, E., and Pomerantz, M.: 1983, *Solar Phys.* **82**, 55.
Gurman, J. B., Leibacher, J. W., Shine, R. A., Woodgate, B. E., and Henze, W.: 1982, *Astrophys. J.* **253**, 939.
Hill, F.: 1982, Ph. D. Thesis, Univ. of Colorado.
Hill, F., Toomre, J., and November, L. J.: 1983, *Solar Phys.* **82**, 411.
Hill, H. A.: 1978, in J. A. Eddy (ed.), *The New Solar Physics*, Westview Press, Boulder, Chap. 5.
Hill, H. A.: 1980, *Lecture Notes in Physics* **125**, 174.
Hill, H. A., Bos, R. J., and Goode, P. R.: 1982, *Phys. Rev. Lett.* **49**, 1794.
Hill, H. A. and Caudell, T. P.: 1979, *Monthly Notices Roy. Astron. Soc.* **186**, 327.
Hill, H A. and Rosenwald, R. D.: 1978, in J. A. Eddv (ed.), *The New Solar Physics*, Westview Press, Boulder, p. 203.
Hill, H. A., Rosenwald, R. D., and Caudell, T. P.: 1978, *Astrophys. J.* **225**, 304.
Hill, H. A. and Stebbins, R. T.: 1975, *Ann. N. Y. Acad. Sci.* **262**, 472.
Hill, H. A., Stebbins, R. T., and Brown, T. M.: 1976, in J. H. Sanders and A. H. Wapstra (eds), *Atomic Masses and Fundamental Constants*, Plenum, New York, p. 622.
Hines, C. O.: 1960, *Can, J. Phys.* **38**, 1441.
Howard, R. F. and La Bonte, B. J.: 1980, *Astrophys. J.* **239**, L33.
Isaak, G. R.: 1982, *Nature* **296**, 130.
Jordan, S. D.: 1977, *Solar Phys.* **51**, 51.
Keeley, D. A.: 1977, in *Proc. Symposium on Large-Scale Motions on the Sun*. Sacramento Peak Observatory, Sunspot, N. M., p. 24.
Keeley, D. A.: 1980, *Lecture Notes in Physics,* **125**, 245.
KenKnight, C., Gatewood, G. D., Kipp, S. L., and Black, R.: 1977, *Astron. Astrophys.* **59**, L27.
Knapp, J., Hill, H. A., and Caudell, T. P.: 1980, *Lecture Notes in Physics* **125**, 394.
Kopal, Z.: 1949, *Astrophys. J.* **109**, 509.

Kosovichev, A. G. and Severny, A. B.: 1983, *Solar Phys.* **82**, 231.
Kotov, V. A. and Koutchmy, S.: 1979, *Uspehi Fiz. Nauk (USSR)* **128**, 730.
Kotov, V. A., Severny, A. B., and Tsap, T. T.: 1977, *Trans. Int. Astron. Union* **163**, 244.
Kotov, V. A., Severny, A. B., and Tsap, T. T.: 1978, *Monthly Notices Roy, Astron Soc.* **183**, 61.
Ledoux, P.: 1951, *Astrophys. J.* **114**, 373.
Ledoux, P.: 1974, *IAU Symp.* **59**, 135.
Ledoux, P. and Walraven, Th.: 1958, in S. Flugge (ed.), *Handbuch der Physik* **51**, 353.
Leibacher, J. W. and Stein, R. F.: 1971, *Astrophys. Lett.* **7**, 191.
Leibacher, J. W. and Stein, R. F.: 1981, in S. Jordan (ed.), *The Sun as a Star*, NASA SP-450, Washington, D.C., p. 263.
Leighton, R. B., Noyes, R. W., and Simon, G. W.: 1962, *Astrophys. J.* **135**, 474.
Lighthill, M. J., 1952, *Proc. Roy. Soc. London* **A211**, 564.
Lighthill, M. J., 1967, *IAU Symp.* **28**, 429.
Lites, B. W. and Chipman, E. G.: 1979, *Astrophys. J.* **231**, 570.
Lites, B. W., Chipman, E. G., and White, O. R.: 1982a, *Astrophys. J.* **253**, 367.
Lites, B. W., White, O. R., and Packman, D.: 1982b, *Astrophys. J.* **253**, 386.
Liu, S.-Y.: 1974, *Astrophys. J.* **189**, 359.
Lubow, S. L., Rhodes, E. J., Jr, and Ulrich, R. K.: 1980, *Lecture Notes in Physics* **125**, 300.
Lynden-Bell, D. and Ostriker, J. P.: 1967, *Monthly Notices Roy. Astron. Soc.* **136**, 293.
McLellan, A. and Winterberg, F.: 1968, *Solar Phys.* **4**, 401.
Melrose, D. B. and Simpson, M. A.: 1977, *Aust. J. Phys.* **30**, 647.
Michelitsanos, A. G.: 1973, *Solar Phys.* **30**, 47.
Mihalas, B.: 1979, Ph. D. Thesis, Univ. of Colorado.
Mihalas, B. W. and Toomre, J.: 1981, *Astrophys. J.* **249**, 349.
Mihalas, B. W. and Toomre, J.: 1982, *Astrophys. J.* **263**, 386.
Moore, R. L.: 1973, *Solar Phys.* **30**, 403.
Moore, D. W. and Speigel, E. A.: 1966, *Astrophys. J.* **143**, 871.
Moore, R. L. and Tang, F.: 1975, *Solar Phys.* **41**, 81.
Musman, S. A. and Nye, A. H.: 1977, *Astrophys. J.* **212**, L95.
Musman, S., Nye, A. H., and Thomas, J. H.: 1976, *Astrophys. J.* **206**, L175.
Nakagawa, Y., Priest, E. R., and Wellck, R. E.: 1973, *Astrophys. J.* **184**, 931.
Newkirk, G. A.: 1980, *Study of the Solar Cycle from Space: Report of the Science Working Group*, NASA, SCADM #3, Washington, D.C.
Nye, A. H. and Hollweg, J. V.: 1980, *Solar Phys.* **68**, 279.
Nye, A. H. and Thomas, J. H.: 1976, *Astrophys J.* **204**, 582.
Osaki, Y.: 1975, *Publ. Astron. Soc. Japan* **27**, 237.
Parker, E. N.: 1957, *Proc. N.A.S. U.S.A.* **43**, 8.
Parker, E. N.: 1974a, *Solar Phys.* **36**, 249.
Parker, E. N.: 1974b, *Solar Phys.* **37**, 127.
Parker, E. N.: 1979, *Astrophys. J.* **233**, 1005.
Parker, R. L.: 1977, *Ann. Rev. Earth Planet. Sci.* **5**, 35.
Petrukhin, N. S. and Fainshtein, S. M.: 1976, *Sov. Astron.* **20**, 713.
Plumpton, C.: 1957, *Astrophys. J.* **125**, 494.
Plumpton, C. and Ferraro, V.C.A.: 1953, *Monthly Notices Roy. Astron. Soc.* **113**, 647.
Rhodes, E. J., Jr: 1977, Ph. D. Dissertation, U.C.L.A., Univ. Microfilms, Ann Arbor.
Rhodes, E. J., Jr: 1979, in *Study of the Solar Cycle from Space*, NASA CP-2098, Washington, D.C., p. 159.
Rhodes, E. J., Jr, Deubner, F.-L., and Ulrich, R. K.: 1979, *Astrophys. J.* **227**, 629.
Rhodes, E. J., Jr, Harvey, J. W., and Duvall, T. L., Jr: 1983, *Solar Phys.* **82**, 111.
Rhodes, E. J., Jr, Howard, R. J., Ulrich, R. K., and Smith, E. J.: 1981a, in R. B. Dunn (ed.), *Solar Instrumentation: What's Next?*, Sunspot, p. 102.
Rhodes, E. J., Jr, Howard, R. J., Ulrich, R. K., and Smith, E. J.: 1983, *Solar Phys.* **82**, 245.
Rhodes, E. J., Jr and Ulrich, R. K.: 1977, *Astrophys. J.* **218**, 521.
Rhodes, E. J., Jr and Ulrich, R. K.: 1982, in J. P. Cox and C. J. Hansen (eds), *Proc. Conf. Pulsations in Classical and Cataclysmic Variable Stars*, J.I.L.A., Boulder, p. 147.

Rhodes, E. J., Jr, Ulrich, R. K., Harvey, J. W., and Duvall, T. L., Jr: 1981b, in R. B. Dunn (ed.), *Solar Instrumentation: What's Next?*, Sacramento Peak Observatory, Sunspot, p. 37.

Rhodes, E. J., Jr, Ulrich, R. K., and Simon, G. W.: 1977a, *Astrophys. J.* **218**, 901.

Rhodes, E. J., Jr, Ulrich, R. K., and Simon, G. W.: 1977b, in E. Hansen and S. Schaffner (eds), *Proc. Nov. 1977 OSO-8 Workshop*, Univ. of Colorado, Boulder, p. 365.

Rice, J. B. and Gaizauskas, V.: 1973, *Solar Phys,* **32**, 421.

Robe, H.: 1968, *Ann Astrophys.* **31**, 475.

Rosenwald, R. D. and Hill, H. A.: 1980, *Lecture Notes in Physics* **125**, 404.

Saio, H.: 1980, *Astrophys. J.* **240**, 685.

Scherrer, P. H.: 1978, in *Proc. Symposium of Large-Scale Motions on the Sun*, Sacramento Peak Observatory, Sunspot, p. 12.

Scherrer, P. H. and Wilcox, J. M.: 1983, *Solar Phys.* **82**, 37.

Scherrer, P. H., Wilcox, J. M., Christensen-Dalsgaard, J., and Gough, D. O.: 1982, *Nature* **297**, 312.

Scherrer, P. H., Wilcox, J. M., Kotov, V. A., Severny, A. B., and Tsap, T. T.: 1979, *Nature* **277**, 635.

Scherrer, P. H., Wilcox, J. M., Severny, A. B., Kotov, V. A., and Tsap, T. T.: 1980, *Astrophys. J.* **237**, L97.

Scheuer, M. A. and Thomas, J. H.: 1981, *Solar Phys.* **71**, 21.

Schmieder, B.: 1977, *Solar Phys.* **57**, 245.

Schultz, R. B.: 1974, Ph. D. Thesis, Univ. of Colorado.

Schwartz, R. A. and Stein, R. F.: 1975, *Astrophys J.* **200**, 499.

Schwartzschild, M.: 1948, *Astrophys. J.* **107**, 1.

Scuflaire, R.: 1974, *Astron. Astrophys.* **36**, 107.

Scuflaire, R., Gabriel, M., and Noels, A.: 1981, *Astron. Astrophys.* **99**, 39.

Scuflaire, R., Gabriel, M., Noels, A., and Boury, A.: 1975, *Astron. Astrophys.* **45**, 15.

Severny, A. B., Kotov, V. A., and Tsap, T. T.: 1976, *Nature* **259**, 87.

Severny, A. B., Kotov, V. A., and Tsap, T. T.: 1978, in S. Dumont and J. Rösch (eds). *Pleins Feux Sur la Physique Solaire*, CNRS, Paris, p. 115.

Severny, A. B., Kotov, V. A., and Tsap, T. T.: 1979, *Soviet Astron.* **23**, 641.

Shibahashi, H.: 1979, *Publ. Astron. Soc. Japan* **31**, 87.

Shibahashi, H., Noels, A., and Gabriel, M.: 1983, *Astron. Astrophys.* **123**, 283.

Shibahashi, H. and Osaki, Y.: 1983, *Solar Phys.* **82**, 231.

Shibahashi, H., Osaki, Y., and Unno, W.: 1975, *Publ. Astron. Soc. Japan* **27**, 401.

Snider, J. L., Eisenstein, J. P., and Otten, G. R.: 1974, *Solar Phys.* **36**, 303.

Snider, J. L., Kearns, M. D., and Tinker, P. A.: 1978, *Nature* **275**, 730.

Souffrin, P.: 1966, *Ann. Astrophys.* **29**, 55.

Spiegel, E. A.: 1957, *Astrophys. J.* **126**, 202.

Spruit, H. C.: 1981, in S. D. Jordan (ed.), *The Sun as a Star*, NASA SP-450, Washington, D.C., p. 385.

Stebbins, R. T.: 1980, *Lecture Notes in Physics* **125**, 191.

Stebbins, R. T. and Wilson, C.: 1983, *Solar Phys.* **82**, 43.

Stein, R. F.: 1968, *Astrophys. J.* **154**, 297.

Stein, R. F.: 1982, *Astron. Astrophys.* **105**, 417.

Stein, R. F. and Leibacher, J. W.: 1974, *Ann. Rev. Astron. Astrophys.* **12**, 407.

Sturrock, P. A. and Shoub, E. C.: 1981, SUIPR Rept No. 824 R.

Tanenbaum, A., Wilcox, J., Frazier, E., and Howard, R.: 1969, *Solar Phys.* **9**, 328.

Tassoul, J. L.: 1978, *The Theory of Rotating Stars,* Princeton Univ. Press, Princeton.

Tassoul, M.: 1980, *Astrophys. J. Supp.* **43**, 469.

Thomas, J. H.: 1978, *Astrophys. J.* **225**, 275.

Thomas, J. H.: 1983, *Ann. Rev. Fluid Mech.* **15**, 321.

Thomas, J. H., Clark, P. A., and Clark, A., Jr: 1971, *Solar Phys.* **16**, 51.

Thomas, J. H. and Scheuer, M. A.: 1982, *Solar Phys.* **68**, 229.

Title, A. M. and Ramsey, H. E.: 1980, *Appl. Optics* **19**, 2046.

Title, A. and Rosenberg, W.: 1980, in R. B. Dunn (ed.), *Solar Instrumentation: What's Next?*, Sacramento Peak Observatory, Sunspot, p. 326.

Toth, P.: 1977, *Nature* **270**, 159.

Uchida, Y. and Sakurai, T.: 1975, *Publ Astron. Soc. Japan* **27**, 259.

Ulmschneider, P.: 1971, *Astron. Astrophys.* **14**, 275.
Ulmschneider, P.: 1974, *Astron. Astrophys.* **39**, 327.
Ulmschneider, P., Schmitz, F., Kalkofen, W., and Bohn, H. U.: 1978, *Astron. Astrophys.* **70**, 487.
Ulrich, R. K.: 1970, *Astrophys. J.* **162**, 933.
Ulrich, R. K.: 1974, *Astrophys. J.* **188**, 369.
Ulrich, R. K. and Rhodes, E. J. Jr: 1977, *Astrophys. J.* **218**, 521.
Ulrich, R. K. and Rhodes, E. J. Jr: 1983, *Astrophys. J.* **265**, 551.
Ulrich, R. K., Rhodes, E. J. Jr, and Deubner, F.-L.: 1979, *Astrophys J.* **227**, 638.
Ulrich, R. K., Rhodes, E. J. Jr, Tomczyk, S., Dumont, P. J., and Brunish, W. M.: 1983, Preprint.
Unno, W.: 1967, *Publ. Astron. Soc. Japan* **19**, 140.
Unno, W., Osaki, Y., Ando, H. and Shibahashi, H.: 1981, *Nonradial Oscillations of Stars*, Univ. of Tokyo Press.
Unno, W.: 1975, *Publ. Astron. Soc. Japan* **27**, 81.
Unno, W. and Spiegel, E. A.: 1966, *Publ. Astron. Soc. Japan* **18**, 85.
Vandakurov, Y. U.: 1967a, *Astro. Zh.* **44**, 786.
Vandakurov, Y. U.: 1967b, *Astrophys. J.* **149**, 435.
Wentzel, D. G.: 1978, *Solar Phys.* **58**, 307.
White, O. R. and Athay, R. G.: 1979, *Astrophys. J. Suppl.* **39**, 347.
Willson, R. C.: 1979, *J. Appl. Opt.* **18**, 179.
Wilson, P. R.: 1978a, *Astrophys. J.* **221**, 672.
Wilson, P. R.: 1978b, *Astrophys. J.* **225**, 1058.
Wolff, C. L.: 1972a, *Astrophys. J.* **176**, 833.
Wolff, C. L.: 1972b, *Astrophys. J.* **177**, L87.
Wolff, C. L.: 1979, *Astrophys. J.* **227**, 943.
Woodard, M. and Hudson, H.: 1983, *Solar Phys.* **82**, 67.
Yoshimura, H.: 1975, *Astrophys. J.* **201**, 740.
Yoshimura, H.: 1981, *Astrophys. J.* **247**, 1102.
Zirin, H. and Stein, A.: 1972, *Astrophys. J.* **173**, L85.

Timothy M. Brown & Barbara Weibel Mihalas,
High Altitude Observatory,
National Center for Atmospheric Research,
Boulder, CO 80307,
U.S.A.

Edward J. Rhodes, Jr,
Dept of Astronomy and Space Sciences Center,
University of Southern California,
Los Angeles, CA 90007,
U.S.A.

INDEX